BARRON'S
MATH WORKBOOK
for the
NEW SAT*

6TH EDITION

Lawrence S. Leff, M.S.
Former Assistant Principal, Mathematics Supervision
Franklin D. Roosevelt High School
Brooklyn, New York

BARRON'S

*SAT is a registered trademark of the College Entrance Examination Board, which was not involved in the production of, and does not endorse, this book.

Dedication

To Rhona:
For the understanding,
for the sacrifices,
and
for the love

All inquiries should be addressed to:
Barron's Educational Series, Inc.
250 Wireless Boulevard
Hauppauge, New York 11788
www.barronseduc.com

ISBN: 978-1-4380-0621-5

PCN: 2015914881

PRINTED IN THE UNITED STATES OF AMERICA

9 8 7 6 5 4 3 2 1

10%
POST-CONSUMER
WASTE
Paper contains a minimum
of 10% post-consumer
waste (PCW). Paper used
in this book was derived
from certified, sustainable
forestlands.

Contents

Preface

This new edition of the Barron's SAT Math Workbook is based on the redesigned 2016 SAT. It is organized around a simple, easy-to-follow, and proven four-step study plan:

> STEP 1. Know what to expect on test day.
> STEP 2. Become testwise.
> STEP 3. Review SAT Math topics and SAT-type questions.
> STEP 4. Take practice exams under test conditions.

STEP 1 KNOW WHAT TO EXPECT ON TEST DAY

Chapter 1 gets you familiar with the format of the test, types of math questions, and special directions that will appear on the SAT you will take. This information will save you valuable testing time when you take the SAT. It will also help build your confidence and prevent errors that may arise from not understanding the directions on test day.

STEP 2 BECOME TESTWISE

By paying attention to the test-taking tips and SAT Math facts that are strategically placed throughout the book, you will improve your speed and accuracy, which will lead to higher test scores. Chapter 2 is a critically important chapter that discusses essential SAT Math strategies while also introducing some of the newer math topics that are tested by the redesigned SAT.

STEP 3 REVIEW SAT MATH TOPICS AND SAT-TYPE QUESTIONS

The SAT test redesigned for 2016 and beyond places greater emphasis on your knowing the topics that matter most from your college preparatory high school mathematics courses. Chapters 3 to 6 serve as a math refresher of the mathematics you are expected to know and are organized around the four key SAT Math content areas: Heart of Algebra, Problem Solving and Data Analysis, Passport to Advanced Math, and Additional Topics in Math (geometric and trigonometric relationships). These chapters also feature a large number and variety of SAT-type math questions organized by lesson topic. The easy-to-follow topic and lesson organization makes this book ideal for either independent study or use in a formal SAT preparation class. Answers and worked-out solutions are provided for all practice problems and sample tests.

STEP 4   TAKE PRACTICE EXAMS UNDER TEST CONDITIONS

Practice makes perfect! At the end of the book, you will find two full-length SAT Math practice tests with answer keys and detailed explanations of answers. Taking these exams under test conditions will help you better manage your time when you take the actual test. It will also help you identify and correct any remaining weak spots in your test preparation.

Lawrence S. Leff

LEARNING ABOUT SAT MATH

Know What You're Up Against

1

This chapter introduces you to the test format, question types, and the mathematics topics you need to know for the redesigned 2016 SAT. Compared to prior editions of the SAT, the *new* SAT

- Places a greater emphasis on algebra: forming and interpreting linear and exponential models; analyzing scatterplots, and two-way tables.
- Includes two math test sections: in one section you can use a calculator and in the other section a calculator is *not* allowed.
- Does *not* deduct points for wrong answers.

LESSONS IN THIS CHAPTER

Lesson 1-1 Getting Acquainted with the Redesigned SAT

Lesson 1-2 Multiple-Choice Questions

Lesson 1-3 Grid-In Questions

OVERVIEW

The March 2016 SAT test date marks the first administration of a redesigned SAT. The mathematics content of the new version of the test will be more closely aligned to what you studied in your high school math classes. The redesigned SAT is a timed exam lasting 3 hours (or 3 hours and 50 minutes with an optional essay).

What Does the SAT Measure?

The math sections of the new SAT seek to measure a student's understanding of and ability to apply those mathematics concepts and skills that are most closely related to successfully pursuing college study and career training.

Why Do Colleges Require the SAT?

College admissions officers know that the students who apply to their colleges come from a wide variety of high schools that may have different grading systems, curricula, and academic standards. SAT scores make it possible for colleges to compare the course preparation and the performances of applicants by using a common academic yardstick. Your SAT score, together with your high school grades and other information you or your high school may be asked to provide, helps college admission officers to predict your chances of success in the college courses you will take.

How Have the SAT Math Sections Changed?

Here are five key differences between the math sections of the SAT given before 2016 and the SAT for 2016 and beyond:

- There is no penalty for wrong answers.
- Multiple-choice questions have four (A to D) rather than five (A to E) answer choices.
- Calculators are permitted on only one of the two math sections.
- There is less emphasis on arithmetic reasoning and a greater emphasis on algebraic reasoning with more questions based on real-life scenarios and data.

New Math Topics

Beginning with the 2016 SAT, these additional math topics will now be required:

- Manipulating more complicated algebraic expressions including *completing the square* within a quadratic expression. For example, the circle equation $x^2 + y^2 + 4x - 10y = 7$ can be rewritten in the more convenient center-radius form as $(x + 2)^2 + (y - 5)^2 = 36$ by completing the square for both variables.
- Performing operations involving the imaginary unit i where $i = \sqrt{-1}$.

- Solving more complex equations including quadratic equations with a leading coefficient greater than 1 as well as nonfactorable quadratic equations.
- Working with trigonometric functions of general angles measured in radians as well as degrees.

Table 1.1 summarizes the major differences between the math sections of the previous and newly redesigned SATs.

Table 1.1 Comparing Old and New SAT Math

Test Feature	Old SAT Math (before 2016)	Redesigned SAT Math (2016 and after)
Test Time	70 minutes	80 minutes
Number of sections	Three	Two: one 55-minute calculator section and one 25-minute no-calculator section
Number of questions	54 = 44 multiple-choice + 10 grid-in	58 = 45 multiple-choice + 13 grid-in
Calculators	Allowed for each math section	Permitted for longer math section only
Point penalty for a wrong answer?	Yes	No
Multiple-choice questions	5 answer choices (A to E)	4 answer choices (A to D)
Point value	Each question counts as 1 point.	Each question counts as 1 point.
Math content	■ Topics from arithmetic, algebra, and geometry ■ Only a few algebra 2 topics ■ Not aligned with college-bound high school mathematics curricula	■ Greater focus on three key areas: algebra, problem solving and data analysis, and advanced math ■ More algebra 2 and trigonometry topics, more multistep problems, and more problems with real-world settings ■ Stronger connection to college-bound high school mathematics courses

TIP

If you don't know an answer to an SAT Math question, make an educated guess! There is *no* point penalty for a wrong answer on the redesigned SAT. You get points for the questions you answer correctly but do *not* lose points for any wrong answers.

What Math Content Groups Are Tested?

The new test includes math questions drawn from four major content groups:

- Heart of Algebra: linear equations and functions
- Problem Solving and Data Analysis: ratios, proportional relationships, percentages, complex measurements, graphs, data interpretation, and statistical measures
- Passport to Advanced Math: analyzing and working with advanced expressions
- Additional Topics in Math: essential geometric and trigonometric relationships

Table 1.2 summarizes in greater detail what is covered in each of the four math content groups tested by the redesigned SAT.

Table 1.2 The Four SAT Math Content Groups

Math Content Group	Key Topics
Heart of Algebra	■ Solving various types of linear equations ■ Creating equations and inequalities to represent relationships between quantities and to use these to solve problems ■ Polynomials and Factoring ■ Calculating midpoint, distance, and slope in the *xy*-plane ■ Graphing linear equations and inequalities in the *xy*-plane ■ Solving systems of linear equations and inequalities ■ Recognizing linear functions and function notation
Problem Solving and Data Analysis	■ Analyzing and describing relationships using ratios, proportions, percentages, and units of measurement ■ Describing and analyzing data and relationships using graphs, scatter plots, and two-way tables ■ Describing linear and exponential change by interpreting the parts of a linear or exponential model ■ Summarizing numerical data using statistical measures
Passport to Advanced Math	■ Performing more advanced operations involving polynomial rational expressions, and rational exponents ■ Recognizing the relationship between the zeros, factors, and graph of a polynomial function ■ Solving radical, exponential, and fractional equations ■ Completing the square ■ Solving nonfactorable quadratic equations ■ Parabolas and their equations ■ Nonlinear systems of equations ■ Transformations of functions and their graphs
Additional Topics in Math	■ Area and volume measurement ■ Applying geometric relationships and theorems involving lines, angles, and triangles (isosceles, right, and similar). Pythagorean theorem, regular polygons, and circles ■ Equation of a circle and its graph ■ Performing operations with complex numbers ■ Working with trigonometric functions (radian measure, cofunction relationships, unit circle, and the general angle)

What Types of Math Questions Are Asked?

The redesigned SAT includes two types of math questions:

- Multiple-choice (MC) questions with four possible answer choices for each question.
- Student-produced response questions (grid-ins) which do not come with answer choices. Instead, you must work out the solution to the problem and then "grid-in" the answer you arrived at on a special four-column grid.

How Are the Math Sections Set Up?

The redesigned SAT has two math sections: a section in which a calculator is permitted and a shorter section in which a calculator is **not** allowed.

- The 55-minute calculator section contains 38 questions. Not all questions in the calculator section require or benefit from using a calculator.
- The 25-minute no-calculator section has 20 questions.

Table 1.3 Breaking Down the Two Math Sections

The Two Types of Math Sections		
Calculator math section	55 minutes	30 MC + 8 grid-ins = 38 questions
No-calculator math section	25 minutes	15 MC + 5 grid-ins = 20 questions

Table 1.4 summarizes how the four math content areas are represented in each of the math sections.

Table 1.4 Number of Questions by Math Content Area

Math Content Area	Calculator Section	No-Calculator Section	Total
Heart of Algebra	11	8	19
Problem Solving and Data Analysis	17	—	17
Passport to Advanced Math	7	9	16
Additional Topics in Math	3	3	6
Total	38	20	58

How Are the SAT Math Scores Reported?

When you receive your SAT Math score, you will find that your raw math test score has been converted to a scaled score that ranges from 200 to 800, with 500 representing the average SAT Math score. In addition, three math test subscores will be reported for the following areas: (1) Heart of Algebra, (2) Problem Solving and Data Analysis, and (3) Advanced Math.

The Difficulty Levels of the Questions

As you work your way through each math section, questions of the same type (multiple-choice or grid-in) gradually become more difficult. Expect easier questions at the beginning of each section and harder questions at the end. You should, therefore, concentrate on getting as many of the earlier questions right as possible as each correct answer counts the same.

TIPS FOR BOOSTING YOUR SCORE

- If a question near the beginning of a math section seems very hard, then you are probably not approaching it in the best way. Reread the problem, and try solving it again, as problems near the beginning of a math section tend to have easier, more straightforward solutions.
- If a question near the end of a math section seems easy, beware—you may have fallen into a trap or misread the question.
- Read each question carefully, and make sure you understand what is being asked. Keep in mind that when creating the multiple-choice questions, the test makers tried to anticipate common student errors and included these among the answer choices.
- When you take the actual SAT, don't panic or become discouraged if you do not know how to solve a problem. Very few test takers are able to answer all of the questions in a section correctly.
- Since easy and hard questions count the same, don't spend a lot of time trying to answer a question near the end of a test section that seems very difficult. Instead, go back and try to answer the easier questions in the same test section that you may have skipped over.

OVERVIEW

Almost 80 percent of the SAT's math questions are standard multiple-choice questions with four possible answer choices labeled from (A) to (D). After figuring out the correct answer, you must fill in the corresponding circle on a machine-readable answer form. If you are choosing choice (B) as your answer for question 8, then on the separate answer form you would locate item number 8 for that test section and use your pencil to completely fill the circle that contains the letter B, as in

The Most Common Type of SAT Math Question

Forty-five of the 58 math questions that appear on the SAT are regular multiple-choice questions. Using a No. 2 pencil, you must fill in the circle on the answer form that contains the same letter as the correct choice. Since answer forms are machine scored, be sure to completely fill in the circle you choose as your answer. When filling in a circle, be careful not to go beyond its borders. If you need to erase, do so completely without leaving any stray pencil marks. Figure 1.1 shows the correct way to fill in an oval when the correct answer is choice (B).

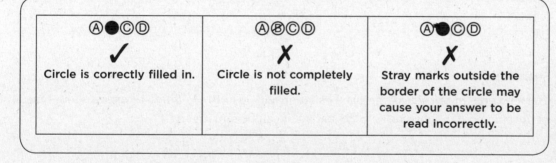

Figure 1.1 Correcting filling in the circle with your answer.

TIP

If you don't know the answer to a multiple-choice question, try to eliminate as many of the answer choices as you can. Then guess from the remaining choices. Since there is no penalty for a wrong answer, it is always to your advantage to guess rather than to omit an answer to a question.

➥ **Example :: No-Calculator Section :: Multiple-Choice**

If $3x - 2y = 13$ and $x + y = 1$, then $xy =$

(A) −6

(B) −3

(C) 3

(D) 6

Solution

Begin by solving the second equation for y.

- Since $x + y = 1$, $y = 1 - x$. Substitute $1 - x$ for y in the first equation gives $3x - 2(1 - x) = 13$, which simplifies to $5x - 2 = 13$ so $5x = 15$ and $x = \dfrac{15}{5} = 3$.

- In $x + y = 1$, replace x with 3, which gives $3 + y = 1$ so $y = -2$.

- Since $x = 3$ and $y = -2$, $xy = (3)(-2) = -6$.

Fill in circle **A** on the answer form: ●ⒷⒸⒹ

➥ **Example :: No-Calculator Section :: Multiple-Choice**

If $2 \cdot 4^x + 3 \cdot 4^x + 5 \cdot 4^x + 6 \cdot 4^x = 4^{12} + 4^{12} + 4^{12} + 4^{12}$, then $x =$

(A) 16

(B) 14

(C) 12

(D) 11

Solution

The terms on the left side of the given equation add up to $16 \cdot 4^x$. Since 4^{12} appears four times in the sum on the right side of the equation, it can be replaced by $4 \cdot 4^{12}$:

$$16 \cdot 4^x = 4 \cdot 4^{12}$$

$$4^x = \frac{\overset{1}{\cancel{4}} \cdot 4^{12}}{\underset{4}{\cancel{16}}}$$

$$= \frac{4^{12}}{4^1}$$

$$4^x = 4^{11} \quad \text{so} \quad x = 11$$

Fill in circle **D** on the answer form: ⒶⒷⒸ●

➥ Example :: No-Calculator Section :: Multiple-Choice

If k is a positive constant, which of the following could represent the graph of $k(y + 1) = x - k$?

(A)

(C)

(B)

(D)

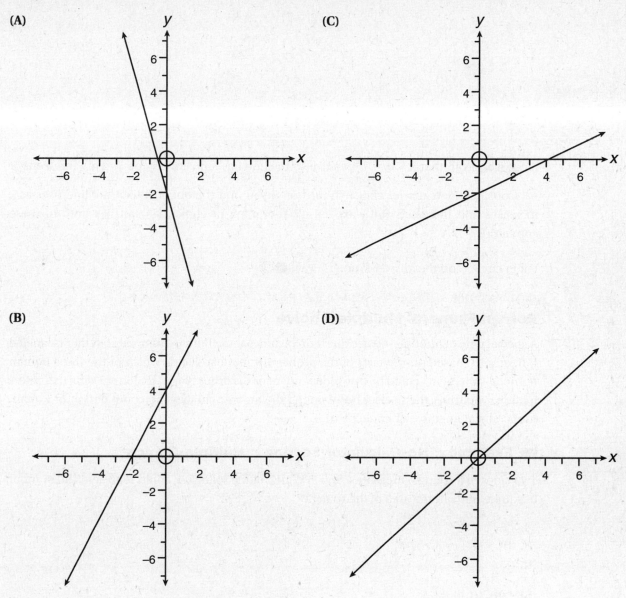

Solution

Write the given equation in $y = mx + b$ slope-intercept form where m, the coefficient of x, is the slope of the line and b is the y-intercept:

$$k(y+1) = x - k$$
$$ky + k = x - k$$
$$ky = x - 2k$$
$$\frac{\cancel{k}y}{\cancel{k}} = \frac{x}{k} - \frac{2\cancel{k}}{\cancel{k}}$$
$$y = \frac{1}{k}x - 2$$

Since it is given that $k > 0$, $\dfrac{1}{k}$ is positive so the line has a positive slope and a y-intercept of -2. Consider each answer choice in turn until you find the one in which the line rises as x increases and intersects the y-axis at -2. The graph in choice (C) satisfies both of these conditions.

Fill in circle **C** on the answer form: Ⓐ Ⓑ ● Ⓓ

Roman Numeral Multiple-Choice

A special type of multiple-choice question includes three Roman numeral statements labeled I, II, and III. Based on the facts of the problem, you must decide which of the three Roman numeral statements could be true independent of the other two statements. Using that information, you must then select from among the answer choices the combination of Roman numeral statements that could be true.

➥ Example :: No-Calculator Section :: Multiple-Choice

Two sides of a triangle measure 4 and 9. Which of the following could represent the number of square units in the area of the triangle?

  I.   6
 II. 18
III. 20

(A) I only
(B) II only
(C) I and II only
(D) I, II, and III

Solution

- The maximum area of the triangle occurs when the two given sides form a right angle:

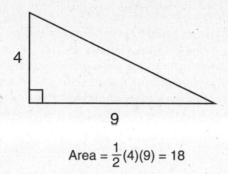

$$\text{Area} = \frac{1}{2}(4)(9) = 18$$

Since 18 square units is the *maximum* area of the triangle, the area of the triangle can be any positive number 18 or less.

- Determine whether each of the Roman numeral statements are True (T) or False (F). Then select the answer choice that contains the correct combination of statements:

I. 6 T	(A) I only	✗
II. 18 T	(B) II only	✗
III. 20 F	(C) I and II only	✓
	(D) I, II, and III	✗

- Since Roman numeral statements I and II are true while statement III is false, only choice (C) gives the correct combination of true statements.

The correct choice is **(C)**.

TIPS FOR BOOSTING YOUR SCORE

1. **Don't keep moving back and forth from the question page to the answer sheet.** Instead, record the answer next to each question. After you accumulate a few answers, transfer them to the answer sheet at the same time. This strategy will save you time.

2. After you record a group of answers, check to make sure that you didn't accidentally skip a line and enter the answer to question 3, for instance, in the space for question 4. This will save you from a possible disaster!

3. On the answer sheet, be sure to fill one oval for each question you answer. If you need to change an answer, erase it completely. If the machine that scans your answer sheet "reads" what looks like two marks for the same question, the question will not be scored.

4. Do not try to do all your reasoning and calculations in your head. Freely use the blank areas of the test booklet as a scratch pad.

5. Write a question mark (**?**) to the left of a question that you skip over. If the problem seems much too difficult or time consuming for you to solve, write a cross mark (**X**) instead of a question mark. This will allow you to set priorities for the questions that you need to come back to and retry, if time permits.

OVERVIEW

Although most of the SAT Math questions are multiple-choice, student produced response questions, also called **grid-ins**, account for the about 20 percent of the math questions. Instead of selecting your answer from a list of four possible answer choices, you must come up with your own answer and then enter it in a four-column grid provided on a separate answer sheet. By learning the rules for gridding-in answers before you take the test, you will boost your confidence and save valuable time when you take the SAT.

Know How to Grid-In an Answer

When you figure out your answer to a grid-in question, you will need to record it on a four-column answer grid like the one shown in Figure 1.2. The answer grid can accommodate whole numbers from 0 to 9999, as well as fractions and decimals. If the answer is 0, grid it in column 2 since zero is not included in the first column. Be sure you check the accuracy of your gridding by making certain that no more than one circle in any column is filled in. If you need to erase, do so completely. Otherwise, an incomplete or sloppy erasure may be incorrectly interpreted by the scoring machine as your actual answer.

TIP

- The answer grid does not contain a negative sign so your answer can never be a negative number or include special symbols such as a dollar sign ($) or a percent symbol (%).
- Unless a problem states otherwise, answers can be entered in the grid as a decimal or as a fraction. All mixed numbers must be changed to an improper fraction or an equivalent decimal before they can be entered in the grid. For example, if your answer is $3\frac{1}{2}$, grid-in 7/2 or 3.5.
- Always fill the grid with the most accurate value of an answer that the grid can accommodate. If the answer is $\frac{2}{3}$, grid-in 2/3, .666, or .667, but *not* 0.66.

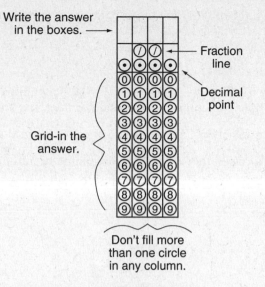

Write the answer in the boxes. →

Fraction line

Decimal point

Grid-in the answer.

Don't fill more than one circle in any column.

Figure 1.2 An Answer Grid

To grid-in an answer:

- Write the answer in the top row of the column boxes of the grid. A decimal point or fraction bar (slash) requires a separate column. Although writing the answer in the column boxes is not required, it will help guide you when you grid the answer in the matching circles below the column boxes.
- Fill the circles that match the answer you wrote in the top row of column boxes. Make sure that no more than one circle is filled in any column. Columns that are not needed should be left blank. If you forget to fill in the circles, the answer that appears in the column boxes will *NOT* be scored.

Here are some examples:

Answer: 0.237   Answer: 23.7   Answer: $\frac{23}{7}$

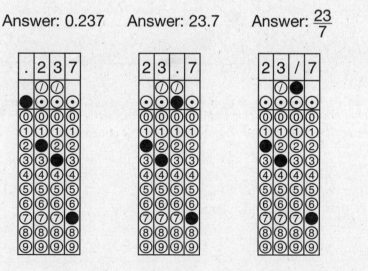

- If the answer is a fraction that fits the grid, don't try to reduce it, as may lead to a careless error. If the answer is $\frac{52}{4}$, don't try to reduce it, because it can be gridded in as:

- If the answer is a fraction that needs more than four columns, reduce the fraction, if possible, or enter the decimal form of the fraction. For example, since the fractional answer $\frac{17}{25}$ does not fit the grid and cannot be reduced, use a calculator to divide the denominator into the numerator. Then enter the decimal value that results. Since $17 \div 25 = 0.68$, grid-in .68 for $\frac{17}{25}$.

Answer: $\frac{17}{25}$

➡ Example :: No-Calculator Section :: Grid-In

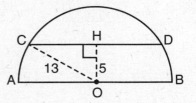

In semicircle O above, chord $\overline{CD}$ is parallel to diameter $\overline{AB}$, $AB = 26$, and the distance of $\overline{CD}$ from the center of the circle is 5. What is the length of chord $\overline{CD}$?

Solution

- From O, draw $\overline{OH}$ perpendicular to $\overline{CD}$ so $OH = 5$:

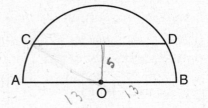

- Draw radius $\overline{OC}$. Since the diameter of the circle is 26, $OC = \dfrac{1}{2} \times 26 = 13$. The lengths of the sides of right triangle OHC form a 5–*12*–13 Pythagorean triple with $CH = 12$.
- A line through the center of a circle and perpendicular to a chord bisects the chord. Hence, $DH = 12$ so $CD = 12 + 12 = 24$.

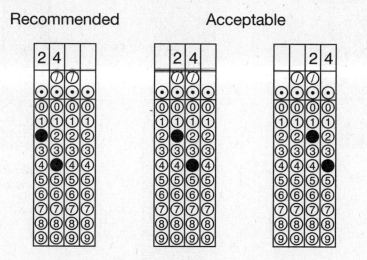

Recommended   Acceptable

Start All Answers, Except 0, in the First Column

If you get into the habit of always starting answers in the first column of the answer grid, you won't waste time thinking about where a particular answer should begin. Note, however, that the first column of the answer grid does not contain 0. Therefore a zero answer can be entered in any column after the first. If your answer is a decimal number less than 1, don't bother writing the answer with a 0 in front of the decimal point. For example, if your answer is 0.126, grid .126 in the four column boxes on the answer grid.

Enter Mixed Numbers as Fractions or as Decimals

The answer grid cannot accept a mixed number such as $1\dfrac{3}{5}$. You *must* change a mixed number into an improper fraction or a decimal before you grid it in. For example, to enter a value of $1\dfrac{3}{5}$, grid-in either 8/5 or 1.6.

Answer: $\frac{8}{5}$ Answer: 1.6   Answer: $1\frac{3}{5}$

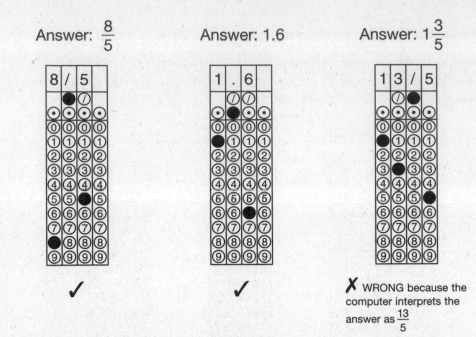

✓   ✓   ✗ WRONG because the computer interprets the answer as $\frac{13}{5}$

TIP

Time Saver

If your answer fits the grid, don't change its form. If you get a fraction as an answer and it fits the grid, then do not waste time and risk making a careless error by trying to reduce it or change it into a decimal number.

Entering Long Decimal Answers

If your answer is a decimal number with more digits than can fit in the grid, it may either be rounded or the extra digits deleted (truncated), provided the decimal number that you enter as your final answer fills the entire grid. Here are some examples:

■ If you get the repeating decimal 0.6666 … as your answer, you may enter it in any of the following correct forms:

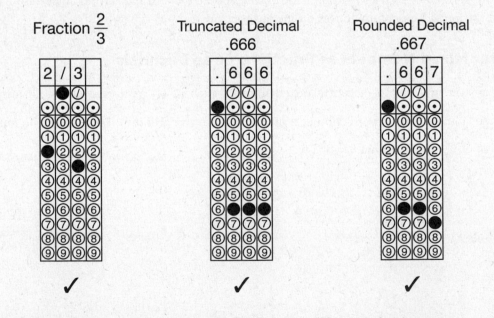

Fraction $\frac{2}{3}$ Truncated Decimal .666   Rounded Decimal .667

✓   ✓   ✓

Entering less accurate answers such as .66 or .67 will be scored as incorrect.

- If your answer is $\frac{25}{19}$, then you must convert the mixed number to a decimal since it does not fit the grid. Using a calculator, $\frac{25}{19} = 1.315789474$. The answer can be entered in either truncated or rounded form. To truncate 1.315789474, simply delete the extra digits so that the final answer fills the entire four-column grid:

1.315789474 = 1.31

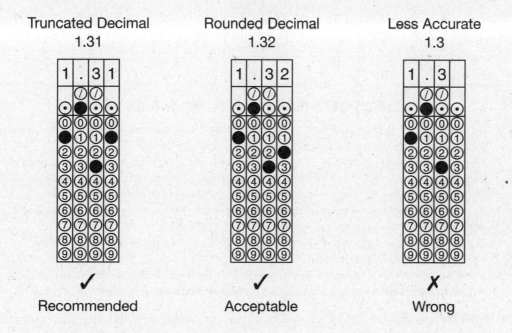

Truncated Decimal 1.31	Rounded Decimal 1.32	Less Accurate 1.3
✓	✓	✗
Recommended	Acceptable	Wrong

Grid-In Questions May Have More than One Correct Answer

When a grid-in question has more than one correct answer, enter only one.

➡ Example :: No-Calculator Section :: Grid-In

If $4 < |2 - x| < 5$ and $x < 0$, what is one possible value of $|x|$?

Solution

- The expression $4 < |2 - x| < 5$ states that the value of $|2 - x|$ is between 4 and 5. Pick any such number, say 4.5, and then solve $|2 - x| = 4.5$ for x.
- By inspection, the solutions of $|2 - x| = 4.5$ are $x = 6.5$ or $x = -2.5$.
- Since it is given that $x < 0$, use $x = -2.5$ so $|x| = |-2.5| = 2.5$.

Grid-in **2.5**

TIP

To avoid possible rounding errors, truncate rather than round off long decimal answers. Make sure your final answer fills the entire grid.

TIPS FOR BOOSTING YOUR SCORE

1. An answer to a grid-in question can never contain a negative sign, be greater than 9999, or include a special symbol, as these cannot be entered in the grid.

2. Write your answer at the top of the grid to help reduce gridding errors. Because handwritten answers at the top of the grid are not scored, make sure you then enter your answer by completely filling in the ovals in the four-column grid.

3. Review the accuracy of your gridding. Make certain that a decimal point or slash is entered in its own column.

4. Do not grid zeros before the decimal point. Grid-in .55 rather than 0.55. Begin nonzero answers in the leftmost column of the grid.

5. Enter an answer in its original form, fraction or decimal, provided it fits the grid. Mixed numbers must be changed to a fraction or decimal. For example, if your answer is $1\frac{1}{8}$, grid-in $\frac{9}{8}$.

6. If a fraction does not fit the grid, enter it as a decimal. For example, if your answer is $\frac{23}{10}$, grid in 2.3. Truncate long decimal answers that do not fit the grid while making certain that your final answer completely fills the grid.

SAT Math Strategies

2

One of the most powerful general problem-solving strategies in mathematics is the ability to *reason analogously*—to recognize that an unfamiliar problem you are confronted with can be likened to a more familiar yet related type of problem that you already know how to solve. As you work through the problems in this chapter, as well as the chapters that follow, pay close attention to the solution approaches and strategies used as well as to the mechanical details of the solution. Whenever possible, try to make connections between problems that may look different at first glance but that have similar solution strategies. By so doing, you will be developing your ability to reason analogously.

This chapter, in particular, offers 20 key mathematics strategies that will help you maximize your SAT Math score. The problems used to illustrate these strategies will not only expose you to types of problems that you may not have encountered in your regular mathematics classes but will also serve to introduce some of the new mathematics topics that you are expected to know beginning with the 2016 SAT.

LESSONS IN THIS CHAPTER

OVERVIEW

This lesson features strategies that will help you answer mathematics test questions that may seem unfamiliar or too complicated at first glance. Strategies that begin "Know How To . . ." not only will show you how to efficiently solve certain types of SAT problems, but will also help to further guide you in your preparation for the mathematics part of the SAT.

STRATEGY 1: HAVE A STUDY PLAN

To maximize your score on the SAT, you need to have an organized and realistic study plan that takes into account your individual learning style and study habits. Here are some suggestions:

START EARLY

Rather than cram shortly before you take the test, it is better to start practicing a little each day beginning at least three months before test day and, as the test day gets closer, to gradually increase the amount of study time each week. Set up and keep to a regular study schedule instead of trying to fit in your study sessions whenever you feel time permits.

KNOW YOUR WEAKNESSES

As you work through this book from beginning to end, you will become aware of those areas in which you may need to spend additional study time. Consider this an opportunity to correct any weaknesses. If you get a practice problem wrong, don't ignore it. Instead, review it until you understand it. Mark any troublesome problems so you know which problems you need to revisit. As part of your next study session, redo those problems you marked. Repeat this process as needed. This will help ensure that you remember and internalize the necessary concepts and skills.

WATCH THE CLOCK

As you become more comfortable solving SAT-type math problems, work on increasing your speed as well as your accuracy.

PRACTICE MAKES PERFECT

The more problems you try and work out correctly, the larger the arsenal of math tools and problem-solving skills you will have when you take the actual test. By working conscientiously and systematically through this chapter and then the remainder of this book, you will gain the knowledge, skills, and confidence that will help you approach and solve actual SAT problems that might otherwise seem too difficult.

DON'T SKIP STUDY SESSIONS

During the months before the test, make your SAT preparation both a priority and a routine practice—keep to your schedule. A good study plan has value only if you stick to it.

TAKE PRACTICE TESTS

During the last 10 days of your SAT preparation, take at least two SAT practice tests under timed test conditions. This will help you pace yourself when you take the actual test. Whether taking a practice test or the real test, don't waste time by getting stuck on a hard question. Remember that easy questions and hard questions are worth the same. As a last resort, guess after eliminating as many choices as possible, and then move on. Unless you are aiming for a perfect or near perfect score, you should not become discouraged if you cannot answer every question correctly. When taking a practice test, enter your answer to each of the grid-in questions on the special four-column answer grid. This will help you remember the rules for gridding in answers.

BE PREPARED TO SOLVE UNFAMILIAR WORD PROBLEMS

Don't be discouraged if you encounter unfamiliar types of word problems either in your practice sessions or when you take the actual SAT. Most of the word problems you will encounter on the SAT do not have predictable, formula type solutions. Instead, they will require some creative analysis. Start by reading the word problem carefully so that you have a clear understanding of what is given and what you are being asked to find. Underline key phrases and circle numerical values. Pay attention to units of measurement. Write down the key relationships and what the variables represent. Draw a diagram to help organize the information or to help visualize the conditions of the problem. If a diagram is provided with the problem, mark it up with everything you know about it. When appropriate, translate the conditions of the problem into an algebraic equation. If you get stuck, get into the habit of trying one of these strategies or a combination of them:

- Reason analogously—does the problem remind you of a related but simpler problem you already know how to solve? If so, decide how you can use the same or a similar solution approach.
- Plug easy numbers into the problem and then work through the problem using these numbers. This may help you discover an underlying relationship or detect a pattern that will help you solve the problem.
- Work backwards. If you know what the final result must be and need to find the beginning value (or better understand that process that led to that end value), try reversing or undoing the steps or process that would have been used to get the end result.
- Break the problem down into smaller parts. Some problems require that you first calculate a value and then use that value to find the final unknown value. The "working backwards" strategy may be helpful in identifying what intermediate calculations need to be performed.

STRATEGY 2: LOOK AT A SPECIFIC CASE

If a problem does not give specific numbers or dimensions, make up a simple example using easy numbers.

➥ Example :: No-Calculator Section :: Multiple-Choice

The perimeter of a rectangle is 10 times as great as its width. The length of the rectangle is how many times as great as the width of the rectangle?

(A) One-half
(B) Two
(C) Three
(D) Four

Solution

- Consider a specific rectangle whose width is 1. If the perimeter of the rectangle is 10 times as great as the width, the perimeter of the rectangle is 10.

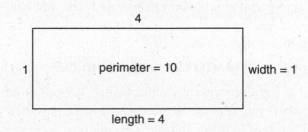

- Because $(2 \times length) + 2 \times 1 = 10$, $2 \times length = 8$, so the length of the rectangle is 4.
- Since length = 4 and width = 1, the length is 4 times as great as the width.

The correct choice is **(D)**.

If you need to figure out by what fractional amount or percent a quantity changes when its starting value is not given, consider a specific case by picking any starting value that makes the arithmetic easy. Do the calculations using this value. Then compare the final answer with the starting value you chose.

➥ Example :: No-Calculator Section :: Multiple-Choice

The current value of a stock is 20% less than its value when it was purchased. By what percent must the current value of the stock rise in order for the stock to have its original value?

(A) 20%
(B) 25%
(C) 30%
(D) 40%

Solution

Choose a convenient starting value of the stock. When working with percents, 100 is usually a good starting value.

- Assume the original value of the stock was $100.
- Find the current value of the stock. Since the current value is 20% less than the original value, the current value is

$$\$100 - 0.20(\$100) = \$100 - \$20 = \$80$$

- Find the amount by which the current value of the stock must increase in order to regain the original value. The value must rise from \$80 to \$100, which is a change of \$20.
- To find the *percent* by which the current value of the stock must rise in order for the stock to have its original value, find what percent of \$80 is \$20. Since

$$\frac{\$20}{\$80} \times 100\% = \frac{1}{4} \times 100\% = 25\%,$$

the current value must increase 25% in order for the stock to regain its original value.

The correct choice is **(B)**.

➦ Example :: No-Calculator Section :: Multiple-Choice

Fred gives $\frac{1}{3}$ of his DVDs to Andy and then gives $\frac{3}{4}$ of the remaining DVDs to Jerry. Fred now has what fraction of the original number of DVDs?

(A) $\frac{1}{12}$

(B) $\frac{1}{6}$

(C) $\frac{1}{4}$

(D) $\frac{1}{3}$

Solution

Since Fred gives $\frac{1}{3}$ and then $\frac{3}{4}$ of his DVDs away, pick any number that is divisible by both 3 and 4 for the original number of DVDs. Since 12 is the lowest common multiple of 3 and 4, assume that Fred starts with 12 DVDs.

- After Fred gives $4\left(=\frac{1}{3}\times 12\right)$ discs to Andy, he is left with 8 (= 12 − 4) DVDs.
- Since $\frac{3}{4}$ of 8 is $\frac{3}{4}\times 8$ or 6, Fred gives 6 of the remaining DVDs to Jerry. This leaves Fred with 2 (= 8 − 6) of the original 12 DVDs.
- Since $\frac{2}{12}=\frac{1}{6}$, Fred now has $\frac{1}{6}$ of the original number of DVDs.

The correct choice is **(B)**.

STRATEGY 3: DRAW AND MARK UP DIAGRAMS

Feel free to mark up the diagrams in the test booklet. Drawing and labeling a diagram with any information that is given can help organize the facts of a problem while allowing you to see relationships that can guide you in solving the problem. Sometimes you may need to solve a problem by drawing your own lines on the diagram.

The diameter of the base of a right circular cylinder is 6 and the distance from the center of a base to a point on the circumference of the other base is 8. What is the height of the cylinder?

(A) 2

(B) 10

(C) $\sqrt{28}$

(D) $\sqrt{55}$

Solution

Add the altitude to the diagram and draw a segment that represents the distance from the center of the upper base to the circumference of the lower base:

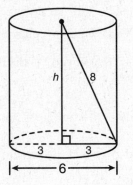

- Since the diameter of the base is 6, the radius is 3.
- The height h of the cylinder is a leg of a right triangle in which the other leg is 3 and the hypotenuse is 8.
- Use the Pythagorean theorem to find h:

$$h^2 + 3^2 = 8^2$$
$$h^2 + 9 = 64$$
$$h^2 = 64 - 9$$
$$h = \sqrt{55}$$

The correct choice is (**D**).

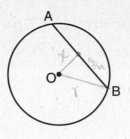

In the figure above, circle O has radius r and $AB = \frac{3}{2}r$. What is the distance in terms of r of $\overline{AB}$ from the center of the circle?

(A) $\dfrac{3}{8}r$

(B) $\dfrac{\sqrt{7}}{4}r$

(C) $\dfrac{\sqrt{5}}{4}r$

(D) $\dfrac{\sqrt{7}r}{8}$

Solution

- To represent the distance of $\overline{AB}$ from the center of the circle, draw the length of a segment from O and perpendicular to $\overline{AB}$. Also draw radius $\overline{OB}$ to form a right triangle:

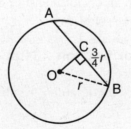

- Since a line through the center of a circle and perpendicular to a chord bisects the chord, $BC = \dfrac{3}{4}r$. Mark off the diagram.

- Use the Pythagorean theorem to find OC:

$$(OC)^2 + \left(\frac{3}{4}r\right)^2 = r^2$$

$$(OC)^2 = r^2 - \frac{9}{16}r^2$$

$$= \frac{7}{16}r^2$$

$$OC = \sqrt{\frac{7}{16}r^2} = \frac{\sqrt{7}}{4}r$$

The correct choice is **(B)**.

STRATEGY 4: PLUG IN NUMBERS TO FIND A PATTERN

➥ Example :: No-Calculator Section :: Grid-In

When a positive integer k is divided by 5, the remainder is 3. What is the remainder when $3k$ is divided by 5?

Solution

List a few positive integers that, when divided by 5, give 3 as a remainder. Any positive integer that is the sum of 5 and 3 or of a multiple of 5 (i.e., 10, 15, etc.) and 3 will have this property. For example, when 8, 13, and 18 are each divided by 5, the remainder is 3:

k	$k \div 5$
8	$8 \div 5 = 1$ remainder 3
13	$13 \div 5 = 2$ remainder 3
18	$18 \div 5 = 3$ remainder 3

Now, using the same values for k, divide $3k$ by 5 and find the remainders.

k	$3k$	$3k \div 5$
8	24	$24 \div 5 = 4$ remainder 4
13	39	$39 \div 5 = 7$ remainder 4
18	54	$54 \div 5 = 10$ remainder 4

The correct answer is **4**.

STRATEGY 5: MAKE AN ORGANIZED LIST OR TABLE

Arranging the facts of a problem in an organized list or table can help you see a pattern or make it easier for you to organize the solution.

➥ Example :: No-Calculator Section :: Multiple-Choice

If p and q are integers such that $6 < q < 17$ and $\dfrac{p}{q} = \dfrac{3}{4}$, how many possible values are there for p?

(A) Two
(B) Three
(C) Four
(D) Five

Solution

If $\dfrac{p}{q} = \dfrac{3}{4}$, then $p = \dfrac{3q}{4}$. Since p is an integer, q must be divisible by 4. Make a list of all integers between 6 and 17:

q	Divisible by 4?
7	No
8	Yes
9	No
10	No
11	No
12	Yes
13	No
14	No
15	No
16	Yes

Since there are three values of q divisible by 4, there are three possible values for p.

The correct choice is **(B)**.

➡ Example :: No-Calculator Section :: Multiple-Choice

Which expression is equivalent to $i + i^{99}$ where $i = \sqrt{-1}$?

(A) $i - 1$

(B) $i + 1$

(C) 0

(D) $2i$

Solution

To simplify i^{99}, list powers of i until you see a pattern:

$$i^0 = 1 \qquad\qquad i^4 = i^2 \cdot i^2 = (-1)(-1) = 1$$
$$i^1 = i \qquad\qquad i^5 = i^4 \cdot i = 1 \cdot i = i$$
$$i^2 = -1 \qquad\qquad i^6 = i^4 \cdot i^2 = 1 \cdot (-1) = -1$$
$$i^3 = i^2 \cdot i = -i \qquad\qquad i^7 = i^4 \cdot i^3 = 1 \cdot (-i) = -i$$

Beginning with $i^0 = 1$, consecutive integer powers of i follow a cyclic pattern of evaluating to 1, i, -1, and $-i$. The pattern repeats when the power of i is a multiple of 4. This suggests that large powers of i can be simplified by dividing the exponent by 4 and using the remainder as the new power of i. To simplify i^{99}, divide 99 by 4, which gives 24 and a remainder of **3** so $i^{99} = i^3 = -i$.

Here is another way of simplifying i^{99}:

- Break down i^{99} as the product of two powers of i such that one of these is the greatest even power of i:

$$i^{99} = i^{98} \times i$$

- Rewrite the even power of i as a power of i^2:

$$i^{99} = (i^2)^{49} \times i$$

- Replace i^2 with -1 and simplify:

$$
\begin{aligned}
i^{99} &= (-1)^{49} \times i \leftarrow -1 \text{ raised to an odd power is } -1 \\
&= -1 \quad\; \times i \\
&= -i
\end{aligned}
$$

Hence, $i + i^{99} = i + (-i) = 0$.

The correct choice is (C).

➥ Example :: Calculator Section :: Grid-In

The first two terms of an ordered sequence of positive integers are 1 and 3. If each number of the sequence after 3 is obtained by adding the two numbers immediately preceding it, how many of the first 1,000 numbers in this sequence are even?

Solution

Write the first several terms of the sequence until you see a pattern:

$$1, 3, 4, 7, 11, 18, 29, 47, 76, \ldots$$

In the first three terms, there is one even number; in the first six terms, there are two even numbers; in the first nine terms of the sequence, there are 3 even numbers. Thus, in the first 999 terms of the sequence, there are $\dfrac{999}{3} = 333$ even numbers. Since the 999th term is even, the 1,000th term is odd. Hence, there are 333 even numbers in the first 1,000 terms of the sequence.

Grid-in **333**

STRATEGY 6: KNOW HOW TO USE PROPORTIONAL REASONING

When the ratio between variable quantities stays the same from one situation to another, set up a proportion to find the value of an unknown quantity that belongs to the proportion.

➥ Example :: Calculator Section :: Grid-In

If 8 ounces of a sports drink contain 110 milligrams of sodium, how many milligrams of sodium are contained in 20 ounces of the same sports drink?

Solution

$$\frac{\text{ounces}}{\text{milligrams}} = \frac{8}{110} = \frac{20}{x}$$

$$8x = 2{,}200$$

$$\frac{8x}{8} = \frac{2{,}200}{8}$$

$$x = 275$$

Grid-in **275**

➡ Example :: Calculator Section :: Grid-In

The trip odometer of an automobile improperly displays only 3 miles for every 4 miles actually driven. If the trip odometer shows 42 miles, how many miles has the automobile actually been driven?

Solution

Form a proportion in which each side represents the rate at which odometer miles translate into actual miles driven. If x represents the number of actual miles driven when the odometer shows 42 miles, then

$$\frac{3\,\text{odometer miles}}{4\,\text{actual miles}} = \frac{42\,\text{odometer miles}}{x\,\text{actual miles}}.$$

Since odometer miles are on top in both fractions and actual miles are on bottom in both fractions, the terms of the proportions have been placed correctly. To solve for x, cross-multiply: $3x = 4 \cdot 42$ so $x = \dfrac{168}{3} = 56$ miles.

Grid-in **56**

➡ Example :: Calculator Section :: Grid-In

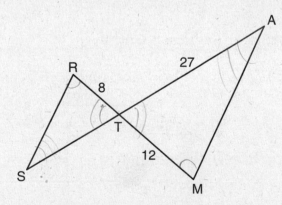

Note: Figure not drawn to scale.

In the figure above, $\overline{RS} \,\|\, \overline{AM}$ and segment SA intersects segment MR at T. What is the length of AS?

Solution

Use the properties of similar triangles to first find the length of segment ST:

- Because alternate interior angles formed by parallel lines have the same measures, triangles RST and MAT have two pairs of congruent angles so that they are similar:

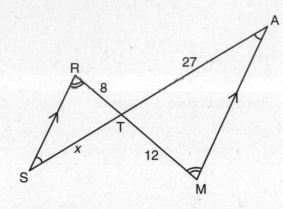

- The lengths of corresponding sides of similar triangles are in proportion:

$$\frac{RT}{MT} = \frac{ST}{AT} \quad \begin{cases} \text{Corresponding sides of similar} \\ \text{triangles face corresponding} \\ \text{congruent angles} \end{cases}$$

$$\frac{8}{12} = \frac{x}{27}$$

$$\frac{2}{3} = \frac{x}{27}$$

$$3x = 54$$

$$x = \frac{54}{3} = 18$$

- Since $AS = AT + ST = 27 + 18 = 45$.

Grid-in **45**

➡ Example :: Calculator Section :: Grid-In

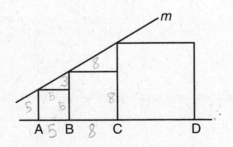

Note: Figure not drawn to scale

Line m intersects the corner points of three adjacent squares as shown in the above figure. If $AB = 5$ and $BC = 8$, what is the length of $\overline{CD}$?

Solution

Label the diagram with the lengths of the sides of the squares where x represents the length of the vertical leg of the second right triangle:

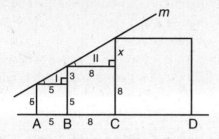

The right triangles formed at the top of the squares are similar. Since the lengths of pairs of corresponding sides of similar triangles have the same ratio:

$$\frac{3}{x} = \frac{5}{8}$$

$$5x = 24$$

$$\frac{5x}{5} = \frac{24}{5}$$

$$x = 4.8$$

Since $x + 8 = 4.8 + 8 = 12.8$, $CD = 12.8$.

Grid-in **12.8**

STRATEGY 7: TEST NUMERICAL ANSWER CHOICES IN THE QUESTION

When each of the answer choices for a multiple-choice question is an easy number, you may be able to find the correct answer by plugging each of the answer choices back into the question until you find the one that works.

➡ Example :: No-Calculator Section :: Multiple-Choice

$$y \le 3x + 1$$
$$x - y > 1$$

Which ordered pair is in the set of all ordered pairs that satisfy the above system of inequalities?

(A) $(-1, -2)$

(B) $(2, -1)$

(C) $(1, 2)$

(D) $(-1, 2)$

Solution

Test each ordered pair in both inequalities until you find the pair that makes both inequalities true:

- (A) $(-1, -2)$: If $x - y > 1$, then $-1 - (-2) > 1$ and $-1 + 2 > 1$ so $1 > 1$, which is false. ✗
- (B) $(2, -1)$: If $y \leq 3x + 1$, then $-1 \leq 3(2) + 1$ so $-1 \leq 7$, which is true. Also, if $x - y > 1$, then $2 - (-1) > 1$ so $2 + 1 > 1$, which is true. ✓
- (C) $(1, 2)$: If $x - y > 1$, then $1 - 2 > 1$ so $-1 > 1$, which is false. ✗
- (D) $(-1, 2)$: If $x - y > 1$, then $-1 - 2 > 1$ so $-3 > 1$, which is false. ✗

The correct choice is **(B)**.

➥ Example :: No-Calculator Section :: Multiple-Choice

When 5 is divided by a number, the result is 3 more than 7 divided by twice the number. What is the number?

(A) $\dfrac{1}{4}$

(B) $\dfrac{1}{2}$

(C) 1

(D) $\dfrac{3}{2}$

Solution

Rather than thinking of an equation and then solving it algebraically, plug in each of the answer choices as the possible correct number until you find the one that works in the problem.

Choice (A): Try $\dfrac{1}{4}$ as the unknown number:

$$\overset{20}{\overbrace{5 \div \dfrac{1}{4}}} \neq \overset{3+14}{\overbrace{3 + 7 \div \dfrac{2}{4}}} \quad ✗$$

Choice (B): Try $\dfrac{1}{2}$ as the unknown number:

$$\overset{10}{\overbrace{5 \div \dfrac{1}{2}}} = \overset{3+7}{\overbrace{3 + 7 \div \dfrac{2}{2}}} \quad ✓$$

The correct choice is **(B)**.

STRATEGY 8: CHANGE VARIABLE ANSWER CHOICES INTO NUMBERS

If you don't know how to find the answer to a regular multiple-choice question in which the answer choices are algebraic expressions, transform each of the answer choices into a number by following these steps:

STEP 1 Work out the problem by choosing numbers that make the arithmetic easy. Circle your answer.

STEP 2 Plug the same numbers into each of the answer choices. Each answer choice should evaluate to a different number. If not, start over and choose different test values.

STEP 3 Compare your circled answer from step 1 with the number obtained for each of the answer choices in step 2. If exactly one answer choice agrees with the answer you calculated in step 1, then that choice is the correct answer. If more than one answer choice agrees, start over choosing different numbers in step 1.

TIP

Don't substitute 0 or 1 as these numbers often lead to more than one choice evaluating to the same number.

➡ Example :: No-Calculator Section :: Multiple-Choice

A yoga studio charges a one-time registration fee of $75 plus a monthly membership fee of $45. If the monthly fee is subject to a sales tax of 6%, which of the following expressions represents the total cost of membership for n months?

(A) $(45 \times 1.06 + 75)n$

(B) $45(1.06)n + 75$

(C) $(75 + 45)(0.06) + 45n$

(D) $(45 + 0.06)n + 75$

Solution

METHOD 1 Change the variable answer choices into numbers.

- The monthly fee is $45 + .06 \times $45 = $47.70.
- If $n = 10$, then the total of the monthly charges is $47.70 \times 10 = $477. Since the registration fee is $75, the total cost of membership is $477 + $75 = $\$552$.
- Evaluate each of the expressions in the four answer choices using $n = 10$. You should verify that only choice (B) gives the correct answer:

$$75 + 45(1.06)n = 75 + 45(1.06)10$$
$$= 75 + (47.70)10$$
$$= 75 + 477$$
$$= \boxed{552}$$

METHOD 2 Translate the conditions of the problem into an algebraic expression.

- The fee for one month is $45 + 0.06 \times 45 = 45(1 + 0.06) = 45(1.06)$.
- Hence, the fee for n months is $45(1.06)n$.
- To get the total cost of membership, add the registration fee:

$$75 + 45(1.06)n$$

The correct choice is **(B)**.

➥ Example :: No-Calculator Section :: Multiple-Choice

The expression $\dfrac{7y-3}{y+2}$ is equivalent to which of the following?

(A) $7 - \dfrac{3}{2y}$

(B) $\dfrac{7-3}{3y}$

(C) $7 - \dfrac{17}{y+2}$

(D) $7 - \dfrac{3}{y+2}$

Solution

Change the given expression and each of the answer choices into numbers by picking a number for y that is easy to work with.

- If $y = 2$,

$$\frac{7y-3}{y+2} = \frac{7(2)-3}{2+2} = \left(\frac{11}{4}\right)$$

- Evaluate each of the answer choices for $y = 2$. You should verify that only choice (C) produces $\dfrac{11}{4}$:

$$7 - \frac{17}{y+2} = 7 - \frac{17}{2+2}$$
$$= \frac{28}{4} - \frac{17}{4}$$
$$= \left(\frac{11}{4}\right)$$

The correct choice is (**C**).

➥ Example :: No-Calculator Section :: Multiple-Choice

Every 8 days a mass of a certain radioactive substance decreases to exactly one-half of its value at the beginning of the 8-day period. If the initial amount of the radioactive substance is 75 grams, which equation gives the number of grams in the mass, M, that remains after d days?

(A) $M = 75\left(\dfrac{d}{16}\right)$

(B) $M = 75\left(\dfrac{8}{d}\right)^2$

(C) $M = 75\left(\dfrac{1}{2}\right)^{8d}$

(D) $M = 75\left(\dfrac{1}{2}\right)^{\frac{d}{8}}$

Solution

- First figure out what the answer must be for 16 days. After the first 8 days, $\frac{1}{2}$ of 75 grams of the substance remains so that after 16 days $\frac{1}{2}\left(\frac{1}{2} \cdot 75\right)$ or $\left(\frac{75}{4}\right)$ grams of the substance are left.

- Evaluate each of the answer choices for $d = 16$. You should verify that only answer choice (D) yields the correct result:

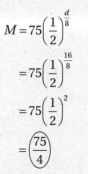

$$M = 75\left(\frac{1}{2}\right)^{\frac{d}{8}}$$

$$= 75\left(\frac{1}{2}\right)^{\frac{16}{8}}$$

$$= 75\left(\frac{1}{2}\right)^{2}$$

$$= \left(\frac{75}{4}\right)$$

The correct choice is **(D)**.

➡ Example :: Calculator Section :: Multiple-Choice

If t ties cost d dollars, how many dollars would $t + 1$ ties cost?

(A) $d + 1$

(B) $\dfrac{dt}{t+1}$

(C) $\dfrac{d+t}{t+1}$

(D) $\dfrac{d(t+1)}{t}$

Solution

Pick easy numbers for t and d, such as $t = 2$ and $d = 10$. If two ties cost \$10, then one tie costs \$5 and three ties $(t + 1)$ cost \$15. Substitute 2 for t and 10 for d in each of the answer choices until you find the one that evaluates to 15.

The correct choice is **(D)**, since

$$\frac{d(t+1)}{t} = \frac{10(2+1)}{2} = \frac{30}{2} = 15$$

> **TIP**
>
> Avoid picking 0 and 1, as these numbers tend to produce more than one "correct" answer choice. If more than one answer choice gives the same correct answer, then start over with different numbers.

STRATEGY 9: UNDERLINE UNITS OF MEASUREMENT

If a problem includes units of measurement, underline them as you read the problem. This will help alert you to whether there is a need to make a units conversion so that all the quantities you are working with are expressed in the same units.

➥ Example :: Calculator Section :: Multiple-Choice

The element copper has a density of 8.9 **grams** per cubic centimeter. What is the number of cubic centimeters in the volume of 3.1 **kilograms** of copper? [Density = Mass divided by Volume]

(A) 287.1
(B) 348.3
(C) 391.4
(D) 418.6

Solution

Notice that there are two different units of measurement so convert kilograms into grams so the units are consistent.

$$\text{Density} = \frac{\text{Mass}}{\text{Volume}}$$

$$8.9\,\text{g/cm}^3 = \frac{3.1 \times 10^3 \,\text{g}}{\text{Volume}}$$

$$\text{Volume} = \frac{3{,}100\,\text{g}}{8.9\,\text{g/cm}^3}$$

$$= 348.3\,\text{cm}^3$$

The correct choice is **(B)**.

➥ Example :: Calculator Section :: Grid-In

Andrea finished her first half-marathon race of 13.1 miles in $2\frac{1}{2}$ **hours**. If she ran the race at a constant rate of speed, how many **minutes** did it take her to run the first 2 miles?

Solution

■ Change $2\frac{1}{2}$ hours to minutes:

$$2\frac{1}{2}\,\text{hours} = 2(60) + \frac{1}{2}(60) = 120 + 30 = 150\,\text{minutes}$$

■ If x represents the number of minutes it takes her to run 2 miles, then

$$\frac{\text{miles}}{\text{min}} = \frac{13.1}{150} = \frac{2}{x}$$

$$13.1x = (150)(2)$$

$$\frac{13.1x}{13.1} = \frac{300}{13.1}$$

$$x = 22.9\,\text{minutes}$$

Grid-in **22.9**

STRATEGY 10: KNOW HOW TO INTERPRET THE PARTS OF A MATHEMATICAL MODEL

You may be presented with an equation that represents or models a real-world situation and asks you to interpret a part of that equation based on the context of the problem.

➡ Example :: No-Calculator Section :: Multiple-Choice

$$C = 65 + 9n$$

The equation above represents the cost in dollars, C, of a cable television subscription that includes n premium high-definition channels. According to the model, what is the meaning of the constant 65 in the equation?

(A) The cost of a cable subscription that includes 9 premium channels.
(B) The cost of a cable subscription that does not include any premium channels.
(C) The cost of adding on 9 premium channels to the cost of the cable subscription.
(D) The amount the cost of a cable subscription increases when one additional premium channel is ordered.

Solution

In general, if a linear equation of the form $y = A + Bx$ is used as a model, the constant A is the starting amount for the quantity that y represents. The constant B is a rate of change—the fixed amount by which y changes each time x changes by 1 unit. The constant 9 is the amount the cost increases each time one additional premium channel is ordered. When no premium channels are ordered, $n = 0$ so $C = 65$.

The correct choice is **(B)**.

➡ Example :: No-Calculator Section :: Multiple-Choice

Connor invests a sum of money at an interest rate of 3% per year compounded annually. The value of the investment, y, after n years can be calculated using the equation $y = A(x)^n$. If Connor invests \$250 and wants to calculate the value of his investment after 10 years, assuming no further deposits or withdrawals, which of the following equations should he use?

(A) $y = 250(0.97)^{10}$
(B) $y = (250 \times 0.03)^{10}$
(C) $y = 250(0.03)^{10}$
(D) $y = 250(1.03)^{10}$

Solution

Compounding is an exponential *growth* process. In the equation $y = A(x)^n$, x represents the growth factor, which is 1 *plus* the interest rate expressed as a decimal. Hence, Connor should use the equation $y = 250(1.03)^{10}$.

The correct choice is **(D)**.

STRATEGY 11: MATCH UP THE SAME TERMS ON OPPOSITE SIDES OF AN IDENTITY

The equation $1 + 1 = 2$ is an example of an *identity*. The equation $2(x + y) = 2x + 2y$ is another example of an identity. An **identity** is an equation that is true regardless of what values are substituted for any of the variables. If $a + 5x = 2 + 5x$ for all values of x, then for this identity we know that a must equal 2. When an identity has opposite sides with exactly the same form, then matching up corresponding terms may help you solve the problem.

➥ Example :: No-Calculator Section :: Grid-In

If $(x + 9)(x + a) = x^2 + (b + 6)x + 45$ is true for all values of x, what is the value of b?

Solution

- The product of the last terms of the binomial factors must be equal to 45. Hence, $9a = 45$ so $a = 5$.
- Use FOIL to multiply $(x + 9)$ and $(x + 5)$ together:

$$(x + 9)(x + 5) = x^2 + (b + 6)x + 45$$

$$x^2 + 5x + 9x + 45 = x^2 + (b + 6)x + 45$$

$$x^2 + 14x + 45 = x^2 + (b + 6)x + 45$$

- Find b by setting matching constants equal to each other:

$$14 = b + 6 \text{ so } b = 8$$

Grid-in **8**

➥ Example :: Calculator Section :: Multiple-Choice

$$p = 3n(5n - 2)$$

$$q = 4(2 - 5n)$$

If $p - q = ax^2 + bx + c$ for all values of x where a, b, and c are constants, what is the value of c?

(A) −8
(B) −4
(C)   4
(D)   8

Solution

- Express $p - q$ in terms of n:

$$p - q = 3n(5n - 2) - 4(2 - 5n)$$
$$= 3n(5n - 2) - 4(-1)(5n - 2)$$
$$= 3n(5n - 2) + 4(5n - 2) \quad \leftarrow \text{Factor out } (5n - 2)$$
$$= (5n - 2)(3n + 4)$$

- Find c:

  Since $(5n - 2)(3n + 4) = ax^2 + bx + c$, c must be equal to the product of the last terms of the binomial factors. Hence, $c = (-2)(4) = -8$.

The correct choice is **(A)**.

➡ Example :: Calculator Section :: Multiple-Choice

$$(ax + 4)(bx - 1) = 21x^2 + kx - 4$$

If the equation above is true for all values of x and $a + b = 10$, what are the possible values for the constant k?

(A) 8 and 14
(B) 5 and 25
(C) 9 and 15
(D) 19 and 31

Solution

- If the two binomials are multiplied together, the product of the coefficients of the x^2-terms must be equal to 21 so $ab = 21$.
- Since it is also given that $a + b = 10$, you are looking for two numbers whose product is 21 and whose sum is 10. The numbers 3 and 7 satisfy these conditions.
- Because the placement of 3 and 7 in the binomial factors matters, there are two possibilities to consider:

$$(7x + 4)(3x - 1) = 21x^2 - 7x + 12x - 4$$
$$= 21x^2 + \underset{k}{5}x - 4 \quad \text{and} \quad$$

$$(3x + 4)(7x - 1) = 21x^2 - 3x + 28x - 4$$
$$= 21x^2 + \underset{k}{25}x - 4$$

Hence, the possible values of k are 5 and 25.

The correct choice is **(B)**.

STRATEGY 12: SOLVE FOR THE QUANTITY ASKED FOR

Some SAT problems may present you with a familiar type of equation and ask you to find the value of a simple expression that includes the same variables. Typically, this involves two steps:

- Solving the equation as usual and
- Evaluating the required expression using the solution.

➡ Example :: No-Calculator Section :: Multiple-Choice

If $3w + 4 = 25$, what is the value of $4w + 3$?

Solution

- Solve for w: $3w + 4 = 25$ so $3w = 21$ and $w = \dfrac{21}{3} = 7$.

- Evaluate $4w + 3$: $4w + 3 = 4(7) + 3 = 28 + 3 = 31$.

Grid-in **31**

➥ Example :: No Calculator Section :: Multiple-Choice

If $6\left(\dfrac{x}{y}\right) = 4$, then $\dfrac{1}{3}\left(\dfrac{y}{x}\right) =$

(A) $\dfrac{1}{2}$

(B) $\dfrac{2}{3}$

(C) $\dfrac{3}{2}$

(D) 2

Solution

If $6\left(\dfrac{x}{y}\right) = 4$, then $\dfrac{x}{y} = \dfrac{4}{6} = \dfrac{2}{3}$ so $\dfrac{y}{x} = \dfrac{3}{2}$ and

$$\dfrac{1}{3}\left(\dfrac{y}{x}\right) = \dfrac{1}{\cancel{3}}\left(\dfrac{\cancel{3}^{\,1}}{2}\right) = \dfrac{1}{2}$$

The correct choice is **(A)**.

➥ Example :: No-Calculator Section :: Multiple-Choice

If $\dfrac{8}{n} = \dfrac{20}{n+9}$, what is the value of $\dfrac{n}{8}$?

(A) $\dfrac{1}{2}$

(B) $\dfrac{3}{4}$

(C) 4

(D) 12

Solution

- Solve for n in the usual way by cross-multiplying:

$$\dfrac{8}{n} = \dfrac{20}{n+9}$$
$$20n = 8(n+9)$$
$$20n = 8n + 72$$
$$12n = 72$$
$$n = \dfrac{72}{12} = 6$$

- Evaluate $\dfrac{n}{8}$: $\dfrac{n}{8} = \dfrac{6}{8} = \dfrac{3}{4}$.

The correct choice is **(B)**.

TIP

Sometimes it may be easier and quicker to manipulate the given equation so that you solve directly for the quantity asked for.

➥ Example: Calculator Section :: Multiple-Choice

If $3x - 7 = 9$, what is the value of $6x + 5$?

(A) 26
(B) 29
(C) 33
(D) 37

Solution

You could begin by solving the given equation for x. However, if you compare the left side of the equation with $6x + 5$ you see that since $6x$ is a multiple of $3x$, it is more efficient to solve directly for the quantity asked for.

$$2(3x - 7) = 2\,(9) \leftarrow \text{To get } 6x \text{ on left side, multiply both sides by 2.}$$
$$6x - 14 = 18 \quad \leftarrow \text{To get 5 on left side, add 19 to both sides.}$$
$$(6x - 14) + 19 = 18 + 19$$
$$6x + 5 = 37$$

The correct choice is **(D)**.

➥ Example: Calculator Section :: Multiple-Choice

If $7 - 2k \leq 3$, what is the least possible value of $2k + 7$?

(A) −3
(B) 10
(C) 11
(D) 13

Solution

Solve for $2k + 7$ directly:

$$7 - 2k \leq 3$$
$$(-1) \times (7 - 2k) \geq (-1) \times (3) \leftarrow \text{Reverse the inequality.}$$
$$2k - 7 \geq -3 \qquad \leftarrow \text{To get +7 on left side, add 14 to both sides.}$$
$$(2k - 7) + 14 \geq (-3) + 14$$
$$2k + 7 \geq 11$$

The least possible value of $2k + 7$ is 11.
   The correct choice is **(D)**.

STRATEGY 13: WRITE AND SOLVE AN EQUATION

Sometimes the most efficient or direct way of solving a word problem is to create and solve an algebraic equation.

➥ Example :: Calculator Section :: Grid-In

Last week, Ben, Kaitlyn, and Emily sent a total of 394 text messages from their cell phones. Kaitlyn sent 50% more text messages than Ben, and Ben sent 30 fewer messages than Emily. How many text messages did Kaitlyn send?

Solution

If x represents the number of text messages that Ben sent, then Kaitlyn sent $x + 0.5x = 1.5x$ or $1.5x$ messages while Emily sent $x + 30$ messages. Since the total number of text messages sent was 394:

$$1.5x + x + (x + 30) = 394$$
$$3.5x = 364$$
$$\frac{3.5x}{3.5} = \frac{364}{3.5}$$
$$x = 104$$
$$1.5x = 156$$

Grid-in **156**

➥ Example :: Calculator Section :: Grid-In

An exterminator needs to dilute a 25% solution of an insecticide with a 15% solution of the same insecticide. How many more liters of the 25% solution than the 15% solution are needed to make a total of 80 liters of a 22% solution of the insecticide?

Solution

If x represents the number of liters needed of the 25% solution, then $80 - x$ represents the number of liters needed of the 15% solution. The sum of the number of liters of pure insecticide in the two solutions that are mixed together must be equal to the number of liters of pure insecticide in the 22% solution.

$$0.25x + 0.15(80 - x) = 0.22(80)$$
$$0.25x + 12 - 0.15x = 17.6$$
$$0.10x = 5.6$$
$$\frac{0.10x}{0.10} = \frac{5.6}{0.10}$$
$$x = 56$$
$$80 - x = 80 - 56 = 24$$

Hence, $56 - 24 = 32$ more liters of the 25% solution than the 15% solution are needed.

Grid-in **32**

STRATEGY 14: CREATIVELY MANIPULATE SYSTEMS OF EQUATIONS

SAT problems involving two or more related equations may seem complicated at first glance. These problems can often be solved easily by eliminating a common variable through:

- Using a straightforward substitution
- Multiplying or dividing both sides of an equation by the same nonzero quantity
- Adding or subtracting corresponding sides of the two equations
- Using some combination of the above methods

➡ Example :: No-Calculator Section :: Grid-In

If $\frac{h}{p} = 28$ and $p = \frac{3}{4}$, what is the value of $\frac{1}{3}h$?

Solution

- Solving for h in the first equation gives $h = 28p$.
- Substitute $\frac{3}{4}$ for p:

$$h = \overset{7}{\cancel{28}} \times \frac{3}{\cancel{4}} = 21$$

- Hence, $\frac{1}{3}h = \frac{1}{3} \times 21 = 7$.

Grid-in **7**

➡ Example :: No-Calculator Section :: Multiple-Choice

$$\frac{x-9}{x+y} = \frac{1}{2}$$

$$\frac{x}{y} - 1 = 3$$

Based on the system of equations above, what is the value of $x + y$?

(A) 18
(B) 24
(C) 27
(D) 30

Solution

- Simplify the first equation by cross-multiplying:

$$2(x - 9) = x + y$$

$$2x - 18 = x + y$$

$$x - 18 = y$$

- Simplify the second equation:

$$\frac{x}{y} - 1 = 3$$

$$x = 4y$$

- Substitute $4y$ for x in the transformed first equation:

$$x - 18 = y$$

$$4y - 18 = y$$

$$\frac{3y}{3} = \frac{18}{3}$$

$$y = 6$$

- Since $y = 6$, $x = 4y = 4(6) = 24$. Hence, $x + y = 24 + 6 = 30$.

The correct choice is **(D)**.

➥ Example :: No-Calculator Section :: Grid-In

$$2x^2 + y^2 = 176$$

$$y + 3x = 0$$

If (x, y) is a solution to the system of equations above, what is the value of y^2?

Solution

- Solve the linear equation for y, which gives $y = -3x$.
- Substitute $-3x$ for y in the first equation:

$$2x^2 + (-3x)^2 = 176$$

$$2x^2 + 9x^2 = 176$$

$$\frac{11x^2}{11} = \frac{176}{11}$$

$$x^2 = 16$$

- Replace x^2 with 16 in the first equation and solve for y^2:

$$2(16) + y^2 = 197$$

$$y^2 = 197 - 32$$

$$= 165$$

Grid-in **165**

➥ Example :: No-Calculator Section :: Grid-In

If $2d - e = 32$ and $2e - d = 10$, what is the average of d and e?

Solution

The average of d and e is $\dfrac{d+e}{2}$. Rather than solving the system of equations for each of the individual variables, think of an easy way of combining the two equations to get a simpler expression. Try adding corresponding sides of the two equations:

$$
\begin{array}{r}
2d - e = 32 \\
-d + 2e = 10 \\
\hline
d + e = 42
\end{array}
$$

This is very close to what you need to find. Now divide each side of $d + e = 42$ by 2:

$$\frac{d+e}{2} = \frac{42}{2} = 21$$

The average of d and e is **21**.

   Grid-in **21**

➥ Example :: No-Calculator Section :: Multiple-Choice

If $x - y = y + 3 = 7$, what is the value of x?

(A) −4
(B)   8
(C) 10
(D) 11

Solution

The extended equation $x - y = y + 3 = 7$ means $x - y = 7$ and $y + 3 = 7$.

- Begin by solving the second equation since it contains only one variable:

$$(y + 3) - 3 = 7 - 3 \text{ so } y = 4.$$

- Replace y with 4 in the first equation, which gives $x - 4 = 7$. Solve for x:

$$(x - 4) + 4 = 7 \text{ so } x = 11.$$

The correct choice is **(D).**

➥ Example :: No-Calculator Section :: Grid-In

If $3u + w = 7$ and $3k - 9u = 2$, what is the value of $w + k$?

Solution

- Eliminate variable u by multiplying both sides of the first equation by -3 and then adding the result to the second equation:

$$
\begin{aligned}
9u + 3w &= 21 \\
-9u + 3k &= 2 \\
\hline
3w + 3k &= 23
\end{aligned}
$$

- Divide both sides of the equation by 3 to get $w + k = \dfrac{23}{3}$.

Grid-in **23/3**

➥ Example :: No Calculator Section :: Grid-In

If $(p + q)^2 = 78$ and $(p - q)^2 = 50$, what is the value of pq?

Solution

Square each binomial and then subtract corresponding sides of the two equations:

$$
\begin{aligned}
(p + q)^2 = p^2 + 2pq + q^2 &= 78 \\
(p - q)^2 = p^2 - 2pq + q^2 &= 50 \\
\hline
4pq &= 28
\end{aligned}
$$

$$
\frac{4pq}{4} = \frac{28}{4}
$$

$$
pq = 7
$$

Grid-in **7**

STRATEGY 15: REWRITE EXPRESSIONS OR EQUATIONS IN EQUIVALENT WAYS

Some SAT Math questions may ask you to find the value of an algebraic expression using an accompanying equation. By rewriting the given expression or equation in an equivalent way, you may be able to make a substitution that will produce its value.

➥ Example :: No-Calculator Section :: Multiple-Choice

If $2x - 3y = 10$, what is the value of $\dfrac{4^x}{8^y}$?

(A) $\left(\dfrac{1}{2}\right)^{10}$

(B) 2^{10}

(C) 4^6

(D) 8^5

Solution

- Write the fraction in an equivalent way by expressing the numerator and denominator as a power of the same base:

$$\frac{4^x}{8^y} = \frac{\left(2^2\right)^x}{\left(2^3\right)^y} = \frac{2^{2x}}{2^{3y}} = 2^{2x-3y}$$

- Since it is given that $2x - 3y = 10$, substitute 10 for the exponent:

$$\frac{4^x}{8^y} = 2^{2x-3y} = 2^{10}$$

The correct choice is (**B**).

➥ Example :: No-Calculator Section :: Grid-In

If $\frac{1}{2}a + \frac{2}{3}b = 5$, what is the value of $3a + 4b$?

Solution

Write the equation in an equivalent way by eliminating its fractional coefficients. Multiply each term on both sides of the equation by 6, the lowest common multiple of 2 and 3:

$$6\left(\frac{1}{2}a\right) + 6\left(\frac{2}{3}b\right) = 6(5)$$

$$3a + 4b = 30$$

Grid-in **30**

➥ Example :: No-Calculator Section :: Grid-In

A population of rabbits doubles every 48 days according to the formula $P = 10(2)^{\frac{t}{48}}$, where P is the population of rabbits on day t. What is the value of t when the population of rabbits is 320?

Solution

If $P = 320$, then $320 = 10(2)^{\frac{t}{48}}$ so $32 = (2)^{\frac{t}{48}}$. Since $2^5 = 32$, $2^5 = (2)^{\frac{t}{48}}$ which means $\frac{t}{48} = 5$ so $t = 48 \times 5 = 240$.

Grid-in **240**

STRATEGY 16: KNOW HOW TO DO THE MATH

Some SAT test questions simply require a straightforward application of some routine algebraic technique, geometric fact, or trigonometric relationship that you studied in your regular mathematics classes.

➥ Example :: No-Calculator Section :: Multiple-Choice

If p and q satisfy $2y^2 + y = 21$ and $p > q$, which of the following is the value of $p - q$?

(A) $\dfrac{15}{2}$

(B) $\dfrac{13}{2}$

(C) $\dfrac{7}{2}$

(D) $\dfrac{1}{2}$

Solution

Rewrite the given quadratic equation as $2y^2 + y - 21 = 0$ and solve it by factoring using the reverse of FOIL:

$$2y^2 + y - 21 = 0$$
$$(2y + ?)(y + ?) = 0$$
$$(2y + 7)(y - 3) = 0$$
$$2y + 7 = 0 \quad or \quad y - 3 = 0$$
$$y = -\frac{7}{2} \quad or \quad y = 3$$

Since $p = 3$ and $q = -\dfrac{7}{2}$,

$$p - q = 3 - \left(-\frac{7}{2}\right)$$
$$= 3 + \frac{7}{2}$$
$$= \frac{6}{2} + \frac{7}{2}$$
$$= \frac{13}{2}$$

The correct choice is **(B)**.

TIP

You are required to know how to factor quadratic trinomials of the form $ax^2 + bx + c$ where the constant a, the coefficient of the x^2-term, is different than 1. You should also know how to solve nonfactorable quadratic equations either by completing the square or by using the quadratic formula.

➡ Example :: No-Calculator Section :: Multiple-Choice

An acute angle of a right triangle measures x radians. Which of the following is equal to $\cos x$ when $x = \dfrac{\pi}{9}$?

(A) $-\cos\left(\dfrac{11\pi}{18}\right)$

(B) $-\sin\left(\dfrac{\pi}{9}\right)$

(C) $\sin\left(\dfrac{8\pi}{9}\right)$

(D) $\sin\left(\dfrac{7\pi}{18}\right)$

Solution

Two angles are complementary if their measures add up to $90°\left(=\dfrac{\pi}{2}\,\text{radians}\right)$. The cosine of an acute angle is equal to the sine of the complement of that angle. In general, for acute angles,

$$\cos x = \sin(90 - x)° \text{ or, equivalently, } \cos x = \sin\left(\dfrac{\pi}{2} - x\right)$$

Substitute $\dfrac{\pi}{9}$ radians for x:

$$\begin{aligned}
\cos\left(\dfrac{\pi}{9}\right) &= \sin\left(\dfrac{\pi}{2} - \dfrac{\pi}{9}\right) \\
&= \sin\left(\dfrac{9\pi}{18} - \dfrac{2\pi}{18}\right) \\
&= \sin\left(\dfrac{7\pi}{18}\right)
\end{aligned}$$

The correct choice is **(D)**.

➡ Example :: No-Calculator Section :: Grid-In

$$\left(2 - \sqrt{-9}\right)\left(1 + \sqrt{-16}\right) = x + yi \quad (\text{Note: } i = \sqrt{-1})$$

In the equation above, what is the value of y?

Solution

- Change the radicals to "i-form":

$$\begin{aligned}
\left(2 - \sqrt{-9}\right)\left(1 + \sqrt{-16}\right) &= \left(2 - \sqrt{9} \cdot \sqrt{-1}\right)\left(1 + \sqrt{16} \cdot \sqrt{-1}\right) \\
&= (2 - 3i)(1 + 4i)
\end{aligned}$$

- Multiply the binomials horizontally using FOIL:

$$(2 - i)(1 + 4i) = \overbrace{2 \cdot 1}^{\mathbf{F}} + \overbrace{2 \cdot 4i}^{\mathbf{O}} + \overbrace{(-i \cdot 1)}^{\mathbf{I}} + \overbrace{(-i)(4i)}^{\mathbf{L}}$$

$$= 2 + 8i - i - 4i^2 \quad \leftarrow i^2 = -1$$

$$= 2 + 7i - 4(-1)$$

$$= 6 + 7i$$

- Since $6 + 7i = x + yi$, $y = 7$.

Grid-in **7**

➡ Example :: No-Calculator Section :: Grid-In

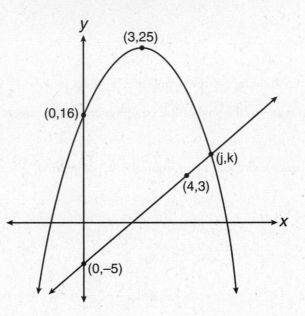

A quadratic function and a linear function are graphed in the *xy*-plane as shown above. The vertex of the graph of the quadratic function is at (3, 25). If the two graphs intersect in the first quadrant at the point (*j*, *k*), what is the value of the product *jk*?

Solution

- The graph of the quadratic function is a parabola. If the vertex of a parabola is at (*h*, *k*), then the equation of the parabola has the form $y = a(x - h)^2 + k$. Since it is given that (*h*, *k*) = (3, 25), $y = a(x - 3)^2 + 25$. From the graph, you know that the *y*-intercept is (0, 16). Plug these coordinates into the equation to find the value of the constant *a*:

$$16 = a(0 - 3)^2 + 25$$

$$16 = 9a + 25$$

$$-9 = 9a$$

$$a = -1$$

Hence, the equation of the parabola is $y = -(x - 3)^2 + 25$.

- Find an equation of the line in $y = mx + b$ slope-intercept form where m is the slope of the line and b is its y-intercept. The y-intercept of the line is -5 so $b = -5$. Using the points $(4, 3)$ and $(0, -5)$, find the slope, m, of the line:

$$m = \frac{\Delta y}{\Delta x} = \frac{\text{change in } y\text{-coordinates}}{\text{change in } x\text{-coordinates}}$$
$$= \frac{-5 - 3}{0 - 4}$$
$$= \frac{-8}{-4}$$
$$= 2$$

Hence, an equation of the line is $y = 2x - 5$.

- To find the coordinates of the point of intersection of the two graphs, solve the linear-quadratic system of equations:

$$y = -(x - 3)^2 + 25$$
$$y = 2x - 5$$

Substitute $2x - 5$ for y in the quadratic equation:

$$-(x - 3)^2 + 25 = 2x - 5$$
$$-(x^2 - 6x + 9) + 25 = 2x - 5$$
$$-x^2 + 6x + 16 = 2x - 5$$
$$-x^2 + 4x + 21 = 0$$
$$x^2 - 4x - 21 = 0$$
$$(x - 7)(x - 3) = 0$$
$$x = 7 \text{ or } x = -3$$

Since (j, k) is in the first quadrant, $j = 7$. Find the corresponding value of y by substituting 7 for x in the linear equation, which gives $y = 2(7) - 5 = 9$. Hence, $k = 9$ so $jk = 7 \times 9 = 63$.

Grid-in **63**

STRATEGY 17: KNOW HOW TO WORK WITH FUNCTIONS

Functions and function notation are commonly used on the SAT. A *function* is simply a mathematical *rule* that describes how to use an input value, say x, to produce an output value, say y. A function may take the form of either a "$y = \ldots$" type of equation, a graph, or a table. If a function named f is described by the equation $y = x^2 - 4$, then in function notation, $f(x) = x^2 - 4$. The notation $f(3)$ represents the "y-value" of function f when $x = 3$ so

$$f(3) = \overbrace{3^2 - 4}^{\text{Replace } x \text{ with } 3} = 5$$

The ordered pair $(3, 5)$ belongs to function f and is a point on its graph.

➥ Example :: No-Calculator Section :: Grid-In

If function h is defined by $h(x) = ax^2 - 7$ and $h(-3) = 29$, what is $h\left(\dfrac{1}{2}\right)$?

(A) −6
(B) −5
(C) 4
(D) 8

Solution

- Use the fact that $h(-3) = 29$ to find the value of the constant a:

$$h(-3) = a(-3)^2 - 7 = 9a - 7 = 29 \text{ so } 9a = 36 \text{ and } a = 4$$

- Evaluate $h\left(\dfrac{1}{2}\right)$:

$$h(x) = 4x^2 - 7$$
$$h\left(\frac{1}{2}\right) = 4\left(\frac{1}{2}\right)^2 - 7$$
$$= 4\left(\frac{1}{4}\right) - 7$$
$$= 1 - 7$$
$$= -6$$

The correct choice is **(A)**.

➥ Example :: No-Calculator Section :: Multiple-Choice

If function f is defined by $f(x) = 3x - 4$, then $f(-2x) =$

(A) $-6x + 4$
(B) $8 - 6x$
(C) $-6x - 4$
(D) $8x + 6x^2$

Solution

$$f(x) = 3x - 4$$
$$f(-2x) = 3(-2x) - 4$$
$$= -6x - 4$$

The correct answer choice is **(C)**.

➥ Example :: No-Calculator Section :: Grid-In

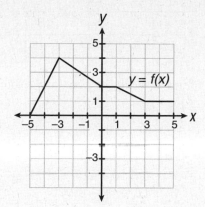

The graph of function f over the interval $-5 \leq x \leq 5$ is shown in the figure above. If $f(w) = 2$ and $w > 0$, what is one possible value of w?

Solution

The notation $f(w) = 2$ represents the point $(w, 2)$ on the graph of function f. Each point whose x-coordinate is between 0 and 1 has a corresponding y-coordinate of 2. Hence, w can be any number between 0 and 1. For example, if $w = \dfrac{1}{3}$, then $f\left(\dfrac{1}{3}\right) = 2$.

Grid-in **1/3**

➥ Example :: No-Calculator Section :: Multiple-Choice

Based on the graph in the previous example, which of the following statements must be true?

  I. $f(5) + f(-5) = 0$.
 II. If $-5 \leq x \leq 5$, the maximum value of function f is 4.
III. The equation $f(x) = 3$ has 3 real solutions.

(A) I only
(B) II only
(C) I and II only
(D) II and III only

Solution

- $f(-5) = 0$ and $f(5) = 1$ so $f(5) + f(-5) \neq 0$. Statement I is false. ✗
- The highest point on the graph is $(-3, 4)$ so the maximum value of function f is 4. Statement II is true. ✓
- The horizontal line $y = 3$ intersects the graph at 2 points so $f(x) = 3$ has 2 rather than 3 real solutions. Statement III is false. ✗

Since only Statement II is true, the correct choice is **(B)**.

➥ **Example :: No-Calculator Section :: Multiple-Choice**

x	−1	1	3	5
$f(x)$	5	−1	−7	−13

The table above shows a few values of the linear function f. Which of the following equations defines f?

(A) $f(x) = 2x - 3$
(B) $f(x) = -2x + 3$
(C) $f(x) = -3x + 2$
(D) $f(x) = 3x - 2$

Solution

The graph of a linear function is a line. From the table, each time x increases by 2 units, y *decreases* by 6 units so the slope of the line is $\dfrac{\Delta y}{\Delta x} = \dfrac{-6}{+2} = -3$. Hence, the coefficient of x for a linear function of the form $f(x) = mx + b$ is −3. Only the function in choice (C) satisfies this condition.

 The correct choice is **(C)**.

➥ **Example :: No-Calculator Section :: Multiple-Choice**

Function g is related to function f by the equation $g(x) = f(x - 1) - 2$. If the point $(4, 3)$ is on the graph of function f, what are the coordinates of the corresponding point on the graph of function g?

(A) $(3, 1)$
(B) $(5, 1)$
(C) $(1, 3)$
(D) $(1, 5)$

Solution

In general, $y = f(x - h) + k$ is the graph of $y = f(x)$ shifted right h units and up k units. For the function $g(x) = f(x - 1) - 2$, $h = 1$ and $k = -2$. Hence, the graph of function g is the graph of function f shifted right 1 unit and down 2 units. Under this transformation, $(4, 3) \rightarrow (4 + 1, 3 - 2) = (5, 1)$.

 The correct choice is **(B)**.

➥ **Example :: No-Calculator Section :: Grid-In**

$$g(x) = 2^x$$
$$h(x) = 8 - x^2$$

If functions g and h are defined in the above equations, what is the value of $g(h(3))$?

Solution

- Evaluate the "inside" function first: $h(3) = 8 - 3^2 = 8 - 9 = -1$.
- Then $g(h(3)) = g(-1) = 2^{-1} = \dfrac{1}{2}$.

Grid-in **1/2**

STRATEGY 18: KNOW KEY RELATIONSHIPS AND FORMULAS

Knowing certain relationships and formulas that are not provided on the reference page of the test booklet can increase your speed and accuracy.

➥ Example :: No-Calculator Section :: Grid-In

A parabola $y = ax^2 + bx + c$ with $a > 0$ passes through the points $(-2, 3)$, (p, q), and $(5, 3)$. If (p, q) is the lowest point on the parabola, what is the value of p?

Solution

Draw a diagram:

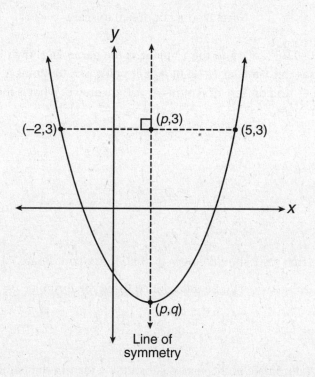

The line of symmetry is the perpendicular bisector of any horizontal segment whose endpoints are on the parabola. Since the points $(-2, 3)$ and $(5, 3)$ are corresponding parabola points on either side of the axis of symmetry, the line of symmetry bisects the segment joining these points and also passes through the vertex of the parabola. Hence, the x-coordinate of (p, q) must be the same as the x-coordinate of the midpoint of this segment:

$$p = \frac{-2+5}{2} = \frac{3}{2}$$

Grid-in **3/2**

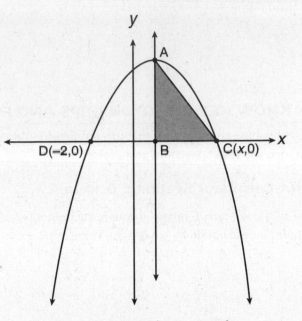

Note: Figure not drawn to scale.

The graph of $f(x) = -0.5x^2 + x + 4$ in the xy-plane is the parabola shown in the figure above. The parabola crosses the x-axis at $D(-2, 0)$ and at point $C(x, 0)$. Point A is the vertex of the parabola. Segment AC and the line of symmetry, $\overline{AB}$, are drawn. What is the number of square units in the area of $\triangle ABC$?

(A)   4.5
(B)   6.25
(C)   6.75
(D) 13.5

Solution

■ A math fact that you should know is that the x-coordinate of the vertex of the parabola $y = ax^2 + bx + c$ can be determined using the formula, $x = -\dfrac{b}{2a}$:

$$x = \frac{-2 + 4}{2} = 1$$

To find the y-coordinate of the vertex, substitute 1 for x in the parabola equation:

$$f(1) = -0.5(1^2) + 1 + 4$$
$$= -0.5 + 5$$
$$= 4.5$$

Hence, $AB = \mathbf{4.5}$.

- The coordinates of point B are $(1, 0)$ so $DB = 1 - (-2) = 3$. Since the line of symmetry is the perpendicular bisector of segment CD, $DB = BC = 3$.
- The area of a right triangle is one-half of the product of the lengths of its legs:

$$\text{Area of } \triangle ABC = \frac{1}{2}(3)(4.5) = 6.75$$

The correct choice is **(C)**.

➡ Example :: No-Calculator Section :: Multiple-Choice

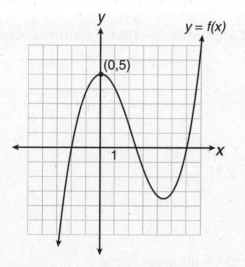

I. $f(x)$ is divisible by $x - 5$.
II. In the interval $0 \leq x \leq 6$, exactly one x-value satisfies the equation $f(x) = 5$.
III. $(x + 2)$ is a factor of $f(x)$.

The diagram above shows the graph of a polynomial function f. Which statement or statements in the box above must be true?

(A) II only
(B) III only
(C) I and II only
(D) II and III only

Solution

The SAT assumes you know some basic facts related to polynomial functions and their graphs. If the graph of a polynomial function $f(x)$ intersects the x-axis at a point whose x-coordinate is c, then the following statements are equivalent and interchangeable:

- $f(c) = 0$.
- c satisfies the equation $f(x) = 0$.
- $x - c$ is a factor of $f(x)$.
- $f(x)$ is divisible by $x - c$.

TIP

If $f(x)$ represents a polynomial and the value of $f(c)$ is r, then r is the remainder when $f(x)$ is divided by $x - c$. If $r = 0$, then $f(x)$ is divisible by $x - c$.

Consider each of the answer choices in turn:

- Since the graph does not intersect the x-axis at $x = 5$, $f(x)$ is *not* divisible by $x - 5$. Statement I is false. ✗
- In the interval $0 \leq x \leq 6$, the line $y = 5$ (not drawn) intersects the graph at only one point so there is exactly *one* value of x that satisfies the equation $f(x) = 5$. Statement II is true. ✓
- Since the graph crosses the x-axis at $x = -2$, $x - (-2)$ or $x + 2$ is a factor of $f(x)$. Statement III is true. ✓

Since only Statements II and III are true, the correct choice is (D).

➡ Example :: No-Calculator Section :: Multiple-Choice

$$p(x) = 4(-x^3 + 11x + 12) - 6(x - c)$$

In the polynomial function $p(x)$ defined above, c is a constant. If $p(x)$ is divisible by x, what is the value of c?

(A) -8

(B) -6

(C) 0

(D) 6

Solution

A polynomial function $p(x)$ has the general form

$$p(x) = a_n x^n + a_{n-1} x^n + a_{n-3} x^{n-3} + \cdots + k$$

If $p(x)$ is divisible by x, then the value of the constant term, k, must be 0. Otherwise, there would be a remainder.

- Write $p(x)$ in standard form by removing the parentheses and collecting like terms:

$$\begin{aligned} p(x) &= 4(-x^3 + 11x + 12) - 6(x - c) \\ &= -4x^3 + 44x + 48 - 6x + 6c \\ &= -4x^3 + 38x + \underbrace{(48 + 6c)}_{= 0} \end{aligned}$$

- Set the constant term equal to 0:

$$48 + 6c = 0 \text{ so } 6c = -48 \text{ and } c = -\frac{48}{6} = -8$$

The correct choice is (A).

➥ Example :: Calculator Section :: Grid-In

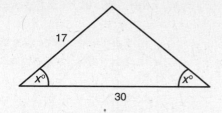

In the figure above, what is the value of $\cos x - \sin x$?

Solution

Since the measures of the base angles are equal, the triangle is isosceles so dropping a perpendicular from the vertex to the opposite side bisects the base and creates two smaller right triangles with side lengths that form a **8-15-17** *Pythagorean triple*:

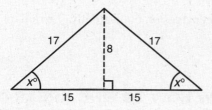

TIP

You should know commonly encountered Pythagorean triples such as 3-4-5, 5-12-13, 8-15-17, and 7-24-25. You are expected to know the right triangle definitions of the sine, cosine, and tangent functions.

Hence,

- $\sin x = \dfrac{\text{length of side opposite } \angle x}{\text{hypotenuse}} = \dfrac{8}{17}$

- $\cos x = \dfrac{\text{length of side adjacent to } \angle x}{\text{hypotenuse}} = \dfrac{15}{17}$

- $\cos x - \sin x = \dfrac{15}{17} - \dfrac{8}{17} = \dfrac{7}{17}$

Grid-in **7/17**

➥ Example :: No-Calculator Section :: Grid-In

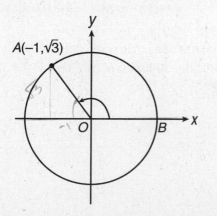

In the *xy*-plane above, O is the center of the circle, and the measure of $\angle AOB$ is $k\pi$ radians. What is a possible value for k?

Solution

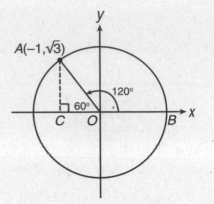

Since $\left(-1,\sqrt{3}\right)=(x,y), \dfrac{y}{x}=-\left(\dfrac{\sqrt{3}}{1}\right)$ so $\triangle AOC$ is a 30°-60° right triangle with m$\angle AOC=60$.

Hence, $\angle AOB$ measures 120 degrees or, equivalently, $\dfrac{2}{3}\pi$ radians.

Grid-in 2/3

TIP

Finding the area of an overlapping shaded region usually involves subtracting the area of one familiar type of figure from a larger one.

STRATEGY 19: KNOW HOW TO FIND AREAS INDIRECTLY

In order to find the area of a particular region, it may be necessary to find it indirectly by finding the difference between the areas of figures that overlap.

➥ **Example :: No-Calculator Section :: Multiple-Choice**

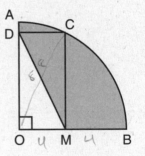

The figure above shows a quarter of a circle with center at O. If the length of diagonal $\overline{DM}$ of rectangle $ODCM$ is 8 and M is the midpoint of $\overline{OB}$, what is the area of the shaded region?

(A) $4\left(4\pi-\sqrt{3}\right)$

(B) $8\left(\pi-\sqrt{3}\right)$

(C) $8\left(2\pi-\sqrt{3}\right)$

(D) $16\left(\pi-\sqrt{3}\right)$

Solution

The area of the shaded region is the difference between the areas of the quarter circle and right triangle *DOM*.

- Since the diagonals of a rectangle have the same length, $OC = DM = 8$:

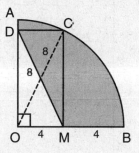

- OC is a radius of the quarter circle. Hence, the area of the quarter circle is $\frac{1}{4}(\pi \times 8^2) = 16\pi$.

- Since segment OB is also a radius, $OB = 8$. It is given that M is the midpoint of $\overline{OB}$ so $OM = MB = 4$. In right $\triangle DOM$, use the Pythagorean theorem to find the length of $\overline{OD}$:

$$(OD)^2 + 4^2 = 8^2$$
$$(OD)^2 = 64 - 16$$
$$OD = \sqrt{48}$$
$$= \sqrt{16} \times \sqrt{3}$$
$$= 4\sqrt{3}$$

- The area of right $\triangle DOM = \frac{1}{2} \times 4\sqrt{3} \times 4 = 8\sqrt{3}$.

Hence,

$$\text{Area of shaded region} = \text{Area of quarter circle} - \text{Area of right } \triangle DOM$$
$$= 16\pi - 8\sqrt{3}$$
$$= 8(2\pi - \sqrt{3})$$

The correct choice is **(C)**.

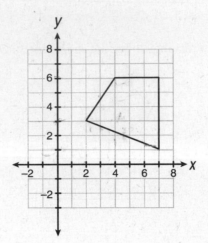

What is the number of square units in the area of the quadrilateral shown in the figure above?

Solution

- Draw line segments to form right triangles at the upper and lower corners of the quadrilateral:

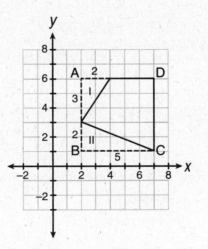

- Find the area of the quadrilateral indirectly by subtracting the sum of the areas of the two right triangles from the area of square *ABCD*:

Area of quadrilateral = Area of square *ABCD* − (Area of right △I + Area of right △II)

$$= 5 \times 5 - \left[\frac{1}{2}(2 \times 3) + \frac{1}{2}(2 \times 5) \right]$$

$$= 25 - 8$$

$$= 17$$

Grid-in **17**

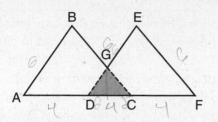

Note: Figure not drawn to scale.

The figure above shows a logo in the shape of overlapping equilateral triangles ABC and DEF. If $AD = DC = CF = 4$, what is the area of the shaded region?

(A) $24\sqrt{3} - 8$

(B) $28\sqrt{3}$

(C) $32\sqrt{3}$

(D) $36\sqrt{3} - 8$

Solution

- $AC = DF = 4 + 4 = 8$
- Since, m$\angle EDF = $ m$\angle BCA = $ m$\angle DGC = 60$, $\triangle DGC$ is equilateral with $DC = 4$.
- The area of the shaded region is the difference between the sum of the areas of the two overlapping equilateral triangles and equilateral $\triangle DGC$. To find the area of the overlapping equilateral triangles, use the formula

$$\text{Area} = \frac{(\text{side})^2}{4}\sqrt{3}$$

Area shaded region $= (\text{Area } \triangle ABC + \text{Area } \triangle DEF) - \text{Area } \triangle DGC$

$$= \frac{8^2}{4}\sqrt{3} + \frac{8^2}{4}\sqrt{3} - \frac{4^2}{4}\sqrt{3}$$

$$= 16\sqrt{3} + 16\sqrt{3} - 4\sqrt{3}$$

$$= 28\sqrt{3}$$

The correct choice is **(B)**.

MATH REFERENCE FACT

Another key formula you should know is the formula for the area, A, of an equilateral triangle with side s:

$$A = \frac{s^2}{4}\sqrt{3}$$

Since a regular hexagon can be divided into six equilateral triangles, the area, A, of a regular hexagon with side s is $A = 6\left(\frac{s^2}{4}\sqrt{3}\right)$.

STRATEGY 20: KNOW HOW TO CHANGE THE FORM OF AN EXPRESSION

Some problems can be simplified by recognizing that an algebraic expression has a familiar structure, whereas in other problems you may also need to change an expression into an equivalent form.

The difference of two perfect squares can be factored as the sum and difference of the two terms that are being squared, as in $x^2 - y^2 = (x + y)(x - y)$. This means that any algebraic expression that you recognize as having the same general form or *structure* as $x^2 - y^2$ can be factored in this way. The expression $2a^6 - 2b^6$ has the same structure as $x^2 - y^2$ with $x = a^3$ and $y = b^3$ so it can be factored in the same way:

$$2a^6 - 2b^6 = 2(a^6 - b^6)$$
$$= 2\left((a^3)^2 - (b^3)^2\right) \quad \leftarrow \text{Think: } x = a^3 \text{ and } y = b^3$$
$$= 2(a^3 + b^3)(a^3 - b^3)$$

To illustrate further, the equation $(2y - 1)^2 - 9(2y - 1) = 0$ can be easily solved by recognizing that it has the same form as $x^2 - 9x = 0$ where $x = 2y - 1$. To solve $(2y - 1)^2 - 9(2y - 1) = 0$,

- Substitute x for $2y - 1$, and solve the simpler equation for x:

$$x^2 - 9x = 0$$
$$x(x - 9) = 0$$
$$x = 0 \quad or \quad x = 9$$

- Substitute $2y - 1$ for x:

$$2y - 1 = 0 \qquad 2y - 1 = 9$$
$$2y = 1 \quad or \quad 2y = 10$$
$$y = \frac{1}{2} \qquad y = \frac{10}{2} = 5$$

➥ Example :: No-Calculator Section :: Grid-In

If $\sqrt[3]{b^5} \cdot \sqrt[4]{b^3} = \sqrt[3]{b^x}$ for all $b \geq 0$, what is the value of x?

Solution

Change the form of each of the radical expressions using the rule that $\sqrt[r]{x^p} = x^{\frac{p}{r}}$:

$$\sqrt[4]{b^3} \cdot \sqrt[6]{b^5} = \sqrt[3]{b^x}$$

$$b^{\frac{3}{4}} \cdot b^{\frac{5}{6}} = b^{\frac{x}{3}}$$

$$b^{\frac{3}{4}+\frac{5}{6}} = b^{\frac{x}{3}}$$

$$\frac{3}{4} + \frac{5}{6} = \frac{x}{3}$$

$$12\left(\frac{3}{4} + \frac{5}{6}\right) = 12\left(\frac{x}{3}\right)$$

$$9 + 10 = 4x$$

$$19 = 4x$$

$$x = \frac{19}{4}$$

Grid-in **19/4**

➡ Example :: No-Calculator Section :: Grid-In

If $\frac{1}{3}r + \frac{1}{4}s = \frac{5}{6}$, what is the value of $4r + 3s$?

Solution

Since the left side of the given equation has fractional coefficients, change its form to eliminate the fractions so that the resulting expression looks like $4r + 3s$. Multiply each term of the given equation by 12, the smallest whole number divisible by 3, 4, and 6:

$$12\left(\frac{1}{3}r\right) + 12\left(\frac{1}{4}s\right) = 12\left(\frac{5}{6}\right)$$

$$\overset{4}{\cancel{12}}\left(\frac{1}{\cancel{3}}r\right) + \overset{3}{\cancel{12}}\left(\frac{1}{\cancel{4}}s\right) = \overset{2}{\cancel{12}}\left(\frac{5}{\cancel{6}}\right)$$

$$4r \quad + \quad 3s \quad = 10$$

Grid-in **10**

➡ Example :: No-Calculator Section :: Grid-In

$$-\frac{12}{5} < 6 - 9y < -\frac{9}{4}$$

In the inequality above, what is one possible value of $3y - 2$?

Solution

You should notice that $3y - 2$ has a similar structure to $6 - 9y$ and can be obtained from it by dividing $6 - 9y$ by -3. Divide each member of the inequality by -3 while keeping in mind that whenever dividing (or multiplying) the terms of an inequality by a negative number, the inequality signs must be reversed:

$$\left(-\frac{12}{5}\right) \div (-3) > \frac{6 - 9y}{-3} > \left(-\frac{9}{4}\right) \div (-3) \qquad \leftarrow \text{Reverse inequality signs}$$

$$\left(\frac{\cancel{-9}^{\,3}}{4}\right) \cdot \left(\frac{1}{\cancel{-3}}\right) < \frac{6 - 9y}{-3} < \left(\frac{\cancel{-12}^{\,4}}{5}\right) \cdot \left(\frac{1}{\cancel{-3}}\right) \qquad \leftarrow \text{Rewrite as an equivalent "less than" inequality}$$

$$\frac{3}{4} < 3y - 2 < \frac{4}{5} \qquad \leftarrow \text{Simplify}$$

$$0.75 < 3y - 2 < 0.80 \qquad \leftarrow \text{Change to decimal form}$$

Grid-in any number between 0.75 and 0.80 such as **.76**.

➥ Example :: Calculator Section :: Multiple-Choice

$$3x + 4y = 7$$
$$kx - 2y = 1$$

For what value of k will the above system of equations have no solution?

(A) -3

(B) $-\dfrac{3}{2}$

(C) 0

(D) $\dfrac{3}{2}$

Solution

A system of linear equations has no solution if the lines the equations represent are parallel and, as a result, have no point in common. Lines are parallel when they have the same slope. Rewrite each equation so they both have the same "$y = mx + b$" slope-intercept form. Then compare their slopes.

■ Solve each equation for y:

$$3x + 4y = 7 \quad \Rightarrow \quad y = \frac{-3}{4}x + \frac{7}{4}$$

$$-2kx + 4y = -2 \quad \Rightarrow \quad y = \frac{k}{2}x - \frac{1}{2}$$

■ The slopes of the two lines are $-\dfrac{3}{4}$ and $\dfrac{k}{2}$, the coefficients of the x-terms. Since the lines must be parallel, $\dfrac{k}{2} = -\dfrac{3}{4}$, so $k = -\dfrac{3}{2}$.

The correct choice is **(B)**.

➥ Example :: No-Calculator Section :: Multiple-Choice

$$f(x) = (x - 8)(x + 2)$$

Which of the following is an equivalent form of the function above in which the minimum value of function f appears in the equation as a constant?

(A) $f(x) = x^2 - 16$
(B) $f(x) = (x + 3)^2 - 16$
(C) $f(x) = (x + 3)^2 - 7$
(D) $f(x) = (x - 3)^2 - 25$

Solution

The minimum (or maximum) value of a quadratic function is the y-coordinate of the vertex of its graph, which appears as the constant k in the vertex form of the equation, $f(x) = a(x - h)^2 + k$.

- Using FOIL, multiply out the binomial factors of $f(x)$, which gives $f(x) = x^2 - 6x - 16$.
- Change to the vertex form of the parabola equation by completing the square, which requires adding and then subtracting $\left(\dfrac{-6}{2}\right)^2 = 9$:

$$f(x) = (x^2 - 6x + 9) - 16 - 9$$
$$= (x - 3)^2 - 25$$

- By comparing $f(x) = (x - 3)^2 - 25$ to $f(x) = a(x - h)^2 + k$, you know that $a = 1$, $h = 3$, and $k = -25$. Since $a > 0$, the vertex of the parabola is a minimum point so that the minimum value of f is −25.

The correct choice is **(D)**.

➥ Example :: No-Calculator Section :: Grid-In

$$x^2 + y^2 + 8x - 10y = 80$$

The equation above represents a circle in the xy-plane. What is the length of the longest chord of the circle?

Solution

Convert the equation of the circle to the more useful center-radius form $(x - h)^2 + (y - k)^2 = r^2$ by completing the square for each variable:

$$x^2 + y^2 + 8x - 10y = 80$$
$$\left(x^2 + 8x + \boxed{?}\right) + \left(y^2 - 10y + \boxed{?}\right) = 80$$

In order to complete the square, add the square of one-half of the coefficient of the x-term and the square of one-half of the coefficient of the y-term to both sides of the equation:

$$\left(x^2 + 8x + 16\right) + \left(y^2 - 10y + 25\right) = 80 + (16 + 25)$$
$$(x + 4)^2 + (y - 5)^2 = 121$$

The equation is now in center-radius form $(x - h)^2 + (y - k)^2 = r^2$ where $(h, k) = (-4, 5)$ and $r^2 = 121$ so $r = \sqrt{121} = 11$. Since the longest chord of a circle is its diameter, the longest chord of the circle is $2 \times 11 = 22$.

Grid-in **22**

TIPS FOR BOOSTING YOUR SCORE

- Before starting your solution, make sure you understand what the question is asking you to find. If the problem involves numbers, the types of numbers or the form of the answer may help you decide on a solution strategy. For example, if there are whole numbers in the question, and decimal or fractions appear as answer choices, the solution may involve division.

- If you think you have solved a multiple-choice problem correctly but do not find your answer among the four answer choices, you may need to change the form of your answer. This may involve writing a fraction in lowest terms; changing from decimal to fractional form or vice versa; rearranging or further simplifying the terms of an expression; or factoring. For example, if you don't find $\dfrac{9}{12}$, look for $\dfrac{3}{4}$; if you don't find $\dfrac{7}{2}$, look for 3.5; if you don't see $y = 4x + 15$, also look for $y = 15 + 4x$; if you don't find $8x + 20$, look for $4(2x + 5)$; if you don't find $\dfrac{a^2 - b^2}{a + b}$, look for $a - b$ since

$$\frac{a^2 - b^2}{a + b} = \frac{(a - b)\,\overset{1}{\cancel{(a + b)}}}{\cancel{a + b}} = a - b$$

- Unless otherwise indicated, figures provided with a question are drawn to scale. If it is stated that the figure is *not* drawn to scale and you are not sure how to solve the problem, try redrawing the figure so that it looks to scale. You may then be able to arrive at the correct answer choice by visually estimating a measurement or by eliminating those answer choices that appear to contradict the revised figure.

OVERVIEW

Since there is no penalty for a wrong answer, you should enter an answer for every question even if it means guessing. Guess smartly by crossing out in your test booklet those choices that you know are unlikely or impossible. Before guessing, however, think through a problem and try to solve it mathematically—guessing should be your last resort.

Make sure you practice with the calculator that you will bring to the exam room. If It takes batteries, replace the batteries with fresh ones the day before the exam.

Ruling Out Answer Choices

If you get stuck on a multiple-choice problem, guess after ruling out answer choices as there is no point deduction for a wrong answer. You can increase your chances of guessing the right answer by first eliminating any answer choices that you know are impossible or unlikely. You may be able to rule out an answer choice by asking questions such as

- Must the answer be a certain type of number: positive or negative? greater than 1 or less than 1? involve a radical?
- Can an accompanying figure be used to estimate the answer? If so, based on the estimate, can you eliminate any of the answer choices?
- Do any of the answer choices look very different from the other three answer choices. For example, suppose the four answer choices to a multiple-choice question are

(A) $\dfrac{1}{2}\pi r$

(B) $\pi\!\left(r - \sqrt{2}\right)$

(C) $\pi\!\left(r + \sqrt{2}\right)$

(D) $\pi(r^2 + 2)$

The test makers often try to disguise the correct answer by making it look similar to other choices. If you need to guess, eliminate choice (A) as its structure looks different than the other three choices, which all include the product of π and a parenthesized expression.

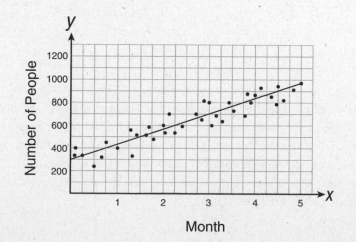

> **TIP**
>
> **Never omit the answer to a question! For questions you are not sure about, guess after after eliminating as many choices as you can.**

➥ Example :: No-Calculator Section :: Multiple-Choice

A new fitness class was started at a chain of fitness clubs owned by the same company. The scatterplot above shows the total number of people attending the class during the first 5 months in which the class was offered. The line of best fit is drawn. If n is the number of the month, which of the following functions could represent the equation of the graph's line of best fit?

(A) $f(n) = 300n + 125$
(B) $f(n) = 300 + 125n$
(C) $f(n) = 400 + 150n$
(D) $f(n) = 200n + 300$

Solution

The equations of the line in the answer choices are in the slope-intercept form $y = mx + b$ where m is the slope of the line and the constant b represents the y-intercept.

- Review the graph to see if you can eliminate any answer choices. You can tell from the graph that the y-intercept is 300, which eliminates choices (A) and (C).
- Pick two convenient data points on the line of best fit where no estimation is needed, such as (0, 300) and (4, 800). Use these points to find the slope, m, of the line:

$$m = \frac{800 - 300}{4 - 0} = \frac{500}{4} = 125$$

- Since $b = 300$ and $m = 125$, the equation of the line in $y = mx + b$ slope-intercept form is $y = 125x + 300$ or, equivalently, $y = 300 + 125x$.

The correct choice is (**B**).

➥ Example :: Calculator Section :: Multiple-Choice

In June, the price of a DVD player that sells for $150 is increased by 10%. In July, the price of the same DVD player is decreased by 10% of its current selling price. What is the new selling price of the DVD player?

(A) $140
(B) $148.50
(C) $150
(D) $152.50

Solution

Is the new selling price equal to the original price of $150, less than the original price, or greater than the original price? Rule out choice (C) since "obvious" answers that do not require any work are rarely correct. In June the price of a DVD player is increased by 10% of $150. In July the price of the DVD player is decreased by 10% of an amount *greater than* $150 (the June selling price). Since the amount of the price decrease was greater than the amount of the price increase, the July price must be *less than* the starting price of $150. You can, therefore, eliminate choices (C) and (D).

This analysis improves your chances of guessing the correct answer. Of course, if you know how to solve the problem without guessing, do so:

STEP 1 June price = \$150 + (10% × \$150) = \$150 + \$15 = \$165
STEP 2 July price = \$165 − (10% × \$165) = \$165.00 − \$16.50 = \$148.50
STEP 3 The correct choice is (**B**).

Calculators

Calculators are permitted on only one of the two math sections. When you take the SAT, you should bring either a scientific or graphing calculator that you are comfortable using and that you used during your practice sessions. Using a calculator wisely and selectively can help you to solve *some* problems in the calculator section more efficiently and with less chance of making a computational error.

When working in the calculator section of the SAT,

- Approach each problem by first deciding how you will use the given information to obtain the desired answer. Then decide whether using a calculator will be helpful.
- Remember that a solution involving many steps with complicated arithmetic is probably not the right way to tackle the problem. Look for another method that involves fewer steps and less complicated computations.
- Use a calculator to help avoid careless arithmetic errors, while keeping in mind that you can often save time by performing very simple arithmetic mentally or by using mathematical reasoning rather than calculator arithmetic.

TIP

Not every problem in the calculator section requires or benefits from using a calculator. Many of the problems do not require a calculator or can be solved faster without using one.

➦ Example :: Calculator Section :: Grid-In

$$v(t) = v_0 - gt \qquad \text{(velocity-time function)}$$
$$h(t) = v_0 t - \frac{1}{2}gt^2 \qquad \text{(position-time function)}$$
$$g = 9.8 \frac{\text{meters}}{\text{sec}^2} \qquad \text{(acceleration due to gravity)}$$

The set of equations above describes projectile motion influenced by gravity after an initial launch velocity of v_0, where t is the number of seconds that have elapsed since the projectile was launched, $v(t)$ is its speed, and $h(t)$ is its height above the ground. If a model rocket is launched upward with an initial velocity of 88 meters per second, what is the maximum height from the ground the rocket will reach correct to *the nearest meter*?

Solution

- When the rocket reaches its maximum height, its velocity is 0. Use the velocity-time function to find the value of t that makes $v(t) = 0$:

$$v(t) = v_0 - gt$$
$$0 = 88 - 9.8t$$
$$9.8t = 88$$
$$\frac{9.8t}{9.8} = \frac{88}{9.8}$$
$$t \approx 9 \text{ seconds}$$

- Substitute 9 for t in the position-time function:

$$h(t) = v_0 t - \frac{1}{2}gt^2$$

$$\left. \begin{aligned} h(9) &= (88)(9) - \frac{1}{2}(9.8)(9)^2 \\ &= 792 - 396.9 \\ &\approx 395 \text{ meters} \end{aligned} \right\} \text{Use calculator to avoid errors.}$$

Grid-in **395**

➥ Example :: Calculator Section :: Multiple-Choice

A certain car is known to depreciate at a rate of 20% per year. The equation $V(n) = p(x)^n$ can be used to calculate the value of the car, V, after n years where p is the purchase price. If the purchase price of the car is \$25,000, to the *nearest dollar*, how much more is the car worth after 2 years than after 3 years?

(A) 1,600
(B) 2,400
(C) 3,200
(D) 4,000

Solution

In the equation $V(n) = p(x)^n$, x represents the exponential decay factor since the car is *losing* value. Hence, x represents 1 *minus* the rate of depreciation.

COMPUTATION METHOD 1

Use your calculator to find the difference between $V(2)$ and $V(3)$, where $p = 25,000$ and $x = 1 - 0.2 = 0.8$:

$$\begin{aligned} V(2) - V(3) &= 25,000(0.8)^2 - 25,000(0.8)^3 \\ &= 25,000(0.64) - 25,000(0.512) \\ &= 16,000 - 12,800 \\ &= 3,200 \end{aligned}$$

COMPUTATION METHOD 2

Avoid the need for a calculator by simplifying the expression before performing the multiplication:

$$V(2) - V(3) = 25{,}000(0.8)^2 - 25{,}000(0.8)^3$$

$$= 25{,}000(0.8)^2[1 - 0.8] \qquad \leftarrow \text{Factor out } 25{,}000(0.8)^2.$$

$$= 25{,}000\left(\frac{4}{5}\right)\left(\frac{4}{5}\right)\left(\frac{1}{5}\right) \qquad \leftarrow \text{Rewrite as fractions.}$$

$$= \frac{\overset{1{,}000}{\cancel{25{,}000}} \times 16}{\cancel{25} \times 5} \qquad \leftarrow \text{Multiply.}$$

$$= \frac{16{,}000}{5}$$

$$= 3{,}200$$

The correct choice is **(C)**.

THE FOUR MATHEMATICS CONTENT AREAS

ANSWER SHEET FOR TUNE-UP EXERCISES
Lesson # _____

ANSWER SHEET

Before you work through the SAT Tune-Up exercises found at the end of each review lesson, you may wish to reproduce the answer form below. This may contain more circles and grids than you will need for a particular exercise section. Mark only those you need and leave the rest blank.

Multiple-Choice

1. Ⓐ Ⓑ Ⓒ Ⓓ   6. Ⓐ Ⓑ Ⓒ Ⓓ   11. Ⓐ Ⓑ Ⓒ Ⓓ   16. Ⓐ Ⓑ Ⓒ Ⓓ

2. Ⓐ Ⓑ Ⓒ Ⓓ   7. Ⓐ Ⓑ Ⓒ Ⓓ   12. Ⓐ Ⓑ Ⓒ Ⓓ   17. Ⓐ Ⓑ Ⓒ Ⓓ

3. Ⓐ Ⓑ Ⓒ Ⓓ   8. Ⓐ Ⓑ Ⓒ Ⓓ   13. Ⓐ Ⓑ Ⓒ Ⓓ   18. Ⓐ Ⓑ Ⓒ Ⓓ

4. Ⓐ Ⓑ Ⓒ Ⓓ   9. Ⓐ Ⓑ Ⓒ Ⓓ   14. Ⓐ Ⓑ Ⓒ Ⓓ   19. Ⓐ Ⓑ Ⓒ Ⓓ

5. Ⓐ Ⓑ Ⓒ Ⓓ   10. Ⓐ Ⓑ Ⓒ Ⓓ   15. Ⓐ Ⓑ Ⓒ Ⓓ   20. Ⓐ Ⓑ Ⓒ Ⓓ

Grid-In

1.   2.   3.   4.   5.

6.   7.   8.   9.   10.

Heart of Algebra

3

"Heart of Algebra" represents the first of the four major mathematics content groups tested by the redesigned SAT. The new SAT places particular emphasis on mastery of the core algebraic concepts and techniques covered in this chapter with more than one-third of the SAT Math questions based on this content area.

LESSONS IN THIS CHAPTER

Lesson 3-1   Some Beginning Math Facts

Lesson 3-2   Solving Linear Equations

Lesson 3-3   Equations with More Than One Variable

Lesson 3-4   Polynomials and Algebraic Fractions

Lesson 3-5   Factoring

Lesson 3-6   Quadratic Equations

Lesson 3-7   Systems of Equations

Lesson 3-8   Algebraic Inequalities

Lesson 3-9   Absolute Value Equations and Inequalities

Lesson 3-10 Graphing in the *xy*-Plane

Lesson 3-11 Graphing Linear Systems

Lesson 3-12 Working with Functions

> ### OVERVIEW
>
> This lesson reviews some prerequisite mathematics terms and skills you should already know including rules for positive integer exponents.

REAL NUMBERS AND THEIR SUBSETS

A **real** number is any number that can be located on a ruler like number line. The set of real numbers includes

- Whole numbers (0, 1, 2, 3, …)
- Integers (… −3, −2, −1, 0, 1, 2, 3, …)
- Rational numbers (fractions with an integer numerator and a nonzero integer denominator)
- Irrational numbers (numbers that are not rational such as π, $\sqrt{3}$, and any nonrepeating, nonending decimal number such as 0.1010010001 …)

CONSTANTS AND VARIABLES

A **constant** is a quantity that always has a fixed value like the number of inches in one foot. A **variable** is a quantity that can change in value such as the number of hours of sleep a person gets from week to week. Variables in algebra are commonly represented by letters such as x and y. A number that multiplies a variable is called its **coefficient**. In the formula $F = \frac{9}{5}C + 32$, which tells how to convert from degrees Celsius to degrees Fahrenheit,

- C is a variable
- $\frac{9}{5}$ is the coefficient of C
- 32 is a constant

COMPARISON SYMBOLS

Table 3.1 summarizes the symbols used to compare two quantities.

Table 3.1. Comparison Symbols

Symbol	Translation	Example
=	is equal to	5 = 5
≠	is *not* equal to	5 ≠ 3
>	is greater than	5 > 3
≥	is greater than *or* equal to	$x \geq 5$ means that x can be 5 *or* any number greater than 5.
<	is less than	3 < 5
≤	is less than *or* equal to	$x \leq 3$ means that x can be 3 *or* any number less than 3.

Since 7 is between 6 and 8, you can write $6 < 7 < 8$. The inequality $6 < 7 < 8$ means that 6 is less than 7 and, at the same time, 7 is less than 8. In general, if x is between a and b with $a < b$, then $a < x < b$. The inequality $a \le x \le b$ means that x is between a and b, or may be equal to either a or b, as shown in Figure 3.1.

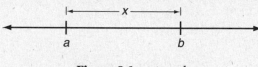

Figure 3.1 $a \le x \le b$

PRODUCTS AND FACTORS

When two or more quantities are multiplied together, the answer is called the **product**. Each of the quantities being multiplied together is called a **factor** of that product. Since $2 \times 3 \times 4 = 24$, the numbers 2, 3, and 4 are each factors of the 24.

EXPONENT RULES

The expression x^n is read as "x raised to the nth power" where the *exponent n* tells how many times the *base x* is used in the repeated multiplication of x:

$$\underbrace{x \cdot x \cdot x \cdots x \cdot x}_{x \text{ appears as a factor } n \text{ times}} = x^n$$

where n is a positive integer and $x^1 = x$. From this definition, there are several useful exponent rules that follow when the bases are the same.

RULES FOR POSITIVE INTEGER EXPONENTS
(*n* and *m* are positive integers and *x* and *y* are real numbers)

Product Rule	Quotient Rule	Power Rule
$x^n \cdot x^m = x^{n+m}$	$x^n \div x^m = x^{n-m}$	$\left(x^n\right)^m = x^{n \times m}$ and $(xy)^m = x^m \cdot y^m$

Here are some examples:

- $\left(2p^4\right)^3 = 2^3 \cdot \left(p^4\right)^3 = 8p^{12}$

- $\left(\dfrac{x}{2}\right)^4 = \dfrac{x^4}{2^4} = \dfrac{x^4}{16}$

- $\dfrac{r^{m+1}}{r^2} = r^{m-1}$

- $\dfrac{x^7 y^2}{x^3 y} = x^{7-3} \cdot y^{2-1} = x^4 y$

➡ Example

If $3x - 2y = 5$, what is the value of $\dfrac{8^x}{4^y}$?

Solution

Change the fraction into an equivalent fraction having the same base in both the numerator and the denominator:

$$\frac{8^x}{4^y} = \frac{\left(2^3\right)^x}{\left(2^2\right)^y}$$

$$= \frac{2^{3x}}{2^{2y}}$$

$$= 2^{3x-2y} \quad \leftarrow \text{Substitute } 5 \text{ for } 3x - 2y$$

$$= 2^5$$

$$= 32$$

➥ Example

If $3^{k+1} - 3^k = m$, what is 3^{k+2} in terms of m?

(A) $\dfrac{4}{3}m$

(B) $3m$

(C) $\dfrac{9}{2}m$

(D) $9m$

Solution

- Rewrite the left side of the given equation in terms of 3^k:

$$3 \cdot 3^k - 3^k = m$$

$$2 \cdot 3^k = m$$

$$3^k = \frac{m}{2}$$

- Multipy 3^k by 3^2 to get 3^{k+2} :

$$3^2\left(3^k\right) = 3^2\left(\frac{m}{2}\right)$$

$$3^{k+2} = \frac{9m}{2}$$

The correct choice is **(C)**.

DISTRIBUTIVE LAW

Two times the sum of x and y can be represented by $2(x + y)$, which must be equivalent to multiplying x and y individually by 2 and then adding the products together, as in $2x + 2y$. Thus, $2(x + y) = 2x + 2y$.

Here are some more examples:

- $3(a - 2b) = 3a - 6b$
- $x(xy + 4) = x^2y + 4x$
- $rs^2\left(\dfrac{1}{r} + \dfrac{1}{s}\right) = \dfrac{rs^2}{r} + \dfrac{rs^2}{s} = s^2 + rs$

The process of reversing the distributive law is called **factoring**:

- $4x + 4y = 4(x + y)$ ← Factor out 4
- $2x + 3x = x(2 + 3)$ ← Factor out x
  $\qquad\quad = 5x$
- $2 - x = -(x - 2)$ ← Factor out −1
- $a^2b - ab^2 = ab(a - b)$ ← Factor out ab

ROOTS

The square root notation $\sqrt{x}$ means one of two equal nonnegative numbers whose product is x as in $\sqrt{16} = 4$ since $4 \times 4 = 16$.

- $\sqrt[2]{x}$ means the same thing as $\sqrt{x}$.
- $\sqrt[3]{x}$ means the cube root of x, which is one of three equal numbers whose product is x as in $\sqrt[3]{8} = 2$ since $2 \times 2 \times 2 = 8$. Also, $\sqrt[4]{x}$ represents the fourth root of x, $\sqrt[5]{x}$ is the fifth root of x, and so on.
- The square root symbol $\sqrt{}$ is called a **radical** and always represents the *positive* or **principal square root** of the number underneath it. Thus, $\sqrt{9} = 3$ but $\sqrt{9} \neq -3$ despite the fact that $(-3) \times (-3) = 9$. However, if $x^2 = 9$, then either $x = -3$ or $x = 3$.

Some Properties of Square Root Radicals

- To multiply square root radicals together, multiply the radicands:

$$\sqrt{3} \times \sqrt{7} = \sqrt{21}$$

and

$$\sqrt{2}\left(\sqrt{8} + \sqrt{5}\right) = \sqrt{16} + \sqrt{10} = 4 + \sqrt{10}$$

- To simplify a square root radical, factor the radicand as the product of two numbers, one of which is the greatest perfect square factor of the radicand. Then write the radical over each factor and simplify:

$$\sqrt{20} = \sqrt{4 \times 5} = \sqrt{4} \times \sqrt{5} = 2\sqrt{5}$$

■ To combine square root radicals, the number or expression underneath the radical sign, called the **radicand**, must be the same:

$$\sqrt{18} + 5\sqrt{2} = \sqrt{9 \cdot 2} + 5\sqrt{2}$$
$$= 3\sqrt{2} + 5\sqrt{2}$$
$$= 8\sqrt{2}$$

■ To eliminate a radical from the denominator of a fraction, multiply the numerator and denominator of the fraction by the radical:

$$\frac{12}{\sqrt{3}} \times \frac{\sqrt{3}}{\sqrt{3}} = \frac{12\sqrt{3}}{3} = 4\sqrt{3}$$

ALGEBRAIC REPRESENTATION

The SAT may include problems that test your ability to use a combination of variables, constants, and arithmetic operations to represent a given situation.

➡ Example

Sam and Jeremy have ages that are consecutive odd integers. The product of their ages is 783. Which equation could be used to find Jeremy's age, j, if he is the younger man?

(A) $j^2 + 2 = 783$
(B) $j^2 - 2 = 783$
(C) $j^2 + 2j = 783$
(D) $j^2 - 2j = 783$

Solution

Since consecutive odd integers differ by 2, if Jeremy's age is represented by j, then $j + 2$ represents Sam's age so $j(j + 2) = 783$. Using the distributive law, $j^2 + 2j = 783$.

The correct choice is (**C**).

➡ Example

A crew of painters are hired to paint a house. The cost of the paint is estimated to be $250 and each of the p painters will get paid $15 per hour. If the crew of painters will work a 7 hour day and the job is estimated to take d days, which expression best represents the total cost of painting the house?

(A) $250 + \dfrac{15p}{7d}$

(B) $250 + 105pd$

(C) $250 + \dfrac{105}{pd}$

(D) $7d(250 + 15p)$

Solution

- The hourly cost of the p painters for each 7 hour day is $(p \times 15) \times 7 = 105p$.
- Since the job is estimated to take d days, the cost of the painter is $105pd$.
- Adding the cost of the paint gives $250 + 105pd$.

The correct choice is **(B)**.

Multiple-Choice

1. If $5 = a^x$, then $\dfrac{5}{a} =$

   (A) a^{x+1}
   (B) a^{x-1}
   (C) a^{1-x}
   (D) $a\dfrac{x}{5}$

2. If $y = 25 - x^2$ and $1 \le x \le 5$, what is the smallest possible value of y?

   (A) 0
   (B) 1
   (C) 5
   (D) 10

3. Given $y = wx^2$ and y is not 0. If the values of x and w are each doubled, then the value of y is multiplied by

   (A) 1
   (B) 2
   (C) 4
   (D) 6

4. If $\dfrac{x^{23}}{x^m} = x^{15}$ and $(x^4)^n = x^{20}$, then $mn =$

   (A) 13
   (B) 24
   (C) 28
   (D) 40

5. If $2 = p^3$, then $8p$ must equal

   (A) p^6
   (B) p^8
   (C) p^{10}
   (D) $8\sqrt{2}$

6. If $10^{k-3} = m$, then $10^k =$

   (A) $1,000m$
   (B) $m + 1,000$
   (C) $\dfrac{m}{1,000}$
   (D) $m - 1,000$

7. If w is a positive number and $w^2 = 2$, then $w^3 =$

   (A) $\sqrt{2}$
   (B) $2\sqrt{2}$
   (C) 4
   (D) $3\sqrt{2}$

8. If $x = \sqrt{6}$ and $y^2 = 12$, then $\dfrac{4}{xy} =$

   (A) $\dfrac{3}{2\sqrt{2}}$
   (B) $\dfrac{\sqrt{2}}{3}$
   (C) $\dfrac{3}{\sqrt{2}}$
   (D) $\dfrac{2\sqrt{2}}{3}$

9. If x is a positive integer such than $x^9 = r$ and $x^5 = w$, which of the following must be equal to x^{13}?

   (A) $rw - 1$
   (B) $r + w - 1$
   (C) $\dfrac{r^2}{w}$
   (D) $r^2 - w$

10. Caitlin has a movie rental card worth $175. After she rents the first movie, the card's value is $172.25. After she rents the second movie, its value is $169.50. After she rents the third movie, the card is worth $166.75. Assuming the pattern continues, which of the following equations define A, the amount of money on the rental card after n rentals?

   (A) $175 - 2.75n$
   (B) $2.75n - 175$
   (C) $(175 - 2.75)n$
   (D) $\dfrac{175}{2.75}n$

11. Three times the sum of a number and four is equal to five times the number, decreased by two. If x represents the number, which equation is a correct translation of the statement?

(A) $3(x + 4) = 5x - 2$
(B) $3(x + 4) = 5(x - 2)$
(C) $3x + 4 = 5x - 2$
(D) $3x + 4 = 5(x - 2)$

12. Owen gets paid $280 per week plus 5% commission on all sales for selling electronic equipment. If he sells d dollars worth of electronic equipment in one week, which algebraic expression represents the amount of money he will earn in w weeks?

(A) $(280d + 5)w$
(B) $280 + 0.05dw$
(C) $(280 + 0.05d)w$
(D) $280w + 0.05d$

13. Which expression represents the number of hours in w weeks and d days?

(A) $7w + 12d$
(B) $84w + 24d$
(C) $168w + 24d$
(D) $168w + 60d$

14. Which verbal expression can be represented by $2(x - 5)$?

(A) 5 less than 2 times x
(B) 2 multiplied by x less 5
(C) twice the difference of x and 5
(D) the product of 2 and x, decreased by 5

15. If k pencils cost c cents, what is the cost in cents of p pencils?

(A) $\dfrac{pc}{k}$

(B) $\dfrac{pk}{c}$

(C) $\dfrac{c}{kp}$

(D) cpk

16. If it takes h hours to paint a rectangular wall that is x feet wide and y feet long, how many *minutes* does it take to paint 1 square foot of the bulletin board?

(A) $\dfrac{xy}{60h}$

(B) $\dfrac{60h}{xy}$

(C) $\dfrac{h}{60}(x + y)$

(D) $\dfrac{hxy}{60}$

17. If Carol earns x dollars a week for 3 weeks and y dollars a week for 4 weeks, what is the average number of dollars per week that she earns?

(A) $\dfrac{1}{7}(x + y)$

(B) $\dfrac{3x + 4y}{7}$

(C) $\dfrac{12xy}{7}$

(D) $\dfrac{x}{3} + \dfrac{y}{4}$

18. Tim bought a skateboard and two helmets for a total of d dollars. If the skateboard costs s dollars, the cost of each helmet could be represented by which of the following expressions?

(A) $2ds$

(B) $\dfrac{ds}{2}$

(C) $\dfrac{d - s}{2}$

(D) $d - \dfrac{s}{2}$

19. The length of a rectangle is three feet less than twice its width. If x represents the width of the rectangle, in feet, which inequality represents the area of the rectangle that is *at most* 30 square feet?

(A) $x(2x - 3) \le 30$
(B) $x(2x - 3) \ge 30$
(C) $x(3 - 2x) \le 30$
(D) $x(3 - 2x) \ge 30$

20. In 1995, the U.S. federal government paid off one-third of its debt. If the original amount of the debt was \$4,920,000,000,000, which expression represents the amount that was *not* paid off?

(A) 1.64×10^4
(B) 1.64×10^{12}
(C) 3.28×10^8
(D) 3.28×10^{12}

$$\frac{\left(b^{2n+1}\right)^3}{b^n \cdot b^{5n+1}}$$

21. The expression above is equivalent to which of the following?

(A) b^2
(B) $\dfrac{1}{b^2}$
(C) b^{2n}
(D) $\dfrac{1}{b^{2n}}$

Grid-In

1. If $2^4 \times 4^2 = 16^x$, then $x =$

2. If $a^7 = 7{,}777$ and $\dfrac{a^6}{b} = 11$, what is the value of ab?

3. If $y = 2^{2p-1}$ and $z = p - 2$, what is the value of $\dfrac{y}{z}$ when $p = 2.5$?

4. If $13 \le k \le 21$, $9 \le p \le 19$, $2 < m < 6$, and k, p, and m are integers, what is the largest possible value of $\dfrac{k - p}{m}$?

OVERVIEW

A **linear equation** is an equation in which each term is either a constant or the product of a constant and the first power of a single variable such as $2x + 1 = 7$. To solve a linear equation, isolate the variable on one side of the equation by doing the same thing on both sides until the resulting equation has the form, *variable* = constant, as in $x = 3$.

BASIC TYPES OF SAT EQUATIONS

Some basic types of equations that appear on the SAT may look a little different than the equations encountered in your regular mathematics classes, but they can be solved using the same procedures.

➡ Example

If $\dfrac{r}{s} = 6$, what is the value of $\dfrac{4s}{r}$?

Solution

If $\dfrac{r}{s} = 6$, then inverting both sides gives $\dfrac{s}{r} = \dfrac{1}{6}$:

$$\frac{4s}{r} = 4\left(\frac{s}{r}\right) = 4\left(\frac{1}{6}\right) = \frac{4}{6} = \frac{2}{3}$$

Grid-in **2/3**

➡ Example

If $kx - 19 = k - 1$ and $k = 3$, what is the value of $x + k$?

Solution

Replace k with 3 and solve the resulting equation in the usual way:

$$3x - 19 = 3 - 1$$
$$3x - 19 = 2$$
$$3x = 21$$
$$\frac{3x}{3} = \frac{21}{3}$$
$$x = 7$$
$$x + k = 7 + 3 = 10$$

Grid-in **10**

TIP

In your regular math class, solving for *x* typically gives you the final answer to the problem. But *x* is not always the final answer on the SAT. When reading a question, circle or underline the quantity that the question asks you to find. After you solve the problem, check that your answer matches what you were required to find by looking back at what you circled or underlined.

➥ Example

If $2y - 7 = 18$, what is the value of $2y + 3$?

(A) 15
(B) 18
(C) 27
(D) 28

Solution

Rather than first solving the equation for y, solve for the quantity asked for directly. To get $2y + 3$ from $2y - 7$, add 10 to both sides of the equation:

$$2(y - 7) + 10 = 18 + 10$$
$$2y + 3 = 28$$

The correct choice is **(D)**.

➥ Example

When 6 times a number x is added to 5, the result is 19. What number results when 5 is subtracted from 3 times x?

(A) -3
(B) $\dfrac{2}{3}$
(C) 2
(D) 7

Solution

According to the conditions of the problem, $6x + 5 = 19$, and you must find the value of $3x - 5$.

- If $6x + 5 = 19$, then $6x = 19 - 5$ so $6x = 14$.
- Dividing each side of $6x = 14$ by 2 gives $3x = 7$, and subtracting 5 from each side yields $3x - 5 = 2$.

The correct choice is **(C)**.

➥ Example

If $\dfrac{5}{x} = \dfrac{9}{x + 12}$, what is the value of $\dfrac{x}{3}$?

Solution

If $\dfrac{5}{x} = \dfrac{9}{x+12}$, then cross-multiplying makes $5(x+12) = 9x$. Remove the parentheses by multiplying each term inside the parentheses by 5:

$$5x + 60 = 9x$$
$$4x = 60 \quad \leftarrow \text{To get } \frac{x}{3} \text{ from } 4x, \text{divide both sides by } 12$$
$$\frac{4x}{12} = \frac{60}{12}$$
$$\frac{x}{3} = 5$$

Grid-in **5**

➡ Example

If $\dfrac{13}{16}x - \dfrac{3}{8}x = \dfrac{2}{5} + \dfrac{3}{10}$, what is the value of x?

(A) $\dfrac{17}{8}$

(B) $\dfrac{9}{4}$

(C) $\dfrac{8}{5}$

(D) $\dfrac{17}{10}$

Solution

Combine fractions on each side of the equation:

$$\frac{13}{16}x - \frac{3}{8}x = \frac{2}{5} + \frac{3}{10}$$
$$\frac{13}{16}x - \frac{6}{16}x = \frac{4}{10} + \frac{3}{10}$$
$$\frac{7}{16}x = \frac{7}{10} \quad \leftarrow \text{Isolate } x \text{ by multiplying both sides}$$
$$\text{by the reciprocal of its coefficient}$$

$$\frac{16}{7}\left(\frac{7}{16}x\right) = \frac{16}{\cancel{7}}\left(\frac{\cancel{7}^{1}}{10}\right)$$
$$1 \cdot x = \frac{16}{10}$$
$$x = \frac{8}{5}$$

The correct choice is **(C)**.

Solving a Linear Equation Using Two Operations

To isolate a variable in a linear equation, it may be necessary to add or subtract and then to multiply or divide. If an equation contains parentheses, remove them by multiplying each term inside the parentheses by the term outside the parentheses.

➡ Example

Solve for b:

$$3(b + 2) + 2b = 21$$

Solution

$$3(b + 2) + 2b = 21$$

- Remove the parentheses by multiplying each term inside the parentheses by 3:
$$3b + 6 + 2b = 21$$
- Combine like terms:
$$5b + 6 = 21$$
- Subtract 6 from each side of the equation:
$$5b + 6 - 6 = 21 - 6$$
- Simplify:
$$5b = 15$$
- Divide each side of the equation by 5:
$$\frac{5b}{5} = \frac{15}{5}$$
$$b = 3$$

➡ Example

$$\frac{7}{3}\left(x + \frac{9}{28}\right) - 3 = 17$$

Which value of x satisfies the equation above?

(A) $\dfrac{33}{4}$

(B) $\dfrac{249}{28}$

(C) $\dfrac{77}{4}$

(D) $\dfrac{45}{28}$

Solution

$$\frac{7}{3}\left(x + \frac{9}{28}\right) = 20$$

$$\frac{7}{3}x + \left(\frac{\cancel{7}}{\cancel{3}}\right)\left(\frac{\cancel{9}^{3}}{\cancel{28}_{4}}\right) = 20$$

$$\frac{7}{3}x = 20 - \frac{3}{4}$$

$$\frac{7}{3}x = \frac{80}{4} - \frac{3}{4}$$

$$\frac{3}{7}\cdot\left(\frac{7}{3}x\right) = \frac{3}{\cancel{7}}\cdot\left(\frac{\cancel{77}^{11}}{4}\right)$$

$$x = \frac{33}{4}$$

The correct choice is **(A)**.

➡ Example ⟍

$$\frac{2(m+4)+13}{5} = \frac{21-(8-3m)}{4}$$

In the equation above, what is the value of m?

Solution

- Simplify the numerators:

$$\frac{2m+21}{5} = \frac{13+3m}{4}$$

- Cross-multiply and simplify:

$$5(13+3m) = 4(2m+21)$$

$$65 + 15m = 8m + 84$$

- Isolate the variable:

$$15m - 8m = 84 - 65$$

$$7m = 19$$

$$m = \frac{19}{7}$$

Grid-in **19/7**

Multiple-Choice

1. If $\dfrac{m-n}{n} = \dfrac{4}{9}$, what is the value of $\dfrac{n}{m}$?

   (A) $\dfrac{9}{13}$

   (B) $\dfrac{7}{4}$

   (C) $\dfrac{9}{5}$

   (D) $\dfrac{13}{7}$

2. If $\dfrac{p+4}{p-4} = 13$, what is the value of p?

   (A) $\dfrac{2}{3}$

   (B) $\dfrac{10}{13}$

   (C) $\dfrac{28}{9}$

   (D) $\dfrac{14}{3}$

3. If $4x + 7 = 12$, what is the value of $8x + 3$?

   (A) 9
   (B) 11
   (C) 13
   (D) 15

4. $\dfrac{6}{x} = \dfrac{4}{x-9}$, what is the value of $\dfrac{x}{18}$?

   (A) 3
   (B) 2
   (C) $\dfrac{1}{2}$
   (D) $\dfrac{3}{2}$

5. If $3j - (k + 5) = 16 - 4k$, what is the value of $j + k$?

   (A) 8
   (B) 7
   (C) 5
   (D) 4

6. If $\dfrac{1}{2}(10p + 2) = p + 7$, then $4p =$

   (A) 6
   (B) $\dfrac{5}{2}$
   (C) 4
   (D) 3

7. If $0.25y + 0.36 = 0.33y - 1.48$, what is the value of $\dfrac{y}{10}$?

   (A) 2.30
   (B) 1.40
   (D) 0.75
   (C) 0.64

8. If $\dfrac{4}{7}k = 36$, then $\dfrac{3}{7}k =$

   (A) 21
   (B) 27
   (C) 32
   (D) 35

9. If $\dfrac{1}{2}x + \dfrac{1}{4}x + \dfrac{1}{8}x = 14$, then $x =$

   (A) 4
   (B) 8
   (C) 12
   (D) 16

10. If $\dfrac{2}{x} = 2$, then $x + 2 =$

   (A) $\dfrac{3}{2}$

   (B) $\dfrac{5}{2}$

   (C) 3

   (D) 4

11. If $\dfrac{y-2}{2} = y + 2$, then $y =$

   (A) -6

   (B) -4

   (C) -2

   (D) 4

12. If $\dfrac{2y}{7} = \dfrac{y+3}{4}$, then $y =$

   (A) 5

   (B) 9

   (C) 13

   (D) 21

13. If $\dfrac{y}{3} = 4$, then $3y =$

   (A) 4

   (B) 12

   (C) 24

   (D) 36

14. When the number k is multiplied by 5, the result is the same as when 5 is added to k. What is the value of k?

   (A) $\dfrac{4}{5}$

   (B) 1

   (C) $\dfrac{5}{4}$

   (D) $\dfrac{3}{2}$

$$\frac{8r + 7}{4s} = 11$$

15. If $\dfrac{1}{2}r + 1 = s + 1$, what is the value of $r + s$ for the equation above?

   (A) $\dfrac{1}{2}$

   (B) $\dfrac{3}{4}$

   (C) 1

   (D) $\dfrac{3}{2}$

16. If $m + 1 = \dfrac{5(m-1)}{3}$, then $\dfrac{1}{m} =$

   (A) $\dfrac{1}{4}$

   (B) $\dfrac{3}{8}$

   (C) 2

   (D) $\dfrac{8}{3}$

$$\frac{5(p-1) + 6}{8} = \frac{7 - (3 - 2p)}{12}$$

17. In the equation above, what is the value of p?

   (A) $\dfrac{1}{3}$

   (B) $\dfrac{5}{11}$

   (C) $\dfrac{2}{3}$

   (D) $\dfrac{9}{11}$

18. During the investigation of an archeological dig, a femur bone was found and used to estimate the height of the person it came from using the formula $h = 61.4 + 2.3F$, where h is the height, in *centimeters*, of a person whose femur is F *centimeters* in length. Using this formula, the height of the person was estimated to be 5 feet 8 inches. The length of the femur, in centimeters, was closest to which of the following lengths? [Note: 1 inch = 2.54 centimeters]

(A) 37.3
(B) 48.4
(C) 51.0
(D) 56.3

19. Last month, Sara, Ryan, and Taylor received a total of 882 emails. If Sara received 25% more emails than the sum of the number of emails received by Ryan and Taylor, how many emails did Sara receive?

(A) 448
(B) 486
(C) 490
(D) 504

20. In an election for senior class president, Emily received approximately 25% more votes than Alexis. If Emily received 163 votes, the number of votes Alexis received is closest to

(A) 122
(B) 130
(C) 138
(D) 204

21. At City High School, the sophomore class has 60 more students than the freshman class. The junior class has 50 fewer students than twice the students in the freshman class. The senior class is three times as large as the freshman class. If there are a total of 1,424 students at City High School, how many students are in the freshman class?

(A) 202
(B) 205
(C) 235
(D) 236

Grid-In

1. If $2w - 1 = 2$, what is the value of $w^2 - 1$?

2. If $11 - 3x$ is 4 less than 13, what is the value of $6x$?

3. If $2x + 1 = 8$ and $15 - 3y = 0$, what is value of $\dfrac{x}{y}$?

4. The total score in a football game was 72 points. The winning team scored 12 points more than the losing team. How many points did the winning team score?

5. Max's cell phone plan charges a monthly "pay-as-you-go rate" of $0.13 for each text message sent or received. If Max was charged $7.41 for 15 more sent texts than he received, how many texts did Max send?

6. If $\dfrac{7}{12}x - \dfrac{1}{3}x = \dfrac{1}{2} + \dfrac{3}{8}$, what is the value of x?

$$\frac{5(y-2)}{y} - \frac{1}{3} = 0$$

7. In the equation above, what is the value of y?

$$D = 141 - 0.16p$$
$$S = 64 + 0.28p$$

8. The set of equations above describes how the supply, S, and demand, D, for a computer memory chip depends on market price. If p represents the price in dollars for each lot of 10 memory chips, for what dollar price per memory chip does supply equal demand?

9. If $\dfrac{7(x+9)}{4} - 1 = 41$, what is the value of $x - 9$?

10. Gerald and Jim work at a furniture store. Gerald is paid $185 per week plus 4% of his total sales. Jim is paid $275 per week plus 2.5% of his total sales. What amount of sales will make their weekly pay the same?

11. If 7 quarters and n nickels is equivalent to 380 pennies, what is the value of n?

12. A gardener is planting two types of trees:

   - Type A is three feet tall and grows at a rate of 15 inches per year.
   - Type B is five feet tall and grows at a rate of 9 inches per year.

   How many years will it take for these trees to grow to the same height?

$$\frac{9n - (5n - 3)}{8} = \frac{2(n-1) - (3 - 7n)}{12}$$

13. In the equation above, what is the value of n?

14. An animal shelter spends $2.35 per day to care for each cat and $5.50 per day to care for each dog. If $89.50 is spent caring for a total of 22 cats and dogs, how much was spent on caring for all of the cats?

$$r = 2.24 + 0.06x$$
$$p = 2.89 + 0.10x$$

15. In the equations above, r and p represent the price per gallon, in dollars, of regular and premium grades of gasoline, respectively, x months after January 1 of last year. What was the cost per gallon, in dollars, of premium gasoline for the month in which the per gallon price of premium exceeded the per gallon price of regular by $0.93?

OVERVIEW

An equation that has more than one letter can have many different numerical solutions. For example, in the equation $x + y = 8$, if $x = 1$, then $y = 7$; if $x = 2$, then $y = 6$. Since the value of y depends on the value of x, and x may be any real number, the equation $x + y = 8$ has infinitely many solutions.

The SAT may include questions that ask you to

- Find the value of an expression that is a multiple of one side of an equation. For example, if $x + 2y = 9$, then the value of $2x + 4y$ is 18 since

$$2x + 4y = 2(x + 2y) = 2(9) = 18$$

- Solve for one letter in terms of the other letter(s) of an equation. For example, if $x + y = 8$, then x in terms of y is $x = 8 - y$.

WORKING WITH AN EQUATION IN TWO UNKNOWNS

Some SAT questions can be answered by multiplying or dividing an equation by a suitable number.

➡ Example

If $3x + 3y = 12$, what is the value of $x + y - 6$?

Solution

- Solve the given equation for $x + y$ by dividing each member by 3:

$$\frac{3x}{3} + \frac{3y}{3} = \frac{12}{3}$$
$$x + y = 4$$

- Evaluate the required expression by replacing $x + y$ with 4:

$$x + y - 6 = 4 - 6$$
$$= -2$$

The value of $x + y - 6$ is -2.

➡ Example

If $\frac{t}{s} = \frac{3}{4}$, what is the value of $\frac{3s}{4t}$?

Solution

Since $\frac{t}{s} = \frac{3}{4}$, then $3s = 4t$ $\quad s \cdot \frac{3s}{4t} = 1$.

➡ Example

If $x + 5 = t$, then $2x + 9 =$

(A) $t - 1$
(B) $t + 1$
(C) $2t$
(D) $2t - 1$

Solution

METHOD 1: Use algebraic reasoning:

- Since the coefficient of x in $2x + 9$ is 2, multiply the given equation by 2:

$$2(x + 5) = 2t, \text{ so } 2x + 10 = 2t$$

- Comparing the left side of $2x + 10 = 2t$ with $2x + 9$ suggests that you subtract 1 from both sides of the equation:

$$(2x + 10) - 1 = 2t - 1$$
$$2x + 9 = 2t - 1$$

The correct choice is (**D**).

METHOD 2: Pick easy numbers for x and t that satisfy $x + 5 = t$. Then use these values for x and t to compare $2x + 9$ with each of the answer choices. For example, when $x = 1$, $1 + 5 = t = 6$. Using $x = 1$, you know that $2x + 9 = 2(1) + 9 = 11$. Hence, the correct answer choice is the one that evaluates to 11 when $t = 6$:

$$
\begin{aligned}
\text{Choice (A):} \quad & t - 1 = 6 - 1 && = 5 && ✗ \\
\text{Choice (B):} \quad & t + 1 = 6 + 1 && = 7 && ✗ \\
\text{Choice (C):} \quad & 2t = 2(6) && = 12 && ✗ \\
\text{Choice (D):} \quad & 2t - 1 = 2(6) - 1 && = 11 && ✓
\end{aligned}
$$

➡ Example

If $\dfrac{4}{x+1} = \dfrac{8}{y-2}$ where $x \neq -1$ and $y \neq 2$, what is y in terms of x?

(A) $y = 2x + 1$
(B) $y = 2(x + 2)$
(C) $y = 2x$
(D) $y = 4x + 2$

Solution

$$\frac{4}{x+1} = \frac{8}{y-2}$$
$$4(y-2) = 8(x+1)$$
$$4y - 8 = 8x + 8$$
$$4y = 8x + 16$$
$$\frac{4y}{4} = \frac{8x}{4} + \frac{16}{4}$$
$$y = 2x + 4$$
$$y = 2(x+2)$$

The correct choice is **(B)**.

➡ Example

$$3(r + 2s) = r + s$$

If (r, s) is a solution to the equation above and $s \neq 0$, what is the value of $\frac{r}{s}$?

(A) $-\frac{5}{2}$

(B) $-\frac{2}{5}$

(C) $\frac{2}{3}$

(D) $\frac{3}{2}$

Solution

Removing the parentheses gives $3r + 6s = r + s$. Collecting like terms on the same side of the equation makes $2r = -5s$ so $r = -\frac{5}{2}s$ and $\frac{r}{s} = -\frac{5}{2}$.

The correct choice is **(A)**.

Multiple-Choice

1. If $V = \frac{1}{3}Bh$, what is h expressed in terms of B and V?

   (A) $\frac{1}{3}VB$

   (B) $\frac{V}{3B}$

   (C) $\frac{3V}{B}$

   (D) $3VB$

2. If $F = \frac{kmM}{r^2}$, then $m =$

   (A) $\frac{Fr^2}{kM}$

   (B) $\frac{kFr^2}{M}$

   (C) $\frac{kM}{Fr^2}$

   (D) $F(r^2 + kM)$

3. If $P = 2(L + W)$, what is W in terms of P and L?

   (A) $P - 2L$

   (B) $\frac{P - 2L}{2}$

   (C) $\frac{2L - P}{2}$

   (D) $\frac{1}{2}(P - L)$

4. If $A = \frac{1}{2}h(x + y)$, what is y in terms of A, h, and x?

   (A) $\frac{2A - hx}{h}$

   (B) $\frac{A - hx}{2h}$

   (C) $2Ah - x$

   (D) $2A - hx$

5. If $s = \frac{2x + t}{r}$, then $x =$

   (A) $\frac{rs - t}{2}$

   (B) $\frac{rs + 1}{2}$

   (C) $2rs - t$

   (D) $rs - 2t$

6. If $x = x_0 + \frac{1}{2}(v + v_0)t$, what is v in terms of the other variables?

   (A) $\frac{2(x - x_0)}{v_0 t}$

   (B) $\frac{2(x - x_0)}{t} - v_0$

   (C) $\frac{t(x - x_0)}{2v_0}$

   (D) $v_0 t - \frac{2(x - x_0)}{t}$

7. If $2s - 3t = 3t - s$, what is s in terms of t?

   (A) $\frac{t}{2}$

   (B) $2t$

   (C) $t + 2$

   (D) $\frac{t}{2} + 1$

8. If $xy + z = y$, what is x in terms of y and z?

   (A) $\frac{y + z}{y}$

   (B) $\frac{y - z}{z}$

   (C) $\frac{y - z}{y}$

   (D) $1 - z$

9. If $b(x + 2y) = 60$ and $by = 15$, what is the value of bx?

(A) 15
(B) 20
(C) 25
(D) 30

10. If $\dfrac{a-b}{b} = \dfrac{2}{3}$, what is the value of $\dfrac{a}{b}$?

(A) $\dfrac{1}{2}$

(B) $\dfrac{3}{5}$

(C) $\dfrac{3}{2}$

(D) $\dfrac{5}{3}$

11. If $s + 3s$ is 2 more than $t + 3t$, then $s - t =$

(A) -2

(B) $-\dfrac{1}{2}$

(C) $\dfrac{1}{2}$

(D) $\dfrac{3}{4}$

12. If $\dfrac{1}{p+q} = r$ and $p \neq -q$, what is p in terms of r and q?

(A) $\dfrac{rq-1}{q}$

(B) $\dfrac{1+rq}{q}$

(C) $\dfrac{r}{1+rq}$

(D) $\dfrac{1-rq}{r}$

13. If $\dfrac{a+b+c}{3} = \dfrac{a+b}{2}$ then $c =$

(A) $\dfrac{a-b}{2}$

(B) $\dfrac{a+b}{2}$

(C) $5a + 5b$

(D) $\dfrac{a+b}{5}$

14. If the value of n nickels plus d dimes is c cents, what is n in terms of d?

(A) $\dfrac{c}{5} - 2d$

(B) $5c - 2d$

(C) $\dfrac{c-d}{10}$

(D) $\dfrac{cd}{10}$

15. If $\dfrac{c}{d} - \dfrac{a}{b} = x$, $a = 2c$, and $b = 5d$, what is the value of $\dfrac{c}{d}$ in terms of x?

(A) $\dfrac{2}{3}x$

(B) $\dfrac{3}{4}x$

(C) $\dfrac{4}{3}x$

(D) $\dfrac{5}{3}x$

$$\frac{4}{t-1} = \frac{2}{w-1}$$

16. If in the equation above $t \neq 1$ and $w \neq 1$, then $t =$

(A) $2w - 1$
(B) $2(w - 1)$
(C) $w - 2$
(D) $2w$

Grid-In

1. If $16 \times a^2 \times 64 = (4 \times b)^2$ and a and b are positive integers, then b is how many times greater than a?

2. If $3a - c = 5b$ and $3a + 3b - c = 40$, what is the value of b?

3. If $a = 2x + 3$ and $b = 4x - 7$, for what value of x is $3b = 5a$?

$$\frac{x}{8} + \frac{y}{5} = \frac{31}{40}$$

4. In the equation above, if x and y are positive integers, what is the value of $x + y$?

5. If $\frac{1}{8}x + \frac{1}{8}y = y - 2x$, then what is the value of $\frac{x}{y}$?

OVERVIEW

A **polynomial** is a single term or the sum or difference of two or more unlike terms. For example, the polynomial $a + 2b + 3c$ represents the sum of the three unlike terms a, $2b$, and $3c$. Since polynomials represent real numbers, they can be added, subtracted, multiplied, and divided using the laws of arithmetic.

Whenever a letter appears in the denominator of a fraction, you may assume that it cannot represent a number that makes the denominator of the fraction equal to 0.

CLASSIFYING POLYNOMIALS

A polynomial can be classified according to the number of terms it contains.

- A polynomial with one term, as in $3x^2$, is called a **monomial**.
- A polynomial with two unlike terms, as in $2x + 3y$, is called a **binomial**.
- A polynomial with three unlike terms, as in $x^2 + 3x - 5$, is called a **trinomial**.

OPERATIONS WITH POLYNOMIALS

Polynomials may be added, subtracted, multiplied, and divided.

- To add polynomials, write one polynomial on top of the other one so that like terms are aligned in the same vertical columns. Then combine like terms. For example:

$$\begin{array}{r} 2x^2 - 3x + 7 \\ +\ \ x^2 + 5x - 9 \\ \hline 3x^2 + 2x - 2 \end{array}$$

- To subtract polynomials, take the opposite of each term of the polynomial that is being subtracted. Then add the two polynomials. For instance, the difference

$$(7x - 3y - 9z) - (5x + y - 4z)$$

can be changed into a sum by adding the opposite of each term of the second polynomial to the first polynomial:

$$\begin{array}{r} 7x - 3y - 9z \\ +\ \ -5x - y + 4z \\ \hline 2x - 4y - 5z \end{array}$$

- To multiply monomials, multiply their numerical coefficients and multiply *like* variable factors by *adding* their exponents. For example:

$$\begin{aligned} (-2a^2b)(4a^3b^2) &= (-2)(4)(a^2 \cdot a^3)(b \cdot b^2) \\ &= -8(a^{2+3})(b^{1+2}) \\ &= -8a^5b^3 \end{aligned}$$

- To divide monomials, divide their numerical coefficients and divide *like* variable factors by *subtracting* their exponents. For example:

$$\frac{14x^5y^2}{21x^2y^2} = \left(\frac{14}{21}\right)\left(\frac{x^5}{x^2}\right)\left(\frac{y^2}{y^2}\right)$$

$$= \left(\frac{2}{3}\right)(x^{5-2})(1)$$

$$= \frac{2}{3}x^3$$

- To multiply a polynomial by a monomial, multiply each term of the polynomial by the monomial and add the resulting products. For example:

$$5x(3x - y + 2) = 5x(3x) + 5x(-y) + 5x(2)$$

$$= 15x^2 - 5xy + 10x$$

The expression $-(a - b)$ can be interpreted as "take the opposite of whatever is inside the parentheses." The result is $b - a$ since

$$-(a - b) = -1(a - b)$$

$$= (-1)a + (-1)(-b)$$

$$= -a + b \text{ or } b - a$$

- To divide a polynomial by a monomial, divide each term of the polynomial by the monomial and add the resulting quotients. For example:

$$\frac{6x + 15}{3} = \frac{6x}{3} + \frac{15}{3} = 2x + 5$$

MULTIPLYING BINOMIALS USING FOIL

To find the product of two binomials, write them next to each other and then add the products of their *F*irst, *O*uter, *I*nner, and *L*ast pairs of terms.

➡ Example

Multiply $(2x + 1)$ by $(x - 5)$.

Solution

For the product $(2x + 1)(x - 5)$, $2x$ and x are the *F*irst pair of terms; $2x$ and -5 are the *O*utermost pairs of terms; 1 and x are the *I*nnermost terms; 1 and -5 are the *L*ast terms of the binomials. Thus:

$$
\begin{array}{cccc}
\textbf{F} & \textbf{O} & \textbf{I} & \textbf{L} \\
\end{array}
$$
$$(2x + 1)(x - 5) = (2x)(x) + (2x)(-5) + (1)(x) + (1)(-5)$$
$$= 2x^2 \quad + [-10x + x] \quad\quad - 5$$
$$= 2x^2 \quad - \quad 9x \quad\quad\quad - 5$$

PRODUCTS OF SPECIAL PAIRS OF BINOMIALS

SAT problems may involve these special products: $(a - b)(a + b)$, $(a + b)^2$, and $(a - b)^2$.

TIP

Time Saver

You can save some time if you memorize and learn to recognize when these special multiplication rules can be applied.

- $(a - b)(a + b) = a^2 - b^2$
- $(a + b)^2 = (a + b)(a + b) = a^2 + 2ab + b^2$
- $(a - b)^2 = (a - b)(a - b) = a^2 - 2ab - b^2$

➡ **Example**

Express $(x + 3)(x - 3)$ as a binomial.

Solution

$$(x + 3)(x - 3) = (x)^2 - (3)^2$$
$$= x^2 - 9$$

➡ **Example**

Express $(2y - 1)(2y + 1)$ as a binomial.

Solution

$$(2y - 1)(2y + 1) = (2y)^2 - (1)^2$$
$$= 4y^2 - 1$$

➡ **Example**

If $(x + y)^2 - (x - y)^2 = 28$, what is the value of xy?

Solution

- Square each binomial:

$$(x + y)^2 - (x - y)^2 = 28$$
$$(x^2 + 2xy + y^2) - (x^2 - 2xy + y^2) = 28$$

- Write the first squared binomial without the parentheses. Then remove the second set of parentheses by changing the sign of each term inside the parentheses to its opposite.

$$x^2 + 2xy + y^2 - x^2 + 2xy - y^2 = 28$$

- Combine like terms. Adding x^2 and $-x^2$ gives 0, as does adding y^2 and $-y^2$. The sum of $2xy$ and $2xy$ is $4xy$. The result is

$$4xy = 28$$
$$xy = \frac{28}{4} = 7$$

COMBINING ALGEBRAIC FRACTIONS

Algebraic fractions are combined in much the same way as fractions in arithmetic.

➡ **Example**

Write $\dfrac{w}{2} - \dfrac{w}{3}$ as a single fraction.

Solution

The LCD of 2 and 3 is 6. Change each fraction into an equivalent fraction that has 6 as its denominator.

$$\frac{w}{2} - \frac{w}{3} = \frac{3}{3}\left(\frac{w}{2}\right) - \frac{2}{2}\left(\frac{w}{3}\right)$$
$$= \frac{3w}{6} - \frac{2w}{6}$$
$$= \frac{3w - 2w}{6}$$
$$= \frac{w}{6}$$

➡ Example

If $h = \dfrac{y}{x - y}$, what is $h + 1$ in terms of x and y?

Solution

- Add 1 on both sides of the given equation:

$$h + 1 = \frac{y}{x - y} + 1$$

- On the right side of the equation, replace 1 with $\dfrac{x - y}{x - y}$:

$$h + 1 = \frac{y}{x - y} + \frac{x - y}{x - y}$$

- Write the sum of the numerators over the common denominator:

$$h + 1 = \frac{y + x - y}{x - y} = \frac{x}{x - y}$$

THE SUM AND DIFFERENCE OF TWO RECIPROCALS

The formulas for the sum and the difference of the reciprocals of two nonzero numbers, x and y, are worth remembering:

➡ Example

If $t = \dfrac{1}{r} + \dfrac{1}{s}$, then what is $\dfrac{1}{t}$ in terms of r and s?

Solution

Since $t = \dfrac{1}{r} + \dfrac{1}{s} = \dfrac{r + s}{rs}$, then $\dfrac{1}{t} = \dfrac{rs}{r + s}$.

TIP

Reciprocal Rules

If x and y are not 0, then
$$\frac{1}{x} + \frac{1}{y} = \frac{x + y}{xy}$$
and
$$\frac{1}{x} - \frac{1}{y} = \frac{y - x}{xy}.$$

SOLVING EQUATIONS WITH FRACTIONS

To solve an equation that contains fractions, eliminate the fractions by multiplying each member of the equation by the lowest common multiple of all of the denominators.

➡ **Example**

$$\frac{2x-9}{7} + \frac{2x}{3} = \frac{8x-1}{21}$$

What value of x is the solution of the above equation?

Solution

The smallest number that 3, 7, and 21 divide evenly into is 21. Eliminate the fractions by multiplying each member of the equation by 21:

$$\overset{3}{\cancel{21}}\left(\frac{2x-9}{\cancel{7}}\right) + \overset{7}{\cancel{21}}\left(\frac{2x}{\cancel{3}}\right) = \overset{1}{\cancel{21}}\left(\frac{8x-1}{\cancel{21}}\right)$$

$$3(2x-9) + 7(2x) = 8x - 1$$

$$6x - 27 + 14x = 8x - 1$$

$$20x - 8x = 27 - 1$$

$$12x = 26$$

$$\frac{12x}{12} = \frac{26}{12}$$

$$x = \frac{13}{6}$$

Multiple-Choice

1. $\dfrac{20b^3 - 8b}{4b} =$

   (A) $5b^2 - 2b$
   (B) $5b^3 - 2$
   (C) $5b^2 - 8b$
   (D) $5b^2 - 2$

$$\frac{3}{w} - \frac{4}{3} = \frac{5w}{10w^2}$$

2. In the equation above, what is the value of w?

   (A) $\dfrac{15}{8}$

   (B) $\dfrac{18}{11}$

   (C) $\dfrac{23}{12}$

   (D) $\dfrac{13}{6}$

$$P = 4x - z + 3y$$
$$Q = -x + 4z + 3y$$

3. Using the definitions above for P and Q, what is $2P - Q$?

   (A) $7x - 6z + 3y$
   (B) $9x + 2z + 9y$
   (C) $9x - 6z + 3y$
   (D) $7x - 6z + 9y$

4. If $(x - y)^2 = 50$ and $xy = 7$, what is the value of $x^2 + y^2$?
   (A) 8
   (B) 36
   (C) 43
   (D) 64

5. If $p = \dfrac{a}{a - b}$ and $a \neq b$, then, in terms of a and b, $1 - p =$

   (A) $\dfrac{a}{b - a}$

   (B) $\dfrac{b}{b - a}$

   (C) $\dfrac{a}{a - b}$

   (D) $\dfrac{b}{a - b}$

6. If $(a - b)^2 + (a + b)^2 = 24$, then $a^2 + b^2 =$

   (A) 4
   (B) 12
   (C) 16
   (D) 18

$$\frac{2}{p} - \frac{1}{2p} = \frac{p^2 + 1}{p^2 + 1}$$

7. In the equation above, what is the value of $\dfrac{1}{p}$?

   (A) $\dfrac{1}{3}$

   (B) $\dfrac{2}{3}$

   (C) $\dfrac{3}{2}$

   (D) 3

8. If $r = t + 2$ and $s + 2 = t$, then $rs =$

   (A) t^2
   (B) 4
   (C) $t^2 - 4$
   (D) $t^2 - 4t + 4$

9. Which statement is true for all real values of x and y?

(A) $(x + y)^2 = x^2 + y^2$

(B) $x^2 + x^2 = x^4$

(C) $\dfrac{2^{x+2}}{2^x} = 4$

(D) $(3x)^2 = 6x^2$

10. If $(p - q)^2 = 25$ and $pq = 14$, what is the value of $(p + q)^2$?

(A) 25

(B) 36

(C) 53

(D) 81

11. If $\dfrac{a}{2} - \dfrac{b}{3} = 1$, what is $2a + 3b$ in terms of b?

(A) $\dfrac{7b}{3} + 1$

(B) $\dfrac{13b}{3} + 4$

(C) $\dfrac{13b + 1}{3}$

(D) $\dfrac{17b}{3}$

12. If $(x + 5)(x + p) = x^2 + 2x + k$, then

(A) $p = 3$ and $k = 5$

(B) $p = -3$ and $k = 15$

(C) $p = 3$ and $k = -15$

(D) $p = -3$ and $k = -15$

13. For what value of p is $(x - 2)(x + 2) = x(x - p)$?

(A) -4

(B) 0

(C) $\dfrac{2}{x}$

(D) $\dfrac{4}{x}$

$$\frac{m}{2} - \frac{3(m - 4)}{5} = \frac{5(3 - m)}{6}$$

14. What value of m makes the equation above a true statement?

(A) $\dfrac{8}{27}$

(B) $\dfrac{3}{22}$

(C) $\dfrac{62}{27}$

(D) $\dfrac{147}{22}$

15. If $\left(k + \dfrac{1}{k}\right)^2 = 16$, then $k^2 + \dfrac{1}{k^2} =$

(A) 4

(B) 8

(C) 12

(D) 14

16. If $\dfrac{a}{b} = 1 - \dfrac{x}{y}$, then $\dfrac{b}{a} =$

(A) $\dfrac{x}{y - x}$

(B) $\dfrac{y}{x} - 1$

(C) $\dfrac{y}{x - y}$

(D) $\dfrac{y}{y - x}$

$$\frac{11 + s}{12r} = \frac{1}{6} + \frac{1 - 3s}{4r}$$

17. In the equation above, what is r in terms of s?

(A) $r = 5s + 4$

(B) $r = \dfrac{s - 2}{3}$

(C) $r = 4s + 5$

(D) $r = \dfrac{s + 3}{2}$

Grid-In

1. If $(3y - 1)(2y + k) = ay^2 + by - 5$ for all values of y, what is the value of $a + b$?

2. If $4x^2 + 20x + r = (2x + s)^2$ for all values of x, what is the value of $r - s$?

$$\frac{5}{8} = \frac{-(11 - 7y)}{4y} + \frac{1}{2y} - \frac{1}{8}$$

3. What value of y makes the equation above a true statement?

OVERVIEW

Factoring reverses multiplication.

Operation	Example
Multiplication	$2(x + 3y) = 2x + 6y$
Factoring	$2x + 6y = 2(x + 3y)$

There are three basic types of factoring that you need to know for the SAT:

- Factoring out a common monomial factor, as in

$$4x^2 - 6x = 2x(x - 3) \quad \text{and} \quad ay - by = y(a - b)$$

- Factoring a quadratic trinomial using the reverse of FOIL, as in

$$x^2 - x - 6 = (x - 3)(x + 2) \text{ and}$$
$$x^2 - 2x + 1 = (x - 1)(x - 1) = (x - 1)^2$$

- Factoring the difference between two squares using the rule

$$a^2 - b^2 = (a + b)(a - b)$$

FACTORING A POLYNOMIAL BY REMOVING A COMMON FACTOR

If all the terms of a polynomial have factors in common, the polynomial can be factored by using the reverse of the distributive law to remove these common factors. For example, in

$$24x^3 + 16x = 8x(3x^2 + 2)$$

$8x$ is the *Greatest Common Factor* (GCF) of $24x^3$ and $16x$ since 8 is the GCF of 24 and 16, and x is the greatest power of that variable that is contained in both $24x^3$ and $16x$. The factor that corresponds to $8x$ can be obtained by dividing $24x^3 + 16x$ by $8x$:

$$\frac{24x^3 + 16x}{8x} = \frac{24x^3}{8x} + \frac{16x}{8x} = 3x^2 + 2$$

You can check that the factorization is correct by multiplying $3x^2 + 2$ by $8x$ and verifying that the product is $24x^3 + 16x$.

USING FACTORING TO ISOLATE VARIABLES IN EQUATIONS

It may be necessary to use factoring to help isolate a variable in an equation in which terms involving the variable cannot be combined into a single term.

➠ **Example**

If $ax - c = bx + d$, what is x in terms of a, b, c, and d?

Solution

Isolate terms involving x on the same side of the equation.

- On each side of the equation add c and subtract bx:

$$ax - bx = c + d$$

- Factor out x from the left side of the equation:

$$x(a - b) = c + d$$

- Divide both sides of the new equation by the coefficient of x:

$$\frac{x\cancel{(a - b)}^{1}}{\cancel{a - b}} = \frac{c + d}{a - b}$$

$$x = \frac{c + d}{a - b}$$

FACTORING A QUADRATIC TRINOMIAL: $x^2 + bx + c$

A quadratic trinomial like $x^2 - 7x + 12$ contains x^2 as well as x. Quadratic trinomials that appear on the SAT can be factored as the product of two binomials by reversing the FOIL multiplication process.

- Think: "What two integers when multiplied together give +12 and when added together give −7?"
- Recall that, since the product of these integers is *positive* 12, the two integers must have the same sign. Hence, the integers are limited to the following pairs of factors of +12:

$$1 \text{ and } 12; \quad -1 \text{ and } -12$$

$$2 \text{ and } 6; \quad -2 \text{ and } -6$$

$$3 \text{ and } 4; \quad -3 \text{ and } -4$$

- Choose −3 and −4 as the factors of 12 since they add up to −7. Thus,

$$x^2 - 7x + 12 = (x - 3)(x - 4)$$

- Use FOIL to check that the product $(x - 3)(x - 4)$ is $x^2 - 7x + 12$.

➡ Example

Factor $n^2 - 5n - 14$.

Solution

Find two integers that when multiplied together give −14 and when added together give −5. The two factors of −14 must have different signs since their product is negative. Since $(+2)(-7) = -14$ and $2 + (-7) = -5$, the factors of −14 you are looking for are +2 and −7. Thus:

$$n^2 - 5n - 14 = (n + 2)(n - 7)$$

FACTORING $ax^2 + bx + c$ WHEN $a > 1$

Factoring a quadratic trinomial becomes more complicated when the coefficient of the x^2-term is greater than 1. To factor $3x + 10x + 8$:

- Factor the x^2-term and set up the binomial factors:

$$3x^2 + 10x + 8 = \left(3x + \boxed{?}\right)\left(x + \boxed{?}\right)$$

- Identify possibilities for the unknown pair of integers in the binomial factors. The product of these integers must be +8, the last term of $3x^2 + 10x + 8$. The possibilities are: 1 and 8; –1 and –8; 2 and 4; and –2 and –4.

- Use trial and elimination to determine the correct pair of factors of 8 and their proper placement in the binomial factors. The factors of +8 must be chosen and placed so that the sum of the outer and inner products of the terms of the binomial factors is equal to +10x, the middle term of $3x^2 + 10x + 8$:

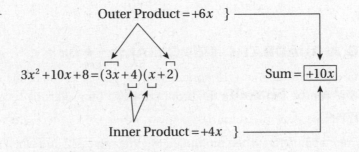

- Check that the factors work. The placement of the factors of 8 matters. Although $(3x + 2)(x + 4)$ contains the correct factors of 8, they are not placed correctly since the sum of the outer and inner products is $12x + 2x = 14x$ rather than $10x$.

Factoring the Difference Between Two Squares

Since $(a + b)(a - b) = a^2 - b^2$, any binomial of the form $a^2 - b^2$ can be rewritten as $(a + b)$ times $(a - b)$.

$$a^2 - b^2 = (a + b)(a - b)$$

This means that the difference between two squares can be factored as the product of the sum and difference of the quantities that are being squared. Here are some examples in which this factoring rule is used:

- $y^2 - 16 = (y + 4)(y - 4)$
- $x^2 - 0.25 = (x + 0.5)(x - 0.5)$
- $100 - x^4 = (10 - x^2)(10 + x^2)$

➡ Example

If $\dfrac{x^{p^2}}{x^{q^2}} = \left(x^{50}\right)^2$ with $x > 1$ and $p - q = 4$, what is the value of $p + q$?

(A) 25
(B) 50
(C) 64
(D) 96

Solution

- Since $\dfrac{x^{p^2}}{x^{q^2}} = x^{p^2-q^2}$ and $\left(x^{50}\right)^2 = x^{100}$, $x^{p^2-q^2} = x^{100}$. Since the bases are the same, the exponents must be the same:

$$p^2 - q^2 = 100$$

- Factoring $p^2 - q^2$ makes $(p-q)(p+q) = 100$.

- Substituting 4 for $p - q$ gives $4(p+q) = 100$ so $p + q = \dfrac{100}{4} = 25$.

The correct choice is **(A)**.

Factoring Completely

To factor a polynomial into factors that cannot be further factored, it may be necessary to use more than one factoring technique. For example:

- $3t^2 - 75 = 3(t^2 - 25) = 3(t+5)(t-5)$
- $t^3 - 6t^2 + 9t = t(t^2 - 6t + 9) = t(t-3)(t-3)$ or $t(t-3)^2$

Using Factoring to Simplify Algebraic Fractions

To simplify an algebraic fraction, factor the numerator and the denominator. Then cancel any factor that is in both the numerator and the denominator of the fraction since any nonzero quantity divided by itself is 1.

➡ **Example**

Simplify $\dfrac{2b - 2a}{a^2 - b^2}$.

Solution

- Factor the numerator and the denominator:

$$\dfrac{2b - 2a}{a^2 - b^2} = \dfrac{2(b-a)}{(a+b)(a-b)}$$

- Rewrite $b - a$ as $-(a-b)$:

$$= \dfrac{2[-(a-b)]}{(a+b)(a-b)}$$

- Cancel $a - b$ since it is a factor of the numerator and a factor of the denominator:

$$= \dfrac{-2\,\overset{1}{\cancel{(a-b)}}}{(a+b)\,\cancel{(a-b)}}$$

$$= \dfrac{-2}{a+b}$$

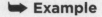 **Example**

$$\frac{x^2 - 4x - 5}{(x-1)^2 - 4(x-1)}$$

The expression above is equivalent to

(A) $x + 1$

(B) $\dfrac{x+5}{x-1}$

(C) $\dfrac{x+1}{x-1}$

(D) $\dfrac{x+1}{x-4}$

TIP

If you get stuck on an algebraic solution, plug in an easy number. If $x = 3$, then

$$\frac{x^2 - 4x - 5}{(x-1)^2 - 4(x-1)}$$

evaluates to 2. After plugging 3 into *each* of the four answer choices, you should verify that only choice (C) evaluates to 2.

Solution

Factor the numerator as the product of two binomials and factor out the common binomial term of $(x - 1)$ in the denominator:

$$\frac{x^2 - 4x - 5}{(x-1)^2 - 4(x-1)} = \frac{(x+1)(x-5)}{(x-1)[(x-1)-4]}$$

$$= \frac{(x+1)\overset{1}{\cancel{(x-5)}}}{(x-1)\cancel{(x-5)}}$$

$$= \frac{x+1}{x-1}$$

The correct choice is **(C)**.

➥ **Example**

$$\frac{3x}{2x-6} + \frac{9}{6-2x}$$

Which of the following is equivalent to the expression above?

(A) $-\dfrac{3}{2}$

(B) $\dfrac{-6x^2 + 36x - 54}{(2x-6)(6-2x)}$

(C) $\dfrac{3x+9}{2x-6}$

(D) $\dfrac{3}{2}$

Solution

$$\frac{3x}{2x-6} + \frac{9}{6-2x} = \frac{3x}{2x-6} + \frac{9}{-(2x-6)}$$

$$= \frac{3x}{2x-6} - \frac{9}{2x-6}$$

$$= \frac{3x-9}{2x-6}$$

$$= \frac{3\cancel{(x-3)}}{2\cancel{(x-3)}}$$

$$= \frac{3}{2}$$

The correct choice is **(D)**.

Multiple-Choice

$$q = \frac{d}{d+n}$$

1. On a manufacturer's assembly line, d parts are found to be defective and n parts are nondefective. The formula above is used to calculate a quality-of-parts ratio. What is d expressed in terms of the other two variables?

   (A) $\dfrac{n}{1-q}$

   (B) $\dfrac{nq}{1-q}$

   (C) $\dfrac{n}{q-1}$

   (D) $\dfrac{nq}{q-1}$

2. The sum of $\dfrac{a}{a^2-b^2}$ and $\dfrac{b}{a^2-b^2}$ is

   (A) $\dfrac{1}{a-b}$

   (B) $\dfrac{a}{a-b}$

   (C) $\dfrac{b}{a-b}$

   (D) $\dfrac{a+b}{a-b}$

3. If $ax + x^2 = y^2 - ay$, what is a in terms of x and y?

   (A) $y - x$
   (B) $x - y$
   (C) $x + y$
   (D) $\dfrac{x^2 + y^2}{x - y}$

4. If $\dfrac{xy}{x+y} = 1$ and $x \neq -y$, what is x in terms of y?

   (A) $\dfrac{y+1}{y-1}$

   (B) $\dfrac{y+1}{y}$

   (C) $\dfrac{y}{y-1}$

   (D) $\dfrac{y}{y+1}$

5. What is the sum of $\dfrac{4x}{x-1}$ and $\dfrac{4x+4}{x^2-1}$ expressed in simplest form?

   (A) $\dfrac{4x+1}{x-1}$

   (B) $\dfrac{4(x+1)}{x-1}$

   (C) $\dfrac{4(x^2+4x+1)}{x^2-1}$

   (D) $\dfrac{4(x+2)}{x^2-1}$

6. If $h = \dfrac{x^2-1}{x+1} + \dfrac{x^2-1}{x-1}$, what is x in terms of h?

   (A) $\dfrac{h}{2}$

   (B) $2h + 1$

   (C) $2h - 1$

   (D) $\sqrt{\dfrac{h}{2}}$

7. If $ax^2 - bx = ay^2 + by$, then $\dfrac{a}{b} =$

   (A) $\dfrac{1}{x - y}$

   (B) $\dfrac{1}{x + y}$

   (C) $\dfrac{x - y}{x + y}$

   (D) $\dfrac{x + y}{x - y}$

8. If $a \neq b$ and $\dfrac{a^2 - b^2}{9} = a + b$, then what is the value of $a - b$?

   (A) $\dfrac{1}{3}$

   (B) 3

   (C) 9

   (D) 12

9. If $\dfrac{r + s}{x - y} = \dfrac{3}{4}$, then $\dfrac{8r + 8s}{15x - 15y} =$

   (A) $\dfrac{32}{45}$

   (B) $\dfrac{8}{15}$

   (C) $\dfrac{7}{16}$

   (D) $\dfrac{2}{5}$

10. If $x^2 = k + 1$, then $\dfrac{x^4 - 1}{x^2 + 1} =$

    (A) k

    (B) k^2

    (C) $k + 2$

    (D) $k - 2$

11. If $p = x(3x + 5) - 28$, then p is divisible by which of the following expressions?

    (A) $3x + 4$

    (B) $x - 4$

    (C) $x + 7$

    (D) $3x - 7$

12. If $(x + p)$ is a factor of both $x^2 + 16x + 64$ and $4x^2 + 37x + k$, where p and k are nonzero integer constants, what could be the value of k?

    (A) 9

    (B) 24

    (C) 40

    (D) 63

OVERVIEW

A *quadratic equation* is an equation in which the greatest exponent of the variable is 2, as in $x^2 + 3x - 10 = 0$. A quadratic equation has two roots, which can be found by breaking down the quadratic equation into two first-degree equations.

ZERO-PRODUCT RULE

If the product of two or more numbers is 0, at least one of these numbers is 0.

➡ Example

For what values of x is $(x - 1)(x + 3) = 0$?

Solution

Since $(x - 1)(x + 3) = 0$, either $x - 1 = 0$ or $x + 3 = 0$.

- If $x - 1 = 0$, then $x = 1$.
- If $x + 3 = 0$, then $x = -3$.

The possible values of x are 1 and -3.

SOLVING A QUADRATIC EQUATION BY FACTORING

The two roots of a quadratic equation may or may not be equal. For example, the equation $(x - 3)^2 = 0$ has a double root of $x = 3$. The quadratic equation $x^2 = 9$, however, has two unequal roots, $x = 3$ and $x = -3$. More complicated quadratic equations on the SAT can be solved by factoring the quadratic expression.

➡ Example

Solve $x^2 + 2x = 0$ for x.

Solution

- Factor the left side of the quadratic equation:

$$x^2 + 2x = 0$$
$$x(x + 2) = 0$$

- Form two first-degree equations by setting each factor equal to 0:

$$x = 0 \quad \text{or} \quad x + 2 = 0$$

- Solve each first-degree equation:

$$x = 0 \quad \text{or} \quad x = -2$$

  The two roots are 0 and -2.

If a quadratic equation does not have all its nonzero terms on the same side of the equation, you must put the equation into this form before factoring.

➡ Example

Solve $x^2 + 3x = 10$ for x.

Solution

To rewrite the quadratic equation so that all the nonzero terms are on the same side:

- Subtract 10 from both sides of $x^2 + 3x = 10$: $\qquad\qquad\qquad\qquad x^2 + 3x - 10 = 0$
- Factor the quadratic polynomial: $\qquad\qquad\qquad\qquad\qquad (x + 5)(x - 2) = 0$
- Set each factor equal to 0: $\qquad\qquad\qquad x + 5 = 0 \quad$ or $\quad x - 2 = 0$
- Solve each equation: $\qquad\qquad\qquad\qquad\quad x = -5 \quad$ or $\qquad x = 2$

The two roots are -5 and 2.

You can check that $x = -5$ and $x = 2$ are the roots by plugging each value into $x^2 + 3x = 10$ and verifying that the left side then equals 10, the right side.

➥ Example

$$2y^2 = 3(y + 9)$$

If j and k represent solutions of the equation above and $j > k$, what is the value of $j - k$?

Solution

- Rewrite the equation in standard $ay^2 + by + c = 0$ form:

$$2y^2 - 3(y + 9) = 0$$
$$2y^2 - 3y - 27 = 0$$

- Solve by factoring:

$$(2y + ?)(y + ?) = 0 \leftarrow \text{Missing terms are the factors of } -27$$
$$(2y - 9)(y + 3) = 0 \leftarrow \text{Verify the factors and their placement}$$
$$\text{by multiplying out using } FOIL$$

$$2y - 9 = 0 \quad or \quad y + 3 = 0$$
$$y = \frac{9}{2} \quad or \qquad y = -3$$

- Set $j = \dfrac{9}{2}$ and $k = -3$ so

$$j - k = \frac{9}{2} - (-3)$$
$$= \frac{9}{2} + 3$$
$$= \frac{9}{2} + \frac{6}{2}$$
$$= \frac{15}{2}$$

➥ Example

If 4 is a root of $x^2 - x - w = 0$, what is the value of w?

Solution

Since 4 is a root of the given equation, replacing x with 4 in that equation gives an equation that can be used to solve for w:

$$4^2 - 4 - w = 0$$
$$16 - 4 - w = 0$$
$$12 - w = 0$$
$$12 = w$$

Hence, $w = \mathbf{12}$.

QUADRATIC EQUATIONS WITH EQUAL ROOTS

A quadratic equation may have equal roots. If $x^2 - 2x + 1 = 0$, then

$$(x - 1)\,(x - 1) = 0$$
$$x - 1 = 0 \quad or \quad x - 1 = 0$$
$$x = 1 \qquad\qquad x = 1$$

Thus, the equation $x^2 - 2x + 1 = 0$ has a double root of **1**.

> **TIP**
>
> **Every quadratic equation has two roots, but the roots may be equal.**

Multiple-Choice

1. The fraction $\dfrac{x - 2}{x^2 + 4x - 21}$ is *not* defined when $x =$

   (A) 2
   (B) 7 or –3
   (C) –7 or 3
   (D) –7 or –3

2. If $\dfrac{a^2}{2} = 2a$, then a equals

   (A) 0 or –2
   (B) 0 or 2
   (C) 0 or –4
   (D) 0 or 4

3. If $(s - 3)^2 = 0$, what is the value of $(s + 3)(s + 5)$?

   (A) 48
   (B) 24
   (C) 15
   (D) 0

4. If $k = 7 + \dfrac{8}{k}$, what is the value of $k^2 + \dfrac{64}{k^2}$?

   (A) 33
   (B) 49
   (C) 64
   (D) 65

$$\frac{18 - 3w}{w + 6} = \frac{w^2}{w + 6}$$

5. Which of the following represents the *sum* of all possible solutions to the equation above?

   (A) –9
   (B) –3
   (C) 3
   (D) 9

Equation (1): $2x^2 + 7x = 4$

Equation (2): $(y - 1)^2 = 9$

6. If f is the greater of the two roots of Equation (1) and g is the lesser of the two roots of Equation (2), what is the value of the product $f \times g$?

   (A) –4
   (B) –1
   (C) 2
   (D) 8

$$x^3 - 20x = x^2$$

7. If a, b, and c represent the set of all values of x that satisfy the equation above, what is the value of $(a + b + c) + (abc)$?

   (A) –1
   (B) 0
   (C) 1
   (D) 9

8. If $\dfrac{x^2}{3} = x$, then $x =$

   (A) 0 or –3
   (B) 3 only
   (C) 0 only
   (D) 0 or 3

9. By how much does the sum of the roots of the equation $(x + 1)(x - 3) = 0$ exceed the product of its roots?

   (A) 1
   (B) 2
   (C) 3
   (D) 5

10. If $x^2 - 63x - 64 = 0$ and p and n are integers such that $p^n = x$, which of the following CANNOT be a value for p?

(A) -8

(B) -4

(C) -1

(D) 4

11. If $r > 0$ and $r^t = 6.25r^{t+2}$, then $r =$

(A) $\dfrac{2}{5}$

(B) $\dfrac{4}{9}$

(C) $\dfrac{5}{8}$

(D) $\dfrac{3}{4}$

$$\frac{x}{2x-1} = \frac{2x+1}{x+2}$$

12. If m and n represent the solutions of the equation above, what is the value of $m + n$?

(A) $-\dfrac{4}{3}$

(B) $-\dfrac{3}{4}$

(C) $\dfrac{2}{3}$

(D) $\dfrac{5}{4}$

$$\frac{1}{(t-2)^2} = 6 + \frac{1}{(t-2)}$$

13. If p and q represent the solutions of the equation above, what is the value of $p \times q$?

(A) $-\dfrac{3}{2}$

(B) $\dfrac{7}{2}$

(C) $\dfrac{9}{4}$

(D) $\dfrac{15}{8}$

Grid-In

1. If $(4p + 1)^2 = 81$ and $p > 0$, what is a possible value of p?

2. If $(x - 1)(x - 3) = -1$, what is a possible value of x?

3. By what amount does the sum of the roots exceed the product of the roots of the equation $(x - 5)(x + 2) = 0$?

$$(3k + 14)k = 5$$

4. If r and s represent the solutions of the equation above and $r > s$, what is the value of the difference $r - s$?

$$x^4 + 16 = 10x^2$$

5. If p and q are distinct roots of the equation above and $pq > 0$, what is the value of the product pq?

$$(2a - 5)^2 = (4 - 3a)^2$$

6. What is the sum of the roots of the equation above?

OVERVIEW

A **system of equations** is a set of equations whose solution makes each of the equations true at the same time. SAT questions involving systems of two linear equations with two different variables can usually be solved by

- Substituting the solution of one equation into the other equation to eliminate one of the variables in that equation; or
- Adding or subtracting corresponding sides of the two equations so that an equation with only one variable results.

SOLVING A LINEAR SYSTEM BY SUBSTITUTION

To solve a linear system using the substitution method, pick the equation in which it is easy to solve for one of the variables. After solving the equation for that variable, plug the solution into the other equation. In the linear system,

$$2y - 3x = 29$$
$$y + 2x = 4$$

it is easy to solve the second equation for y, which gives $y = -2x + 4$. Plug the solution into the first equation to eliminate y:

$$2\overbrace{(-2x + 4)}^{y} - 3x = 29$$
$$-4x + 8 - 3x = 29$$
$$-7x + 8 = 29$$
$$-7x = 21$$
$$\frac{-7x}{-7} = \frac{21}{-7}$$
$$x = -3$$

To find the corresponding value for y, substitute -3 for x in the simpler of the two original equations:

$$y + 2\overbrace{(-3)}^{x} = 4$$
$$y - 6 = 4$$
$$y = 10$$

The solution is **(-3, 10)**.

➡ Example

An automobile repair shop wants to mix a solution that is 35% pure antifreeze with another solution that is 75% pure antifreeze. How many liters of each solution must be used in order to produce 80 liters of solution that is 50% pure antifreeze?

Solution

Use x and y to represent the number of liters of the 35% and 75% solutions, respectively, in the mixture. The problems has two conditions:

CONDITION 1: The mixture will contain 80 liters of solution: $x + y = 80$.

CONDITION 2: The sum of the number of liters of pure antifreeze in the two ingredients must be equal to the number of liters of pure antifreeze in the mixture:

$$0.35x + 0.75y = 0.50(80) = 40$$

■ Represent the two conditions by a system of linear equations:

$$x + y = 80$$
$$0.35x + 0.75y = 40$$

■ From the first equation, $y = 80 - x$. Eliminate y in the second equation:

$$0.35x + 0.75(80 - x) = 40$$
$$0.35x + 60 - 0.75x = 40$$
$$-0.40x = 40 - 60$$
$$\frac{-0.40x}{-0.40} = \frac{-20}{-0.40}$$
$$x = 50$$

■ Find y by substituting 50 for x in the first equation:

$$50 + y = 80 \text{ so } y = 80 - 50 = 30$$

Thus, **50 liters** of the 35% solution and **30 liters** of the 75% solution must be used.

   NOTE: The problem could be solved using one variable where x and $80 - x$ represent the number of liters of the 35% and 75% solutions, respectively. This means that $0.35x + 0.75(80 - x) = 40$, which can be solved as before.

SOLVING A LINEAR SYSTEM BY COMBINING EQUATIONS

Sometimes the easiest way of solving a system of linear equations is to rewrite each equation in the standard $Ax + By = C$ form (where A, B, and C are constants). If the same variable in both equations have the opposite (or the same) numerical coefficients, then adding (or subtracting) corresponding sides of the two equations will eliminate one of the variables. In the system

$$x - 2y = 5$$
$$x + 2y = 11$$

the numerical coefficients of y are opposites so adding corresponding sides of the two equations will eliminate it:

$$x - 2y = 5$$
$$\underline{x + 2y = 11}$$
$$2x + 0 = 16$$
$$x = \frac{16}{2} = 8$$

Find y by substituting 8 for x in either of the original equations:

$$8 + 2y = 11$$
$$2y = 3$$
$$y = \frac{3}{2}$$

➥ Example

$$3x = 5y + 19$$
$$2x + y = -x + 7$$

What ordered pair (x, y) represents the solution to the system of equations above?

Solution

- Rewrite each equation in standard $Ax + By = C$ form:

$$3x - 5y = 19$$
$$3x + y = 7$$

- Eliminate x by subtracting the second equation from the first equation:

$$3x - 5y = 19$$
$$-3x + y = 7$$
$$0 - 6y = 12$$
$$y = \frac{12}{-6} = -2$$

- Solve for x in the first equation by substituting -2 for y:

$$3x = 5(-2) + 19$$
$$3x = -10 + 19$$
$$3x = 9$$
$$x = \frac{9}{3} = 3$$

The solution is **(3, –2)**.

SOLVING A LINEAR SYSTEM BY USING MULTIPLIERS

Before combining corresponding sides of the equations in a linear system, it may be necessary to first multiply one or both equations by a number that makes the same variable in both equations have opposite numerical coefficients.

➥ Example

$$2x + 4y = 5x + 11$$
$$6x - 5y = -16$$

What ordered pair (x, y) represents the solution to the system of equations above?

Solution

- Write the first equation in the standard form $Ax + By = C$:

$$-3x + 4y = 11$$
$$\underline{6x - 5y = -16}$$

- Change the coefficient of x in the first equation to -6 by multiplying the equation by 2. Then add the resulting equation and the second equation thereby eliminating x:

$$(2)[-3x + 4y = 11] \left.\right\rbrace \quad \Longrightarrow \quad \begin{array}{r} -6x + 8y = 22 \\ + \ \underline{6x - 5y = -16} \\ 0 + 3y = 6 \\ y = \dfrac{6}{3} = 2 \end{array}$$

- Find x by substituting 2 for y in either of the original equations:

$$6x - 5y = -16$$
$$6x - 5(2) = -16$$
$$6x = -16 + 10$$
$$6x = -6$$
$$\frac{6x}{6} = \frac{-6}{6}$$
$$x = -1$$

The solution is **(−1, 2)**.

➡ Example

$$2x + 5y = 2y - 6$$
$$5x + 2y = 7$$

In the system of equations above, what is the value of the product xy?

Solution

Solve the system to get the values of x and y. Rewrite the first equation in $Ax + By = C$ form:

$$2x + 3y = -6$$
$$\underline{5x + 2y = \ \ 7}$$

- Eliminate y by multiplying the first equation by -2, multiplying the second equation by 3, and then adding corresponding sides of the two resulting equations:

$$-2[2x + 3y = -6] \ \Rightarrow \ -4x - 6y = 12$$
$$3[5x + 2y = 7] \ \ \ \Rightarrow \ \underline{15x + 6y = 21}$$
$$11x + 0 = 33$$
$$x = \frac{33}{11} = 3$$

- Find y by substituting 3 for x in either of the two original equations:

$$5(3) + 2y = 7$$
$$15 + 2y = 7$$
$$2y = -8$$
$$y = \frac{-8}{2} = -4$$

- Since the solution is $(3, -4)$, $xy = (3)(-4) = \mathbf{-12}$.

➥ Example

A store that offers faxing services charges a fixed amount to fax one page and a different amount for faxing each additional page. If the cost of faxing 5 pages is \$3.05 and the cost of faxing 13 pages is \$6.65,

a. What is the cost of faxing one page?

b. What is the total cost of faxing three pages to the same telephone number?

Solution

a. If x represents the cost of faxing the initial page and y the cost of faxing each additional page, then

$$x + 12y = 6.65$$
$$x + 4y = 3.05$$

Solve for x by multiplying the second equation by -3 and then adding the result to the first equation:

$$x + 12y = 6.65$$
$$\underline{-3x - 12y = -9.15}$$
$$-2x + 0 = -2.50$$
$$\frac{-2x}{-2} = \frac{-2.50}{-2}$$
$$x = 1.25$$

The cost of faxing one page is **\$1.25**.

b. Find the cost of faxing each additional page after the first page.

- Solve for y in the second equation:

$$1.25 + 4y = 3.05$$
$$4y = 3.05 - 1.25$$
$$\frac{4y}{4y} = \frac{1.80}{4}$$
$$y = 0.45$$

- The cost of faxing 3 pages is the sum of the charges for the first page and the two additional pages:

$$x + 2y = 1.25 + 2(0.45)$$
$$= 1.25 + 0.90$$
$$= 2.15$$

The cost of faxing 3 pages is **$2.15**.

SOLVING OTHER TYPES OF SYSTEMS OF EQUATIONS

If a system of equations has more variables than equations, you may be asked to solve for some combination of letters.

➡ Example

If $2r = s$ and $24t = 3s$, what is r in terms of t?

Solution

Since the question asks for r in terms of t, work toward eliminating s.

- Substitute $2r$ for s in the second equation:

$$24t = 3s = 3(2r) = 6r$$

- Solve for r in the equation $24t = 6r$:

$$\frac{24t}{6} = \frac{6r}{6}$$
$$4t = r$$

Hence, $r = 4t$.

➥ Example

If $ab - 3 = 12$ and $2bc = 5$, what is the value of $\dfrac{a}{c}$?

Solution

Since the question asks for $\dfrac{a}{c}$, you must eliminate b.

- Find the value of ab in the first equation. Since $ab - 3 = 12$, then $ab = 15$.
- To eliminate b, divide corresponding sides of $ab = 15$ and $2bc = 5$:

$$\frac{ab}{2bc} = \frac{15}{5}$$

$$\frac{ab}{2bc} = 3$$

$$\frac{a}{2c} = 3$$

Solve the resulting equation for $\dfrac{a}{c}$:

$$2\left(\frac{a}{2c}\right) = 2(3)$$

$$\frac{a}{c} = 6$$

The value of $\dfrac{a}{c}$ is **6**.

Multiple-Choice

1. If $2x - 3y = 11$ and $3x + 15 = 0$, what is the value of y?

   (A) -7
   (B) -5
   (C) $\dfrac{1}{3}$
   (D) 3

2. If $2a = 3b$ and $4a + b = 21$, then $b =$

   (A) 1
   (B) 3
   (C) 4
   (D) 7

3. If $2p + q = 11$ and $p + 2q = 13$, then $p + q =$

   (A) 6
   (B) 8
   (C) 9
   (D) 12

$$2(x + y) = 3y + 5$$
$$3x + 2y = -3$$

4. Which equivalent equation could be used to solve the system of equations above?

   (A) $3\left(\dfrac{5 + y}{2}\right) + 2y = -3$
   (B) $3\left(\dfrac{5}{2} - y\right) + 2y = -3$
   (C) $3x + 2(2x - 5) = -3$
   (D) $3x + 2(5 - 2x) = -3$

5. If $x - y = 3$ and $x + y = 5$, what is the value of y?

   (A) -4
   (B) -2
   (C) -1
   (D) 1

6. If $5x + y = 19$ and $x - 3y = 7$, then $x + y =$

   (A) -4
   (B) -1
   (C) 3
   (D) 4

7. If $x - 9 = 2y$ and $x + 3 = 5y$, what is the value of x?

   (A) -2
   (B) 4
   (C) 11
   (D) 17

8. If $\dfrac{1}{x} + \dfrac{1}{y} = \dfrac{1}{4}$ and $\dfrac{1}{x} - \dfrac{1}{y} = \dfrac{3}{4}$, then $x =$

   (A) $\dfrac{1}{4}$
   (B) $\dfrac{1}{2}$
   (C) 2
   (D) 4

9. If $5a + 3b = 35$ and $\dfrac{a}{b} = \dfrac{2}{5}$, what is the value of a?

   (A) $\dfrac{14}{5}$
   (B) $\dfrac{7}{2}$
   (C) 5
   (D) 7

10. If $\dfrac{x}{y} = 6$, $\dfrac{y}{w} = 4$, and $x = 36$, what is the value of w?

(A) $\dfrac{1}{2}$

(B) $\dfrac{3}{2}$

(C) 2

(D) 4

11. If $4r + 7s = 23$ and $r - 2s = 17$ then $3r + 3s =$

(A) 8

(B) 24

(C) 32

(D) 40

12. If $\dfrac{p - q}{2} = 3$ and $rp - rq = 12$, then $r =$

(A) -1

(B) 1

(C) 2

(D) 4

13. If $(a + b)^2 = 9$ and $(a - b)^2 = 49$, what is the value of $a^2 + b^2$?

(A) 17

(B) 20

(C) 29

(D) 58

$$3x - y = 8 - x$$
$$6x + 4y = 2y - 9$$

14. For the system of equations above, what is the value of the product xy?

(A) -3

(B) -2

(C) 2

(D) 3

15. If $3x + y = c$ and $x + y = b$, what is the value of x in terms of c and b?

(A) $\dfrac{c - b}{3}$

(B) $\dfrac{c - b}{2}$

(C) $\dfrac{b - c}{3}$

(D) $\dfrac{b - c}{2}$

16. If $a + b = 11$ and $a - b = 7$, then $ab =$

(A) 6

(B) 8

(C) 10

(D) 18

$$x - z = 7$$
$$x + y = 3$$
$$z - y = 6$$

17. For the above system of three equations, $x =$

(A) 5

(B) 6

(C) 7

(D) 8

$$a = 4c$$
$$c = re$$
$$a = 5e$$

18. For the system of equations above, if $e \neq 0$, what is the value of r?

(A) $\dfrac{1}{20}$

(B) $\dfrac{4}{5}$

(C) $\dfrac{5}{4}$

(D) 1

19. During the next football season, a player's earnings, x, will be 0.005 million dollars more than those of a teammates' earnings, y. The two players will earn a total of 3.95 milion dollars. Which system of equations could be used to determine the amount each player will earn, in millions of dollars?

   (A) $x + y = 3.95$
   $x + 0.005 = y$

   (B) $x - 3.95 = y$
   $y + 0.005 = x$

   (C) $y - 3.95 = x$
   $x + 0.005 = y$

   (D) $x + y = 3.95$
   $y + 0.005 = x$

Food	Protein	Calories
Cereal	5 g	90
Milk	8 g	80

20. The table above shows the number of grams of protein and the number of calories in single servings of bran flakes cereal and milk. How many servings of each are needed to get a total of 35 grams of protein and 470 calories?

   (A) 2 servings of milk; 4 servings of cereal

   (B) $2\frac{1}{2}$ servings of milk; $2\frac{1}{2}$ servings of cereal

   (C) 3 servings of milk; $2\frac{1}{2}$ servings of cereal

   (D) $2\frac{1}{2}$ servings of milk; 3 servings of cereal

Grid-In

1. If 5 sips + 4 gulps = 1 glass and 13 sips + 7 gulps = 2 glasses, how many sips equal a gulp?

2. When Amy exercises in her fitness center for 1 hour she burns a total of 475 calories. If she burns 9 calories a minute jogging on the treadmill and then burns 6.5 calories a minute pedaling on the stationary bicycle, how many minutes of the hour does she spend exercising on the bicycle?

3. John and Sara each bought the same type of pen and notebook in the school bookstore, which does not charge sales tax. John paid $5.55 for two pens and three notebooks, and Sara paid $3.50 for one pen and two notebooks. How much does the school bookstore charge for one notebook?

$$\frac{1}{2}r - \frac{1}{3}s = 8$$

$$\frac{5}{8}r - \frac{1}{4}s = 29$$

4. For the system of equations above, what is the value of $r + s$?

5. During its first week of business, a market sold a total of 108 apples and oranges. The second week, five times the number of apples and three times the number of oranges were sold. A total of 452 apples and oranges were sold during the second week. How many more apples than oranges were sold in the *first* week?

6. Jacob and Zachary go to the movie theater and purchase refreshments for their friends. Jacob spends a total of $18.25 on two bags of popcorn and three drinks. Zachary spends a total of $27.50 for four bags of popcorn and two drinks. What is the cost for purchasing one bag of popcorn and one drink?

For **Questions 7 and 8** refer to the information below.

A mobile phone–based taxi service charges a base fee of $2 plus an amount per minute and an additional amount per mile for the trip. Ariel is charged $16.94 for a ride that takes 14 minutes and travels 10 miles. Victoria is charged $11.30 for a ride that takes 10 minutes and travels 6 miles.

7. What is the per-minute charge?

8. What would be the charge for a ride that takes 8 minutes and travels 5 miles?

LESSON 3-8 ALGEBRAIC INEQUALITIES

OVERVIEW

A linear inequality such as $2x - 3 \leq 7$ is solved in much the same way that a linear equation is solved. There is one exception: multiplying or dividing both sides of an inequality by the same negative quantity *reverses* the direction of the inequality. For example, if $-3x \leq 12$, then dividing both sides by -3 results in the equivalent inequality, $x \geq -4$.

"AT LEAST" AND "AT MOST"

The phrase "is at least" is translated by $\geq$ and the phrase "is at most" is translated by $\leq$.

➡ Example

What is the greatest integer value of x such that $1 - 2x$ is *at least* 6?

Solution

$$1 - 2x \geq 6$$
$$-2x \geq 5$$
$$\frac{-2x}{-2} \leq \frac{5}{-2} \quad \leftarrow \text{Reverse inequality sign}$$
$$x \leq -2.5$$

The greatest integer value of x that satisfies the inequality is **–3**.

➡ Example

$$C = 8n + 522$$

The equation above gives the cost, C, in dollars of manufacturing n items. A profit is made when the total revenue from selling a quantity of items is greater than the total cost of manufacturing the same quantity of items. If each item can be sold for $14, which of the following inequalities gives all possible values of n that will produce a profit?

(A) $n > 87$
(B) $n > 112$
(C) $n > 522$
(D) $n > 609$

Solution

Since a profit will be made when total revenue is greater than total cost,

$$14n > 8n + 522$$

$$14n - 8n > 522$$

$$6n > 522$$

$$\frac{6n}{6} > \frac{522}{6}$$

$$n > 87$$

The correct choice is **(A)**.

➡ Example

If $2x^2 + 3ax - 7 < -17$, what is the smallest possible integer value of a when $x = -1$?

Solution

Replace x with -1 and solve for a:

$$2(-1)^2 + 3a(-1) - 7 < -17$$

$$2 - 3a - 7 < -17$$

$$-3a - 5 < -17$$

$$-3a < -12$$

$$\frac{-3a}{-3} > \frac{-12}{-3} \quad \leftarrow \text{Reverse inequality sign}$$

$$a > 4$$

The smallest integer value of a that satisfies the inequality is **5**.

SOLVING COMBINED INEQUALITIES

To solve an inequality that has the form $c \le ax + b \le d$, where a, b, c, and d stand for numbers, isolate the letter by performing the same operation on each member of the inequality.

➡ Example

If $-2 \le 3x - 7 \le 8$, find x.

Solution

- Add 7 to each member of the inequality:

$$-2 \le \quad 3x - 7 \quad \le 8$$
$$-2 + 7 \le 3x - 7 + 7 \le 8 + 7$$
$$5 \le \quad 3x \quad \le 15$$

- Divide each member of the inequality by 3:

$$\frac{5}{3} \le \frac{3x}{3} \le \frac{15}{3}$$
$$\frac{5}{3} \le x \le 5$$

The solution consists of all real numbers greater than or equal to $\frac{5}{3}$ and less than or equal to 5:

$$\frac{5}{3} \le x \le 5$$

➡ **Example**

If $3 < x + 1 < 8$ and $2 < y < 9$, which of the following best describes the range of values of $y - x$?

(A) $-7 < y - x < 5$
(B) $-5 < y - x < 7$
(C) $0 < y - x < 2$
(D) $2 < y - x < 7$

Solution

First find the upper and lower limits of x and y. If $3 < x + 1 < 8$, then $3 - 1 < x < 8 - 1$ so $2 < x < 7$. This means that the lower limit of x is 2 and the upper limit is 7. Since $2 < y < 9$, the lower limit of y is 2 and the upper limit is 9.

- The upper limit of $y - x$ is obtained by taking the difference between the upper limit of y and the lower limit of x which is $9 - 2 = 7$.
- The lower limit of $y - x$ is obtained by taking the difference between the lower limit of y and the upper limit of x which is $2 - 7 = -5$.
- Hence, $-5 < y - x < 7$.

The correct choice is **(B)**.

TIP

The direction of the inequality sign also gets reversed when comparing the reciprocals of two positive numbers x and y:

If $x > y$,

then $\frac{1}{x} < \frac{1}{y}$.

ORDERING PROPERTIES OF INEQUALITIES

- If $a < b$ and $b < c$, then $a < c$. For example:

$$2 < 3 \text{ and } 3 < 4, \quad \text{so} \quad 2 < 4$$

- If $a < b$ and $x < y$, then $a + x < b + y$. For example:

$$2 < 3 \text{ and } 4 < 5, \quad \text{so} \quad 2 + 4 < 3 + 5$$

- If $a < b$ and $x > y$, then the relationship between $a + x$ and $b + y$ cannot be determined until each of the variables is replaced by a specific number.

Multiple-Choice

1. What is the largest integer value of p that satisfies the inequality $4 + 3p < p + 1$?

   (A) -2
   (B) -1
   (C) 0
   (D) 1
   (E) 2

2. If $-3 < 2x + 5 < 9$, which of the following CANNOT be a possible value of x?

   (A) -2
   (B) -1
   (C) 0
   (D) 2

3. Roger is having a picnic for 78 guests. He plans to serve each guest at least one hot dog. If each package, p, contains eight hot dogs, which inequality could be used to determine the number of packages of hot dogs Roger must buy?

   (A) $\dfrac{p}{8} \geq 78$
   (B) $8p \geq 78$
   (C) $8 + p \geq 78$
   (D) $78 - p \geq 8$

4. Peter begins his kindergarten year able to spell 10 words. He is going to learn to spell 2 new words every day. Which inequality can be used to determine how many days, d, it takes Peter to be able to spell at least 85 words?

   (A) $2d + 10 \geq 85$
   (B) $20d \leq 85$
   (C) $(d + 2) + 10 \geq 85$
   (D) $2d - 10 \leq 85$

5. Which of the following numbers is NOT a solution of the inequality $7 - 5x \leq -3(x - 5)$?

   (A) -5
   (B) -4
   (C) -2
   (D) 1

6. Tamara has a cell phone plan that charges $0.07 per minute plus a monthly fee of $19.00. She budgets $29.50 per month for total cell phone expenses without taxes. What is the maximum number of minutes Tamara could use her phone each month in order to stay within her budget?

   (A) 150
   (B) 271
   (C) 421
   (D) 692

7. What is the solution of $3(2m - 1) \leq 4m + 7$?

   (A) $m \geq 5$
   (B) $m \leq 5$
   (C) $m \geq 4$
   (D) $m \leq 4$

8. An online music club has a one-time registration fee of $13.95 and charges $0.49 to buy each song. If Emma has $50.00 to join the club and buy songs, what is the maximum number of songs she can buy?

   (A) 73
   (B) 74
   (C) 130
   (D) 131

9. The ninth grade class at a local high school needs to purchase a park permit for $250.00 for their upcoming class picnic. Each ninth grader attending the picnic pays $0.75. Each guest pays $1.25. If 200 ninth graders attend the picnic, which inequality can be used to determine the number of guests, x, needed to cover the cost of the permit?

   (A) $0.75x - (1.25)(200) \geq 250.00$
   (B) $0.75x + (1.25)(200) \geq 250.00$
   (C) $(0.75)(200) - 1.25x \geq 250.00$
   (D) $(0.75)(200) + 1.25x \geq 250.00$

10. If $2(x - 4) \geq \dfrac{1}{2}(5 - 3x)$ and x is an integer, what is the smallest possible value of x^2?

   (A) $\dfrac{1}{4}$
   (B) 1
   (C) 4
   (D) 9

11. Edith tutors after school for which she gets paid at a rate of $20 an hour. She has also accepted a job as a library assistant that pays $15 an hour. She will work both jobs, but she is able to work *no more than* a total of 11 hours a week, due to school commitments. Edith wants to earn *at least* $185 a week working a combination of both jobs. Which inequality can be used to represent the situation?

   (A) $20(11 + x) + \dfrac{185}{x} > 15$
   (B) $20x + 15(11 - x) > 185$
   (C) $15(11 - x) + \dfrac{185}{x} > 20$
   (D) $15x + 20(11 + x) > 185$

12. Guy is paid $185 per week plus 3% of his total sales in dollars, and Jim is paid $275 per week plus 2.5% of his total sales in dollars. If d represents the dollar amount of sales for each person, which inequality represents the amount of sales for which Guy is paid more than Jim?

   (A) $d > 18,000$
   (B) $d < 18,000$
   (C) $d > 12,500$
   (D) $d < 12,500$

13. Connor wants to attend the town carnival. The price of admission to the carnival is $4.50, and each ride costs an additional 79 cents. If he can spend at most $16.00 at the carnival, which inequality can be used to solve for r, the number of rides Connor can go on, and what is the maximum number of rides he can go on?

   (A) $0.79 + 4.50r \leq 16.00$; 3 rides
   (B) $0.79 + 4.50r \leq 16.00$; 4 rides
   (C) $4.50 + 0.79r \leq 16.00$; 14 rides
   (D) $4.50 + 0.79r \leq 16.00$; 15 rides

14. For how many integer values of b is $b + 3 > 0$ and $1 > 2b - 9$?

   (A) Four
   (B) Five
   (C) Six
   (D) Seven
   (E) Eight

Grid-In

1. For what integer value of y is $y + 5 > 8$ and $2y - 3 < 7$?

2. If 2 times an integer x is increased by 5, the result is always greater than 16 and less than 29. What is the least value of x?

3. If $2 < 20x - 13 < 3$, what is one possible value for x?

$$\frac{1}{7} + \frac{1}{8} - \frac{1}{9} + \frac{1}{10} < \frac{1}{8} - \frac{1}{9} + \frac{1}{10} + \frac{1}{n}$$

4. For the above inequality, what is the greatest possible positive integer value of n?

5. Chelsea has $45 to spend at an amusement park. She spends $20 on admission and $15 on snacks. She wants to play a game that costs $0.65 per game. What is the maximum number of times she can play the game?

6. Chris rents a booth at a flea market at a cost of $75 for one day. At the flea market Chris sells picture frames each of which costs him $6.00. If Chris sells each picture frame for $13, how many picture frames must he sell to make a profit of *at least* $200 for that day?

7. An online electronics store must sell at least $2,500 worth of printers and monitors per day. Each printer costs $125 and each monitor costs $225. The store can ship a maximum of 15 items per day. What is the maximum number of printers it can ship each day?

$$-\frac{5}{3} < \frac{1}{2} - \frac{1}{3}x < -\frac{3}{2}$$

8. For the inequality above, what is a possible value of $x - 3$?

LESSON 3-9 ABSOLUTE VALUE EQUATIONS AND INEQUALITIES

OVERVIEW

The absolute value of x, written as $|x|$, refers to the quantity x without regard to whether it is positive or negative. Geometrically, $|x|$ represents the distance from 0 to x on the number line. Since -2 and $+2$ are each 2 units from 0, $|2| = 2$ and $|-2| = 2$.

SOLVING ABSOLUTE VALUE EQUATIONS: $|x - a| = b$

To solve an absolute value equation, remove the absolute value sign by accounting for two possibilities:

- The quantity inside the absolute value sign is nonnegative in which case simply remove the absolute value sign:

$$\text{If } x - a \geq 0, \text{ then } |x - a| = x - a.$$

- The quantity inside the absolute value sign is negative in which case remove the absolute value sign by placing a negative sign in front of the quantity:

$$\text{If } x - a < 0, \text{ then } |x - a| = -(x - a).$$

➡ Example

Solve for x: $|x - 1| = 4$.

Solution

Consider the two possibilities:

- If the quantity inside the absolute value sign is nonnegative, then

$$|x - 1| = x - 1 = 4, \text{ so } x = 5$$

- If the quantity inside the absolute value sign is negative, then

$$|x - 1| = -(x - 1) = 4$$
$$-x + 1 = 4$$
$$-x = 3$$
$$x = -3$$

TIP

If c is a negative number, there is *no* value of x that makes $|x| = c$ or $|x| < c$ true.

- The two possible solutions for x are **5** or **−3**. You should verify that both roots satisfy the original absolute value equation.

➡ Example

Solve for x: $|2x + 3| + 4 = 5$.

Solution

If $|2x + 3| + 4 = 5$, then $|2x + 3| = 1$. Thus,

- $2x + 3 = 1$, so $2x = 1 - 3 = -2$ and $x = \dfrac{-2}{2} = -1$; or

- $2x + 3 = -1$, so $2x = -1 - 3 = -4$ and $x = \dfrac{-4}{2} = -2$

You should verify that both roots satisfy the original equation.

➡ **Example**

Solve and check: $|x - 3| = 2x$.

Solution

If $|x - 3| = 2x$, then

- $x - 3 = 2x$ so $-3 = x$.
  <u>Check</u>: If $x = -3$, then

$$|-3 - 3| = 2(-3)$$
$$|-6| \neq -6$$

- $x - 3 = -2x$ so $3x - 3 = 0$, $3x = 3$ and $x = 1$.
  <u>Check</u>: If $x = 1$, then

$$|x - 3| = 2x$$
$$|1 - 3| = 2(1)$$
$$|-2| = 2 \checkmark$$

Hence, $x = \mathbf{1}$ is the only root of the absolute value equation.

INTERPRETING ABSOLUTE VALUE INEQUALITIES

The absolute value inequality $|x - a| < d$ represents the set of all points x that are less than d units from a. For example,

- the inequality $|t - 68°| < 3°$ states that the temperature, t, is less than 3° from 68°, which means that t is between 65° and 71°, as shown in Figure 3.2.

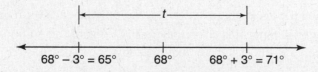

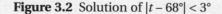

Figure 3.2 Solution of $|t - 68°| < 3°$

- the inequality $|t - 68°| > 3°$ states that the temperature, t, is more than 3° from 68°, which means that t is less than 65° or greater than 71°.

SOLVING ABSOLUTE VALUE INEQUALITIES

To solve an absolute value inequality algebraically, remove the absolute value sign according to the following rules where d is a positive number:

- if $|ax - b| < d$, then $-d < ax - b < d$.
- if $|ax - b| > d$, then $ax - b < -d$ or $ax - b > d$.

➥ Example

Solve and graph the solution set of $|2x - 1| \leq 7$.

Solution

- If $|2x - 1| \leq 7$, then $-7 \leq 2x - 1 \leq 7$.
- Add 1 to each member of the combined inequality:

$$\begin{array}{ccc} -7 \leq 2x - 1 & \leq 7 \\ \underline{+1 \qquad\quad +1 + 1} \\ -6 \leq 2x & \leq 8 \end{array}$$

- Divide each member of the combined inequality by 2:

$$\frac{-6}{2} \leq \frac{2x}{2} \leq \frac{8}{2}$$

$$-3 \leq x \leq 4$$

- Graph the solution set:

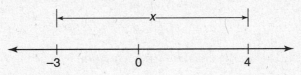

➥ Example

If $5 < |2 - x| < 6$ and $x > 0$, what is one possible value of x?

Solution

Remove the absolute value sign:

- If $2 - x \geq 0$, then $5 < 2 - x < 6$, so $3 < -x < 4$. Multiplying each term of the inequality by -1 gives $-3 > x > -4$. Since it is given that $x > 0$, disregard this solution.
- If $2 - x < 0$, then $5 < -(2 - x) < 6$ or, equivalently, $5 < x - 2 < 6$, so $5 + 2 < x < 6 + 2$ and $7 < x < 8$.
- Therefore, x can be any number between 7 and 8, such as **7.5**.

Multiple-Choice

$$|n - 1| < 4$$

1. How many integers n satisfy the inequality above?

   (A) Two
   (B) Five
   (C) Seven
   (D) Nine

2. If $|x| \leq 2$ and $|y| \leq 1$, then what is the least possible value of $x - y$?

   (A) -3
   (B) -2
   (C) -1
   (D) 0

3. If $\left|\frac{1}{2}x\right| \geq \frac{1}{2}$, then which statement must be true?

   (A) $x \leq -2$ or $x \geq 2$
   (B) $x \leq -1$ or $x \geq 1$
   (C) $x \leq -\frac{1}{2}$ or $x \geq \frac{1}{2}$
   (D) $-1 \leq x \leq 1$

4. If $\frac{1}{2}|x| = 1$ and $|y| = x + 1$, then y^2 could be

   (A) 2
   (B) 3
   (C) 4
   (D) 9

5. In a certain greenhouse for plants, the Fahrenheit temperature, F, is controlled so that it does *not* vary from 79° by more than 7°. Which of the following best expresses the possible range in Fahrenheit temperatures of the greenhouse?

   (A) $|F - 79| \leq 7$
   (B) $|F - 79| > 7$
   (C) $|F - 7| \leq 79$
   (D) $|F - 7| > 79$

6. If $\dfrac{|a + 3|}{2} = 1$ and $2|b + 1| = 6$, then $|a + b|$ could equal any of the following EXCEPT

   (A) 1
   (B) 3
   (C) 5
   (D) 7

7. For what value of x is $|1 + x| = |1 - x|$?

   (A) No value
   (B) 1
   (C) -1
   (D) 0

$$-1 < x < 3$$

8. The inequality above is equivalent to which of the following?

   (A) $|x - 1| < 2$
   (B) $|x + 1| < 2$
   (C) $|x - 2| < 1$
   (D) $|x + 2| < 1$

9. A certain medication must be stored at a temperature, t, that may range between a low of 45° Fahrenheit and a high of 85° Fahrenheit. Which inequality represents the allowable range of Fahrenheit temperatures?

(A) $|t - 65| \leq 20$
(B) $|t + 20| \leq 65$
(C) $|t + 65| \leq 20$
(D) $|t - 20| \leq 85$

10. The inequality $|1.5C - 24| \leq 30$ represents the range of monthly average temperatures, C, in degrees Celsius, during the winter months for a certain city. What was the lowest monthly average temperature, in degrees Celsius, for this city?

(A) −4
(B) 0
(C) 6
(D) 9

Grid-In

$$|t - 7| = 4$$
$$|9 - t| = 2$$

1. What value of t satisfies both of the above equations?

2. If $|-3y + 2| < 1$, what is one possible value of y?

3. If $|x - 16| \leq 4$ and $|y - 6| \leq 2$, what is the greatest possible value of $x - y$?

4. An ocean depth finder shows the number of feet in the depth of water at a certain place. The difference between d, the actual depth of the water, and the depth finder reading, x, is $|d - x|$ and must be less than or equal to $0.05d$. If the depth finder reading is 620 feet, what is the *maximum* value of the actual depth of the water, to the *nearest* foot?

OVERVIEW

The **xy-plane** is the plane formed by two fixed perpendicular lines, called **axes**, intersecting at their 0 points, called the **origin**. Each point in the *xy*-plane can be uniquely represented by an ordered pair of signed numbers of the form (*x*, *y*), called **coordinates**, that represents its distance from the horizontal **x-axis** and the vertical **y-axis**. The *xy*-plane is divided into four quadrants:

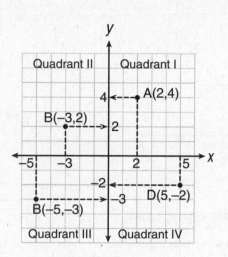

FINDING THE DISTANCE BETWEEN TWO POINTS

The distance formula can be used to find the *distance* between any two points in the *xy*-plane, say *A* and *B*, which is also the length of segment *AB*.

Distance Formula	Example
The distance d between points $A(x_A, y_A)$ and $B(x_B, y_B)$ is $$d = \sqrt{(x_B - x_A)^2 + (y_B - y_A)^2}$$	To find the distance d between points $(4, -1)$ and $(7, 5)$, let $(x_A, y_A) = (4, -1)$ and $(x_B, y_B) = (7, 5)$: $$d = \sqrt{(7-4)^2 + (5-(-1))^2}$$ $$= \sqrt{3^2 + 6^2}$$ $$= \sqrt{45}$$ $$= \sqrt{9} \cdot \sqrt{5}$$ $$= 3\sqrt{5}$$

FINDING THE MIDPOINT OF A SEGMENT

The coordinates of the *midpoint* of a segment are the averages of the corresponding coordinates of the endpoints of the segment.

Midpoint Formula	Example
The coordinates $(\bar{x}, \bar{y})$ of the midpoint of the segment whose endpoints are $A(x_A, y_A)$ and $B(x_B, y_B)$ are $$\bar{x} = \frac{x_A + x_B}{2} \text{ and } \bar{y} = \frac{y_A + y_B}{2}$$	To find the midpoint of a segment whose endpoints are $(4, -1)$ and $(7, 5)$, let $(x_A, y_A) = (4, -1)$ and $(x_B, y_B) = (7, 5)$. Since $$\bar{x} = \frac{4 + 7}{2} = \frac{11}{2}$$ and $$\bar{y} = \frac{(-1) + 5}{2} = 2$$ the midpoint is $\left(\frac{11}{2}, 2\right)$.

FINDING THE SLOPE OF A LINE

The **slope** of a line is a number that represents its steepness. It is calculated by finding the difference of the y-coordinates of any two points on the line (Δy), and dividing it by the difference of the x-coordinates of those two points (Δx) taken in the same order. The letter m is commonly used to represent slope.

Slope Formula	Example
The slope, m, of a nonvertical line that contains $A(x_A, y_A)$ and $B(x_B, y_B)$ is $$m = \frac{\Delta y}{\Delta x} = \frac{y_B - y_A}{x_B - x_A}$$ The order in which the coordinates of the points are subtracted must be the same in the numerator and in the denominator.	To find the slope of the line that contains points $(4, -1)$ and $(7, 5)$, let $(x_A, y_A) = (4, -1)$ and $(x_B, y_B) = (7, 5)$. Then: $$\text{Slope} = \frac{y_B - y_A}{x_B - x_A} = \frac{5 - (-1)}{7 - 4}$$ $$= \frac{5 + 1}{3}$$ $$= 2$$

MATH REFERENCE FACT

If you know the slope and the coordinates of one point on a line, you can find the coordinates of other points on the same line. For example, if the slope of a line is 2 and the point (-1, 3) is on the line, then each time x increases by 1, y increases by 2 so there is another point on the line $(-1 + 1, 3 + 2) = (0, 5)$.

POSITIVE, NEGATIVE, AND ZERO SLOPE

■ A line that rises as x increases has a positive slope. If the line falls as x increases, the line has a negative slope. See Figure 3.3 where the letter m is used to represent the slope of a line.

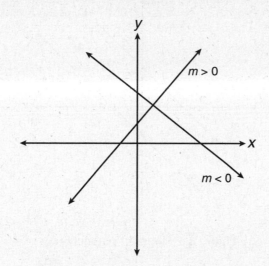

Figure 3.3 Positive vs. negative slope

■ The slope of a horizontal line is 0 and the slope of a vertical line is undefined. See Figure 3.4.

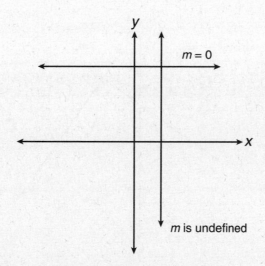

Figure 3.4 Undefined vs. 0 slope

SLOPES OF PARALLEL AND PERPENDICULAR LINES

- Parallel lines have the *same* slope. See Figure 3.5.

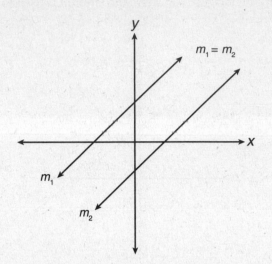

Figure 3.5 Slopes of parallel lines

- Perpendicular lines have slopes that are *negative reciprocals*. See Figure 3.6. If the slope of a line is $\frac{3}{4}$, then the slope of a perpendicular line is $-\frac{4}{3}$. The product of the slopes of perpendicular lines is -1.

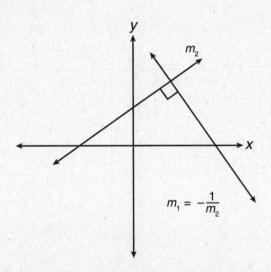

Figure 3.6 Slopes of perpendicular lines

EQUATION OF A LINE: *y = mx + b*

If a nonvertical line has a slope of m and intersects the y-axis at $(0, b)$, then the equation $y = mx + b$ describes the set of all points (x, y) that the line contains. For example:

- If an equation of a line is $y = -5x + 3$, then $m = -5$ and $b = 3$, so the slope of the line is -5 and the line crosses the y-axis at $(0, 3)$.

- If an equation of a line is $y = 2x - 4$, then $m = 2$. The slope of a line parallel to the given line is also 2.
- If an equation of a line is $3x + y = 7$, then solving for y gives $y = -3x + 7$, so $m = -3$ and $b = 7$. The slope of a line perpendicular to the given line is the negative reciprocal of -3, which is $\dfrac{1}{3}$.

TIP

- If a line that passes through the origin contains the point (**a**, **b**), then the slope of the line is $\dfrac{b}{a}$.
- If the slope of line ℓ is m, then the slope of a line parallel to ℓ is also m, and the slope of a line perpendicular to ℓ is $-\dfrac{1}{m}$, provided $m \neq 0$.

➡ Example

Line p contains the points $(-1, 8)$ and $(9, k)$. If line p is parallel to line q whose equation is $3x + 4y = 7$, what is the value of k?

Solution

- Find the slope of $3x + 4y = 7$ by changing the equation to slope-intercept form:

$$3x + 4y = 7$$
$$4y = -3x + 7$$
$$y = -\frac{3}{4}x + \frac{7}{4}$$

The slope of line q is $-\dfrac{3}{4}$.

- Using the slope formula, represent the slope of line p in terms of k:

$$m = \frac{y_B - y_A}{x_B - x_A}$$
$$= \frac{k - 8}{10}$$

$$\Delta y = k - 8$$

$$(-1, 8) \quad \text{and} \quad (9, k)$$

$$\Delta x = 9 - (-1) = 9 + 1 = 10$$

- Since parallel lines have the same slope, set the slope of line p equal to the slope of line q:

$$\frac{k - 8}{10} = \frac{-3}{4}$$
$$4(k - 8) = -3(10)$$
$$4k - 32 = -30$$
$$4k = -30 + 32$$
$$\frac{4k}{4} = \frac{2}{4}$$
$$k = \frac{1}{2}$$

WRITING AN EQUATION OF A LINE

The y-coordinate of the point at which a line crosses the y-axis is called the **y-intercept** of the line. If you know the slope (m) and the y-intercept (b) of a line, you can form its equation by writing $y = mx + b$.

- Suppose you know that the slope of a line is 2 and that the line contains the point $(-1, 1)$. Since it is given that $m = 2$, the equation of the line must look like $y = 2x + b$. Because the line contains the point $(-1, 1)$, when $x = -1$, $y = 1$:

$$y = 2x + b$$
$$1 = 2(-1) + b$$
$$1 = -2 + b$$
$$3 = b$$

Hence, an equation of this line is $y = 2x + 3$.

- If you know that the line contains the points $(4, 0)$ and $(-1, 5)$, you can find its equation by first determining the slope of the line. If Δx represents the difference in the x-coordinates of the two points and Δy stands for the difference in the y-coordinates of these points, then

$$m = \frac{\Delta y}{\Delta x} = \frac{5 - 0}{-1 - 4} = \frac{5}{-5} = -1$$

Since $m = -1$, the equation of the line must look like $y = -x + b$. To find the y-intercept, b, substitute the coordinates of either given point into the equation. Since $y = 0$ when $x = 4$:

$$y = -x + b$$
$$0 = -4 + b$$
$$4 = b$$

Hence, an equation of the line is $y = -x + 4$.

WRITING AN EQUATION OF A LINE FROM ITS GRAPH

You can determine the y-intercept and the slope of a line from its graph.

- To figure out the slope of the line shown in Figure 3.7, form a right triangle by moving 1 unit to the right from any point on the line, say $(0, 3)$, and then moving up or down until the line is reached. The horizontal side of the right triangle is formed by moving 1 unit to the *right*, so $\Delta x = +1$. The vertical side of the right triangle is formed by moving 4 units *up*, so $\Delta y = +4$. Hence, the slope of the line is

$$m = \frac{\Delta y}{\Delta x} = \frac{4}{1} = 4$$

Since $m = 4$ and $b = 3$, an equation of the line is $y = 4x + 3$.

TIP

Memorize the formulas for midpoint, distance, and slope, as these formulas are *not* provided in the math reference section of the actual test. You are expected to be able to recall and apply these formulas, including in problems that involve forming or analyzing the equation of a line.

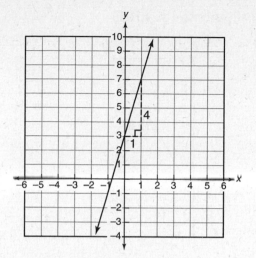

Figure 3.7 Determining the equation $y = 4x + 3$ from its graph

- To figure out the slope of the line shown in Figure 3.8, form a right triangle by moving 1 unit to the right from any point on the line, say $(0, 6)$, and then moving up or down until the line is reached. The horizontal side of the right triangle is formed by moving 1 unit to the *right*, so $\Delta x = +1$. The vertical side of the right triangle is formed by moving 3 units *down*, so $\Delta y = -3$. Hence, the slope of the line is

$$m = \frac{\Delta y}{\Delta x} = \frac{-3}{1} = -3$$

Since $m = -3$ and $b = 6$, an equation of the line is $y = -3x + 6$.

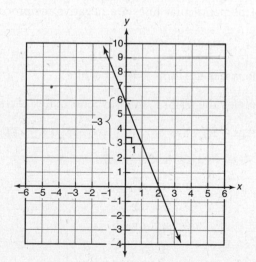

Figure 3.8 Determining the equation $y = -3x + 6$ from its graph

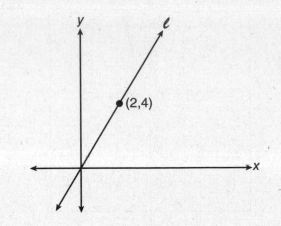

In the accompanying figure, line ℓ passes through the origin and the point (2, 4). Line m (not shown) is perpendicular to line ℓ at (2, 4). Line m intersects the x-axis at which point?

(A) (5, 0)

(B) (6, 0)

(C) (8, 0)

(D) (10, 0)

Solution

The slope of line ℓ is $\dfrac{4-0}{2-0} = 2$.

- Since the slopes of perpendicular lines are negative reciprocals, the slope of line m is $-\dfrac{1}{2}$.

  The equation of line m has the form $y = -\dfrac{1}{2}x + b$.

- Because line m contains the point (2, 4), the coordinates of this point must satisfy its equation:

$$y = -\frac{1}{2}x + b$$

$$4 = -\frac{1}{2}(2) + b$$

$$4 = -1 + b$$

$$5 = b$$

An equation of line m is $y = -\dfrac{1}{2}x + 5$.

■ Line m intersects the x-axis where $y = 0$:

$$0 = -\frac{1}{2}x + 5$$

$$-5 = -\frac{1}{2}(x)$$

$$(-2)(-5) = (-2)\left(-\frac{1}{2}x\right)$$

$$10 = x$$

Since line m intersects the x-axis at (10, 0), the correct choice is (**D**).

➡ Example

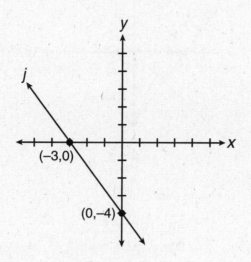

In the figure above, line k (not shown) is perpendicular to line j. If the equation of line k is $y = px$, what is the value of the constant p?

Solution

Use the diagram to determine the slope of line j.

- From the y-intercept to the x-intercept, the vertical "rise" is +4 and the horizontal "run" is –3:

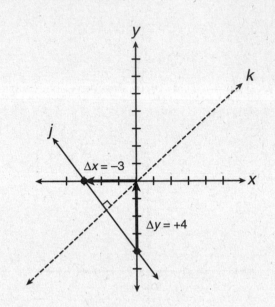

- The slope of line j is "rise" over "run":

$$\frac{\Delta y}{\Delta x} = \frac{\text{rise}}{\text{run}} = \frac{+4}{-3} = -\frac{4}{3}$$

- Since the slopes of perpendicular lines are negative reciprocals, the slope of line k is $\frac{3}{4}$. In the equation $y = px$, p represents the slope of line k so $p = \frac{3}{4}$.

Grid-in **3/4**

➥ **Example**

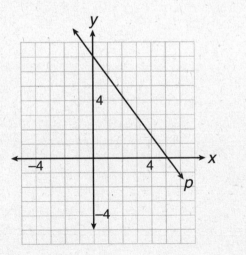

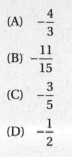

Line p is graphed in the xy-plane above. If line p is translated up 3 units and right 5 units, what is the slope of the resulting line?

(A) $-\dfrac{4}{3}$

(B) $-\dfrac{11}{15}$

(C) $-\dfrac{3}{5}$

(D) $-\dfrac{1}{2}$

Solution

First find the slope of line p by choosing two points on the line with integer coordinates, say $(2, 4)$ and $(5, 0)$. Using these points, the slope of line p is

$$\frac{0-4}{5-2} = -\frac{4}{3}$$

Translating a line shifts all the points on the line the same distance and in the same direction so that the resulting line will be parallel to the original line and, as a result, have the same slope. Hence, the slope of the translated line is $-\dfrac{4}{3}$.

The correct choice is **(A)**.

SYSTEMS OF LINEAR INEQUALITIES

The solution set for a system of two linear equations whose graphs intersect is a single point. The solution set for a system of two linear inequalities is the region over which the solution sets for the two linear inequalities overlap.

Consider the system

$$y \le x + 1$$

$$y > -\frac{1}{2}x + 4$$

■ The solution set for $y \leq x + 1$ consists of the set of points that lie *below* ($\leq$) or on the boundary line $y = x + 1$:

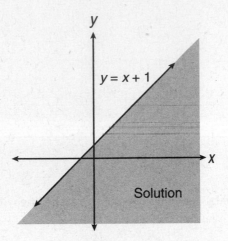

■ The solution set for $y \geq -\dfrac{1}{2}x + 4$ consists of the set of all points that lie *above* ($\geq$) or on the boundary line $y = -\dfrac{1}{2}x + 4$:

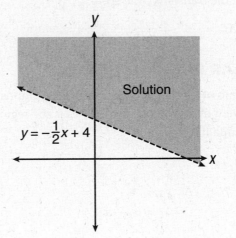

- The solution set for the system lies in the region (labeled I) where the two solution sets overlap:

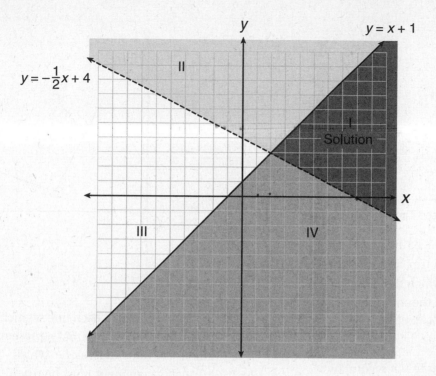

You can also determine which of the four regions above represents the solution set to the system of linear inequalities by picking a point in each region and testing whether it makes both inequalities true at the same time. For example, test the point (6, 2), which lies in region I:

$$\text{Test } (6, 2): \quad y \le x + 1$$
$$2 \le 6 + 1$$
$$2 \le 7 \qquad \textit{True!}$$

$$\text{Test } (6, 2): \quad y \ge -x + 4$$
$$2 \ge -(6) + 4$$
$$2 \ge -2 \qquad \textit{True!}$$

Hence, the bounded region that contains (6, 2) represents the solution set.

> ### MATH FACT
>
> To graph a linear inequality:
>
> - Replace the inequality relation with an "=" sign and graph the boundary line. Draw a solid line for a $\ge$ or $\le$ inequality relation and a broken line if the inequality relation is $>$ or $<$.
> - If after solving for y in terms of x, the inequality relation is $\ge$ or $>$, then the solution region lies *above* the boundary line; if the inequality relation is $\le$ or $<$, the solution region lies *below* the boundary line.

Multiple-Choice

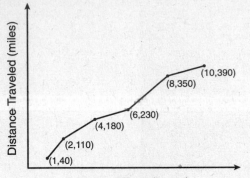

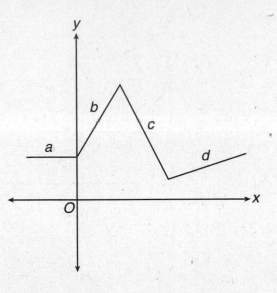

1. A family kept a log of the distance they traveled during a trip, as represented by the graph above in which the points are ordered pairs of the form (hour, distance). During which interval was their average speed the greatest?

    (A) The first hour to the second hour
    (B) The second hour to the fourth hour
    (C) The sixth hour to the eighth hour
    (D) The eighth hour to the tenth hour

2. In the above figure, each line segment is labeled with a variable that represents the numerical value of its slope. Which inequality statement must be true?

    (A) $a < c < b < d$
    (B) $d < b < a < c$
    (C) $d < c < b < a$
    (D) $c < a < d < b$

3. Which of the following represents an equation of the line that is the perpendicular bisector of the segment whose endpoints are (–2, 4) and (8, 4)?

    (A) $x = 3$
    (B) $y = 3$
    (C) $x = 5$
    (D) $y = 5$

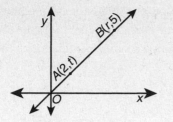

4. In the graph above, what is r in terms of t?

(A) $\dfrac{5}{2}t$

(B) $\dfrac{2}{5}t$

(C) $\dfrac{t}{10}$

(D) $\dfrac{10}{t}$

5. What is the slope of the line $2(x + 2y) = 0$?

(A) $\dfrac{1}{2}$

(B) -2

(C) $-\dfrac{1}{2}$

(D) 0

6. Segments AP and BP have the same length. If the coordinates of A and P are $(-1, 0)$ and $(4, 12)$, respectively, which could be the coordinates of B?

I. $\left(\dfrac{3}{2}, 6\right)$

II. $(9, 24)$

III. $(-8, 7)$

(A) I and II only

(B) II and III only

(C) II only

(D) III only

7. Which of the following is an equation of a line that is parallel to the line $\dfrac{1}{2}y - \dfrac{2}{3}x = 6$ in the xy-plane?

(A) $y = -\dfrac{3}{4}x + 1$

(B) $y = 4\left(\dfrac{x - 1}{3}\right)$

(C) $9x - 6y = 18$

(D) $\dfrac{y}{3} = \dfrac{x - 5}{4}$

8. Which of the following is an equation of a line that is perpendicular to the line $y = -2(x + 1)$?

(A) $x + 2y = 7$

(B) $8x - 4y = 9$

(C) $\dfrac{x - 1}{6} = \dfrac{y}{3}$

(D) $y - 2x = 0$

9. The point whose coordinates are $(4, -2)$ lies on a line whose slope is $\dfrac{3}{2}$. Which of the following are the coordinates of another point on this line?

(A) $(1, 0)$

(B) $(2, 1)$

(C) $(6, 1)$

(D) $(7, 0)$

10. If point $E(5, h)$ is on the line that contains $A(0, 1)$ and $B(-2, -1)$, what is the value of h?

(A) -1

(B) 0

(C) 1

(D) 6

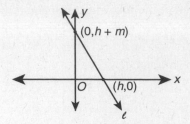

11. In the figure above, if the slope of line ℓ is m, what is m in terms of h?

(A) $\dfrac{h}{1+h}$

(B) $\dfrac{-h}{1+h}$

(C) $\dfrac{h}{1-h}$

(D) $1+h$

12. Which could be the slope of a line that contains $(1, 1)$ and passes between the points $(0, 2)$ and $(0, 3)$?

(A) $-\dfrac{3}{2}$

(B) $-\dfrac{1}{2}$

(C) 0

(D) $\dfrac{1}{2}$

13. The line $y + 2x = b$ is perpendicular to a line that passes through the origin. If the two lines intersect at the point $(k + 2, 2k)$, what is the value of k?

(A) $-\dfrac{3}{2}$

(B) $-\dfrac{2}{3}$

(C) $\dfrac{2}{5}$

(D) $\dfrac{2}{3}$

14. Which of the following is an equation of the line that is parallel to the line $y - 4x = 0$ and has the same y-intercept as the line $y + 3 = x + 1$?

(A) $y = 4x - 2$

(B) $y = 4x + 1$

(C) $y = -\dfrac{1}{4}x + 1$

(D) $y = -\dfrac{1}{4}x - 2$

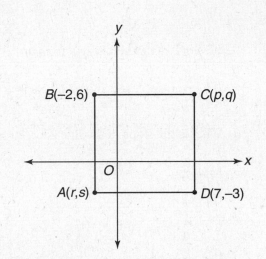

15. Which of the following is an equation of the line that contains diagonal $\overline{AC}$ of square $ABCD$ shown in the accompanying figure?

(A) $y = 2x + 1$

(B) $y = -x + 1$

(C) $y = \dfrac{1}{2}x - 2$

(D) $y = x - 1$

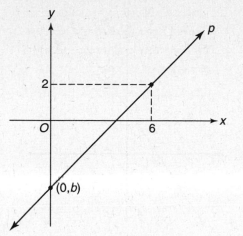

Note: Figure not drawn to scale.

16. If the slope of line p shown in the figure above is $\frac{3}{2}$, what is the value of b?

(A) -8
(B) -7
(C) -5
(D) -3

17. Which of the following graphs shows a line where each value of y is three more than half of x?

(A) Graph (1)
(B) Graph (2)
(C) Graph (3)
(D) Graph (4)

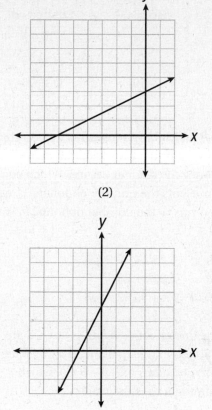

(2)

(3)

(4)

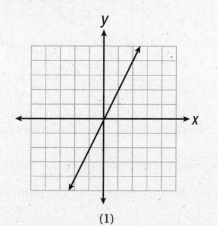

(1)

Number of Hours, h	Dollars Earned, d
8	$70.00
15	$113.75
19	$138.75
30	$207.50

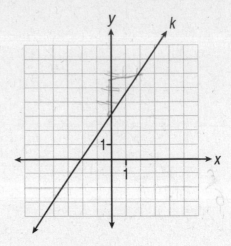

18. The table above represents the number of hours a student worked and the amount of money the student earned. Which equation represents the number of dollars, d, earned in terms of the number of hours, h, worked

(A) $d = 6.25h$
(B) $d = 6.25h + 20$
(C) $d = 5.25h + 28$
(D) $d = 7h + 8.75$

19. The lines $y = ax + b$ and $y = bx + a$ are graphed in the xy-plane. If a and b are non-zero constants and $a + b = 0$, which statement must be true?

(A) The lines are parallel.
(B) The lines intersect at right angles.
(C) The lines have the same x-intercept.
(D) The lines have the same y-intercept.

20. Which of the following is an equation of the line in the xy-plane that is perpendicular to line k in the figure above?

(A) $y = 3\left(1 - \frac{1}{2}x\right)$
(B) $\frac{x}{y} = \frac{2}{3}$
(C) $3y + 2x = 4$
(D) $2y + 3x = -6$

Grid-In

1. A line with a slope of $\frac{3}{14}$ passes through points $(7, 3k)$ and $(0, k)$. What is the value of k?

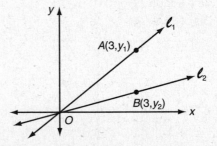

2. In the figure above, the slope of line ℓ_1 is $\frac{5}{6}$ and the slope of line ℓ_2 is $\frac{1}{3}$. What is the distance from point A to point B?

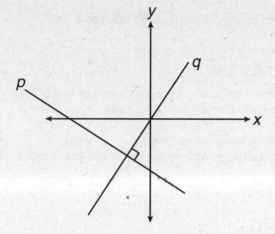

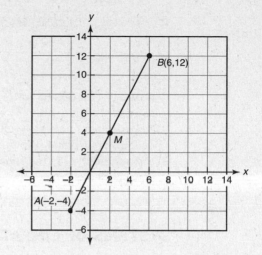

Note: Figure not drawn to scale.

3. In the figure above, lines p and q are perpendicular. Line q passes through the origin and intersects line p at $(-1, -2)$. If $(-20, k)$ is a point on line p, what is the value of k?

4. In the accompanying figure, what is the y-coordinate of the point at which the line that is perpendicular to $\overline{AB}$ (not shown) at point M crosses the y-axis?

5. A line in the xy-plane contains the points $A(c, 40)$ and $B(5, 2c)$. If the line also contains the origin, what is a possible value of c?

> ### OVERVIEW
>
> If two lines intersect in the *xy*-plane, then their point of intersection represents the solution to the system of equations graphed. When a linear inequality is graphed in the *xy*-plane, the set of all points that satisfy the inequality lie on one side of the line called a **half-plane**. When a system of linear inequalities is graphed in the *xy*-plane, the solution set is the region over which the solution half-planes of the individual inequalities overlap.

SYSTEMS OF LINEAR EQUATIONS

If a system of two linear equations is graphed in the *xy*-plane, then there are three possibilities to consider:

(1) The Lines Intersect

The coordinates of the point of intersection represent the solution to the linear system. The linear system $y = 2x + 1$ and $y = -x + 4$ consists of two lines with different slopes and, as a result, intersect at a single point that satisfies both equations. See Figure 3.9.

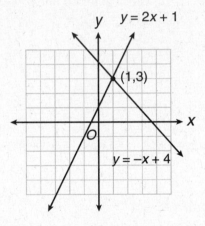

Figure 3.9 Lines with different slopes

(2) The Lines Do *Not* Intersect

If the lines are different and parallel, as in the linear system $y = 2x + 4$ and $y = 2x - 1$ shown in Figure 3.10, the system has no solution since there is no point common to both lines.

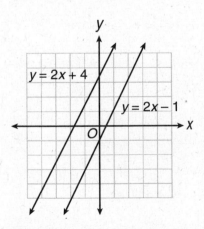

Figure 3.10 Lines with the same slope and different *y*-intercepts

(3) The Lines Coincide

The linear system $x - y = 3$ and $2x - 2y = 6$ consists of two lines that coincide as one equation is a multiple of the other. The lines have the same slope and also have the same y-intercepts. Because each point on the lines is a solution, there are infinitely many solutions.

The three possibilities are summarized in Table 3.2.

To determine the number of solutions of a linear system, compare the equations of the two lines in $y = mx + b$ slope-intercept form.

Table 3.2 Possible Solutions to a Linear System

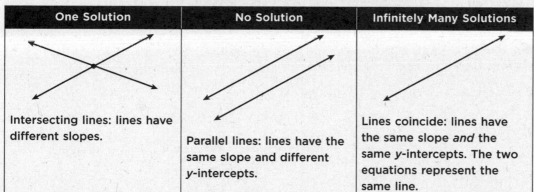

One Solution	No Solution	Infinitely Many Solutions
Intersecting lines: lines have different slopes.	Parallel lines: lines have the same slope and different y-intercepts.	Lines coincide: lines have the same slope *and* the same y-intercepts. The two equations represent the same line.

➡ Example

$$\frac{2}{3}x - \frac{1}{2}y = 7$$
$$kx - 6y = 4$$

In the system of linear equations above, k is a constant. If the system has no solution, what is the value of k?

Solution

Since the system has no solution, the graphs of the equations in the xy-plane are parallel lines so the lines have the same slope. Write each equation in $y = mx + b$ slope-intercept form and then compare their slopes.

- If $\frac{2}{3}x - \frac{1}{2}y = 7$, then $-\frac{1}{2}y = -\frac{2}{3}x + 7$ so $y = \frac{4}{3}x - 14$.

- If $kx - 6y = 4$, then $-6y = -kx + 4$ so $y = \frac{k}{6}x - \frac{2}{3}$.

- When an equation of a line is written in $y = mx + b$ form, the coefficient of the x-term represents the slope of the line. Since parallel lines have equal slopes, $\frac{4}{3} = \frac{k}{6}$ so $3k = 24$ and $k = \frac{24}{3} = 8$.

The value of k is **8**.

GRAPHING A LINEAR INEQUALITY

To graph a linear inequality such as $y + 2x < 6$,

- Replace the inequality relation with an equal sign (=) and graph the boundary line. Draw a solid line for a ≥ or ≤ inequality relation and a broken line if the inequality relation is > or <. See Figure 3.11.

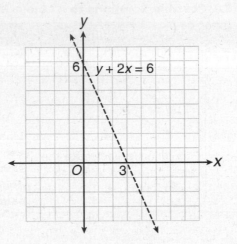

Figure 3.11 Graphing the boundary line of $y + 2x < 6$

- If after solving for y in terms of x, the inequality relation is ≥ or >, then the solution region lies *above* the boundary line; if the inequality relation is ≤ or <, the solution region lies *below* the boundary line. Since $y + 2x < 6$ becomes $y < -2x + 6$ and the inequality is "less than," the solution region lies *below* the line, as shown in Figure 3.12.

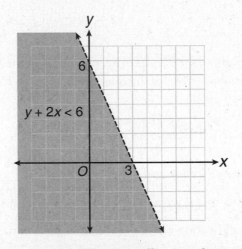

Figure 3.12 Shaded region represents all points that satisfy $y < -2x + 6$

SYSTEMS OF LINEAR INEQUALITIES

The solution set for a system of two linear inequalities is the region over which the solution sets for the two linear inequalities overlap.

Consider the system:

$$y \leq x + 1$$

$$y > -\frac{1}{2}x + 4$$

- The solution set for $y \leq x + 1$ consists of the set of points that lie *below* ($\leq$) or on the boundary line $y = x + 1$. See Figure 3.13.

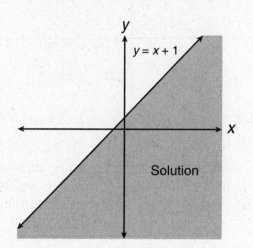

Figure 3.13 Graphing boundary line of $y \leq x + 1$

- The solution set for $y > -\frac{1}{2}x + 4$ consists of the set of all points that lie *above* ($>$) the boundary line $y = -\frac{1}{2}x + 4$ as shown in Figure 3.14.

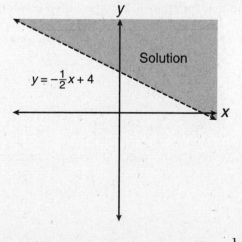

Figure 3.14 Graphing boundary line of $y > -\frac{1}{2}x + 4$

- The solution set for the system lies in the region (labeled I) where the two solution sets overlap, as illustrated in Figure 3.15.

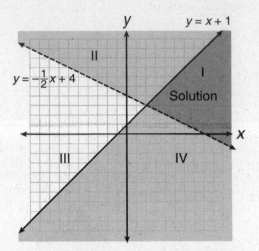

Figure 3.15 Graphing solution of $y \leq x + 1$ and $y > -\frac{1}{2}x + 4$

You can also determine which of the four regions above represents the solution set to the system of linear inequalities by picking a point in each region and testing whether it makes both inequalities true at the same time. For example, test the point (6, 2), which lies in region I:

$$\textbf{Test } (6, 2): y \leq x + 1$$
$$2 \leq 6 + 1$$
$$2 \leq 7 \qquad \textit{True!}$$

$$\textbf{Test } (6, 2): y \geq -x + 4$$
$$2 \geq -(6) + 4$$
$$2 \geq -2 \qquad \textit{True!}$$

Hence, the bounded region that contains (6, 2) represents the solution set.

Multiple-Choice

$$4x + 6y = 12$$
$$y = 8 - kx$$

1. For what value of k does the system of equations above have no solution?

   (A) $-\dfrac{3}{2}$

   (B) 0

   (C) $\dfrac{2}{3}$

   (D) 4

2. Sara correctly solves a system of two linear equations and finds that the system has no solution. If one of the two equations is $\dfrac{y}{6} - \dfrac{x}{4} = 1$, which could be the other equation in this system?

   (A) $y = \dfrac{2}{3}x + 12$

   (B) $y = \dfrac{3}{2}x$

   (C) $y = -\dfrac{3}{2}x$

   (D) $y = \dfrac{3}{2}x + 6$

3. Ben correctly solves a system of two linear equations and finds that the system has an infinite number of solutions. If one of the two equations is $3(x + y) = 6 - x$, which could be the other equation in this system?

   (A) $y = \dfrac{3}{4}x + 2$

   (B) $y = -\dfrac{4}{3}x$

   (C) $y = -\dfrac{4}{3}x + 2$

   (D) $y = -\dfrac{4}{3}x + 6$

4. The graph of the inequality $y \le 2x$ will include all of the points in which quadrant?

   (A) Quadrant I
   (B) Quadrant II
   (C) Quadrant III
   (D) Quadrant IV

$$\dfrac{1}{2}x - \dfrac{5}{6}y = 5$$
$$-2x + ky = 3$$

5. In the system of linear equations above, k is a constant. If the system has no solution, what is the value of k?

   (A) $\dfrac{5}{3}$

   (B) $\dfrac{5}{2}$

   (C) $\dfrac{10}{3}$

   (D) $\dfrac{15}{2}$

6. The graph of a line in the xy-plane has slope $\dfrac{1}{2}$ and contains the point $(0, 7)$. The graph of a second line passes through the points $(0, 0)$ and $(-1, 3)$. If the two lines intersect at the point (r, s), what is the value of $r + s$?

   (A) -3
   (B) -2
   (C) 2
   (D) 4

$$y + x > 2$$

$$y \le 3x - 2$$

7. Which graph shows the solution of the set of inequalities above?

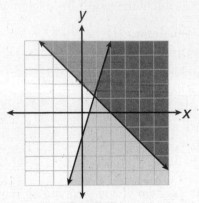

Graph A

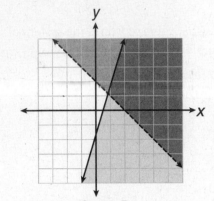

Graph B

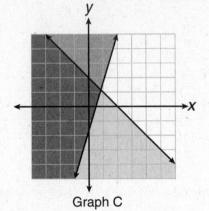

Graph C

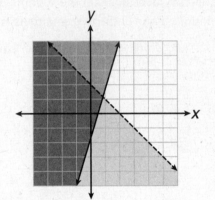

Graph D

(A) Graph A
(B) Graph B
(C) Graph C
(D) Graph D

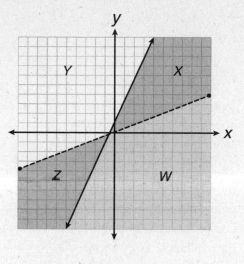

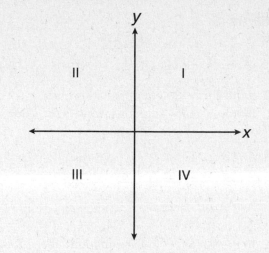

$$2y < x$$

$$y \geq 3x + 1$$

8. A system of inequalities and a graph are shown above. Which region or regions of the graph could represent the set of all ordered pairs that satisfy the system?

(A) Region X
(B) Regions X and Z
(C) Regions X, Y, and W
(D) Region Z

$$3x + 5 = 2y$$

$$\frac{x}{3} + \frac{y}{2} = \frac{2}{3}$$

9. For the system of equations above, which of the following statements is true?

(A) The system has no solution.
(B) The graphs of the equations in the xy-plane intersect at right angles.
(C) The graphs of the equations in the xy-plane intersect but *not* at right angles.
(D) The system has infinitely many solutions.

10. If the system of inequalities $y < 2x + 4$ and $y \geq -x + 1$ is graphed in the xy-plane above which quadrant does not contain any solutions to the system?

(A) Quadrant I
(B) Quadrant II
(C) Quadrant III
(D) Quadrant IV

Grid-In

$$6x + py = 21$$
$$qx + 5y = 7$$

1. If the above system of equations has infinitely many solutions, what is the value of $\frac{p}{q}$?

$$(k-1)x + \frac{1}{3}y = 4$$
$$k(x + 2y) = 7$$

2. In the system of linear equations above, k is a constant. If the system has no solution, what is the value of k?

$$\frac{1}{3}r + 4s = 1$$
$$kr + 6s = -5$$

3. In the system of equations above, k and s are nonzero constants. If the system has no solutions, what is the value of k?

4. The graph of a line in the xy-plane passes through the points $(5, -5)$ and $(1, 3)$. The graph of a second line has a slope of 6 and passes through the point $(-1, 15)$. If the two lines intersect at (p, q), what is the value of $p + q$?

OVERVIEW

A **function** is a *rule* that tells how to pair each member of one set (say the *x*-values) with exactly one member of a second set (say the *y*-values). The rule is typically expressed as either a "*y* = ..." type of equation, a graph, or a table. The set of all possible *x*-values (input) is the **domain** and the resulting set of *y*-values (output) is the **range**.

EQUATIONS AS FUNCTIONS

A function is usually named by a lowercase letter, such as *f* or *g*. The equation $y = 2x + 3$ describes a function, since it gives a rule for pairing any given *x*-value with one particular *y*-value: input any number *x*, multiply it by 2, add 3, and name the result *y*. When $x = 3$, $y = 2(3) + 3 = 9$. If this function is called *f*, then the ordered pair (3, 9) belongs to function *f*. This fact can be abbreviated by writing $f(3) = 9$, which is read as "*f* of three equals nine":

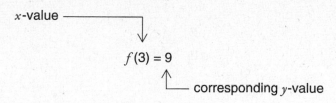

$$f(3) = 9$$

corresponding *y*-value

> **MATH REFERENCE FACT**
>
> The shorthand notation $f(x)$ represents the value of function *f* when *x* has the value inside the parentheses.

Here are some examples of evaluating functions:

- If $h(x) = \dfrac{x}{x^3 + 1}$, then to find $h(2)$, replace *x* with 2:

$$h(2) = \frac{2}{(2)^3 + 1}$$
$$= \frac{2}{8 + 1}$$
$$= \frac{2}{9}$$

- If $g(x) = \dfrac{1}{2}x + 2$, then to find $4g(x) + 1$, multiply the equation that defines function *g* by 4 and then add 1:

$$4g(x) + 1 = 4\overbrace{\left[\frac{1}{2}x + 2\right]}^{g(x)} + 1$$
$$= (2x + 8) + 1$$
$$= 2x + 9$$

- If $f(x) = x^2 + x$, then to find $f(n-1)$, replace each x with $n-1$:

$$f(n-1) = (n-1)^2 \qquad + (n-1)$$
$$= (n-1)(n-1) + (n-1)$$
$$= (n^2 - 2n + 1) + (n-1)$$
$$= n^2 - n$$

➥ Example

If $k(x) = \dfrac{2x - p}{5}$ and $k(7) = 3$, what is the value of p?

Solution

Since $k(7) = 3$, replace x with 7 and set the result equal to 3:

$$k(7) = \frac{2(7) - p}{5}$$
$$3 = \frac{14 - p}{5}$$
$$15 = 14 - p$$
$$p = 14 - 15$$
$$p = -1$$

➥ Example

Function f is defined by $f(x) = 2x + 1$. If $2f(m) = 30$, what is the value of $f(2m)$?

Solution

- If $2f(m) = 30$, then $f(m) = \dfrac{30}{2} = 15$.
- Since $f(m) = 2m + 1 = 15$, $2m = 14$ and $m = 7$.
- Hence, $f(2m) = f(14) = 2(14) + 1 = 29$.

Grid-in **29**

➥ Example

Function f is defined by $f(x) = \dfrac{3}{2}x + c$. If $f(6) = 1$, what is the value of $f(c)$?

(A) −20
(B) −8
(C) 4
(D) 12

Solution

- Find the value of c:

$$f(6) = \frac{3}{2} \times (6) + c = 1$$

$$9 + c = 1$$

$$c = -8$$

- Since $c = -8$, $f(x) = \frac{3}{2}x - 8$.

- Evaluate $f(-8)$:

$$f(x) = \frac{3}{2}x - 8$$

$$f(-8) = \frac{3}{2}(-8) - 8$$

$$= 3(-4) - 8$$

$$= -20$$

The correct choice is **(A)**.

➥ Example

If $g(x) = 3x - 1$ and $f(x) = 4g(x) + 3$, what is $f(2)$?

Solution

- Substitute 2 for x in $f(x) = 4g(x) + 3$:

$$f(2) = 4g(2) + 3$$

- Use $g(x) = 3x - 1$ to find the value of $g(2)$:

$$g(2) = 3(2) - 1 = 5$$

- Hence, $f(2) = 4(5) + 3 = 23$.

➥ Example

If $f(x) = x^2 - 2x$, what is $f(2x + 1)$?

(A) $4x + 2$
(B) $4x^2 - 1$
(C) $4x^2 + 1$
(D) $4x^2 - 4x - 1$

Solution

Since $f(x) = x^2 - 2x$,

$$f(2x + 1) = (2x + 1)^2 - 2(2x + 1)$$

$$= (4x^2 + 4x + 1) - 4x - 2$$

$$= 4x^2 - 1$$

The correct choice is **(B)**.

DETERMINING THE DOMAIN AND RANGE

Unless otherwise indicated, the *domain* of function f is the largest possible set of real numbers x for which $f(x)$ is a real number. There are two key rules to follow when finding the domain of a function:

1. Do **not** divide by zero. Exclude from the domain of a function any value of x that results in division by 0. If $f(x) = \dfrac{x+2}{x-1}$, then $f(1) = \dfrac{1+2}{1-1} = \dfrac{3}{0}$. Since division by 0 is not allowed, x cannot be equal to 1. The domain of function f is the set of all real numbers *except* 1.

2. Do **not** take the square root of a negative number. Since the square root of a negative number is not a real number, the quantity underneath a square root sign must always evaluate to a number that is greater than or equal to 0. If $f(x) = \sqrt{x-3}$, then x must be at least 3, since any lesser value of x will result in the square root of a negative number. For example, $f(1) = \sqrt{1-3} = \sqrt{-2}$, but $\sqrt{-2}$ is not a real number. Thus, the domain of function f is limited to the set of all real numbers greater than or equal to 3.

The *range* of the function $y = f(x)$ is the set of all values that y can have as x takes on each of its possible values. For example, if function f is defined by $f(x) = 1 + \sqrt{x}$, then the smallest possible function value is $f(0) = 1 + \sqrt{0} = 1$.

As x increases without bound, so does $1 + \sqrt{x}$. Therefore, the range of f is the set of all real numbers greater than or equal to 1.

COMPOSITE FUNCTION NOTATION

The output of one function can be used as the input of a second function. The notation $f(g(2))$ means that the output of function g when $x = 2$ is used as the input for function f. If $f(x) = 3x - 1$ and $g(x) = x^2$, then

$$g(2) = 2^2 = 4 \text{ so } f(g(2)) = f(4) = 3(4) - 1 = 11$$

In general, $f(g(x))$ and $g(f(x))$ do *not* necessarily represent the same value. For example, using the same definitions for functions f and g,

$$f(2) = 3(2) - 1 = 5 \text{ so } g(f(2)) = g(5) = 5^5 = 25$$

➡ **Example**

x	0	1	4	5
$f(x)$	–2	4	0	2

x	2	1	3	–4
$g(x)$	0	2	1	5

Some values of functions f and g are given by the tables above. What is the value of $f(g(3))$?

(A) –2

(B) 0

(C) 2

(D) 4

Solution

If $x = 3$, $g(3) = 1$. Then

$$f(g(3)) = f(1) = 4$$

The correct choice is **(D)**.

REPRESENTING A SEQUENCE OF NUMBERS

An ordered sequence of numbers can be represented using function notation. For example, if the first term of an ordered sequence of integers is 4, then $f(1) = 4$. If the nth term of this sequence is given by the function $f(n) = f(n - 1) + n$, then successive terms of the sequence after the first can be obtained by replacing n with 2, 3, 4, and so forth. For example,

- To find the second term of this sequence, set $n = 2$:

$$f(2) = f(2 - 1) + 2 \text{ so } f(2) = f(1) + 2$$

  Since it is given $f(1) = 4$, $f(2) = 4 + 2 = 6$.
- To find the third term of this sequence, set $n = 3$:

$$f(3) = f(3 - 1) + 3 \text{ so } f(3) = f(2) + 3$$

  From the previous step, we know that $f(2) = 6$, so $f(3) = 6 + 3 = 9$.

Hence, the first three terms of the sequence defined by function f are 4, 6, and 9.

> ### MATH REFERENCE FACT
>
> A function may be defined *explicitly* or *recursively*. A function defined in the usual way as a "$y = \ldots$" type of equation is defined explicitly. A function such as $f(n) = f(n - 1) + n$ where $f(1) = 4$ is defined recursively since each of its values after the first is calculated using function values that come before it.

GRAPHS AS FUNCTIONS

Since a graph is a set of ordered pairs located on a coordinate grid, a function may take the form of a graph. If Figure 3.16 shows the complete graph for function g, then you can tell the following from the graph:

- The *domain* of g is $-6 \leq x \leq 5$, since the greatest set of x-values over which the graph extends *horizontally* is from $x = -6$ to $x = 5$, inclusive.
- The *range* of g is $0 \leq y \leq 8$, because the greatest set of y-values over which the graph extends *vertically* is from $y = 0$ to $y = 8$, inclusive.

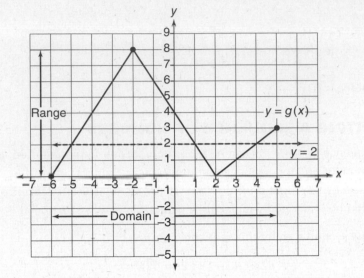

Figure 3.16 Domain and range of function g. The line $y = 2$ intersects the graph at points for which $g(x) = 2$

You can also read specific function values from the graph of function g in Figure 3.16.

TIP

Given the graph of function *f*, you can find those values of *t* that make *f*(*t*) = *k* by drawing a horizontal line through *k* on the *y*-axis. The *x*-coordinates of the points at which the horizontal line intersects the graph, if any, represent the possible values of *t*.

- To find $g(-1)$, determine the y-coordinate of the point on the graph whose x-coordinate is -1. Since the graph contains $(-1, 6)$, $g(-1) = 6$.
- To find the values of t such that $g(t) = 2$, find all points on the graph whose y-coordinates are 2 by drawing a horizontal line through $(0, 2)$. Since the graph passes through $(-5, 2)$, $(1, 2)$, and $(4, 2)$, $g(-5) = 2$, $g(1) = 2$, and $g(4) = 2$. Hence, the possible values of t are -5, 1, and 4.

FINDING THE ZEROS OF A FUNCTION

The zeros of a function f are those values of x, if any, for which $f(x) = 0$. You can determine the zeros of a function from its graph by locating the points at which the graph intersects the x-axis. At each of these points, the y-coordinate is 0, so $f(x) = 0$.

MATH REFERENCE FACT

The *x*-intercepts of the graph of function *f*, if any, correspond to those values of *x* for which *f*(*x*) = 0.

➡ **Example**

Referring to Figure 3.16, which could be the value of s when $g(s) = 0$?

   I. -6

  II. 2

 III. 4

(A) I only
(B) II only
(C) III only
(D) I and II only

Solution

Since the graph of function g has x-intercepts at $x = -6$ and $x = 2$, $g(-6) = 0$ and $g(2) = 0$. Because s can be equal to either -6 or 2, Roman numeral choices I and II are correct. The correct choice is **(D)**.

➡ Example

The zeros of the function $f(x) = 25 - (x + 2)^2$ are

(A) -2 and 5
(B) -3 and 7
(C) -5 and 2
(D) -7 and 3

Solution

The zeros of a function $f(x)$ are those x-values that make the function evaluate to 0. Thus, to find the zeros of function f, set $f(x)$ equal to 0:

$$25 - (x + 2)^2 = 0$$

Since $(-7 + 2)^2 = (3 + 2)^2 = 25$, the equation will be true when $x = -7$ or $x = 3$.

  The correct choice is **(D)**.

➡ Example

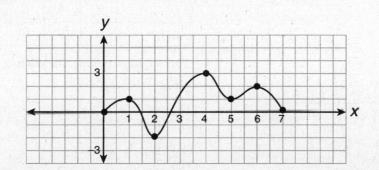

The graph above shows the function $y = f(x)$ over the interval $0 \le x \le 7$. What is the value of $f(f(6))$?

(A) -2
(B) 0
(C) 1
(D) 2

Solution

Reading from the graph, $f(6) = 2$ so $f(f(6)) = f(2) = -2$.
  The correct choice is **(A)**.

Multiple-Choice

1. If the function f is defined by $f(x) = 3x + 2$, and if $f(a) = 17$, what is the value of a?

   (A) 5
   (B) 9
   (C) 10
   (D) 11

2. A function f is defined such that $f(1) = 2$, $f(2) = 5$, and $f(n) = f(n-1) - f(n-2)$ for all integer values of n greater than 2. What is the value of $f(4)$?

   (A) −8
   (B) −2
   (C) 2
   (D) 8

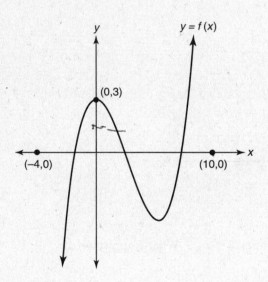

3. The graph of $y = f(x)$ is shown above. If $-4 \leq x \leq 10$, for how many values of x does $f(x) = 2$?

   (A) None
   (B) One
   (C) Two
   (D) Three

4. If function f is defined by $f(x) = 5x + 3$, then which expression represents $2f(x) - 3$?

   (A) $10x - 3$
   (B) $10x + 3$
   (C) $10x$
   (D) 3

5. If the function k is defined by $k(h) = (h + 1)^2$, then $k(x - 2) =$

   (A) $x^2 - x$
   (B) $x^2 - 2x$
   (C) $x^2 - 2x + 1$
   (D) $x^2 + 2x - 1$

x	1	2	3	4	5
$f(x)$	3	4	5	6	7

x	3	4	5	6	7
$g(x)$	4	6	8	10	12

6. The accompanying tables define functions f and g. What is $g(f(3))$?

   (A) 4
   (B) 6
   (C) 8
   (D) 10

7. In 2014, the United States Postal Service charged $0.48 to mail a first-class letter weighing up to 1 oz. and $0.21 for each additional ounce. Based on these rates, which function would determine the cost, in dollars, $c(z)$, of mailing a first-class letter weighing z ounces where z is an integer greater than 1?

(A) $c(z) = 0.48z + 0.21$

(B) $c(z) = 0.21z + 0.48$

(C) $c(z) = 0.48(z − 1) + 0.21$

(D) $c(z) = 0.21(z − 1) + 0.48$

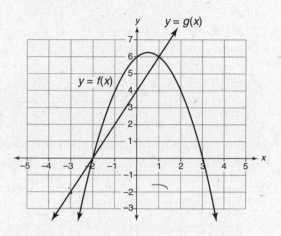

8. Based on the graphs of functions f and g shown in the accompanying figure, for which values of x between −3 and 3 is $f(x) ≥ g(x)$?

I. $−2 ≤ x ≤ 0$

II. $0 ≤ x ≤ 1$

III. $1 ≤ x ≤ 3$

(A) I only

(B) II only

(C) III only

(D) I and II

> ■ $f(2n) = 4f(n)$ for all integers n
> ■ $f(3) = 9$

9. If function f satisfies the above two conditions for all positive integers n, which equation could represent function f?

(A) $f(n) = 9$

(B) $f(n) = n^2$

(C) $f(n) = 3n$

(D) $f(n) = 2n + 3$

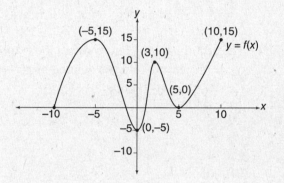

10. If in the accompanying figure (p, q) lies on the graph of $y = f(x)$ and $0 ≤ p ≤ 5$, which of the following represents the set of corresponding values of q?

(A) $−5 ≤ q ≤ 15$

(B) $−5 ≤ q ≤ 10$

(C) $−5 ≤ q ≤ 5$

(D) $5 ≤ q ≤ 10$

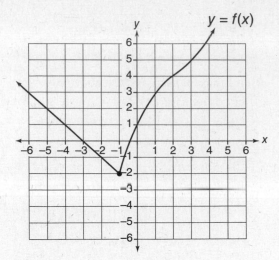

$y = f(x)$

11. The accompanying figure shows the graph of $y = f(x)$. If function g is defined by $g(x) = f(x + 4)$, then $g(-1)$ could be

(A) −2
(B) 3
(C) 4
(D) 5

x	$f(x)$	$g(x)$
1	2	3
2	4	5
3	5	1
4	3	2
5	1	4

Questions 12–13 refer to the accompanying table, which gives the values of functions f and g for integer values of x from 1 to 5, inclusive.

12. According to the table, if $f(5) = p$, what is the value of $g(p)$?

(A) 1
(B) 2
(C) 3
(D) 4

13. Function h is defined by $h(x) = 2f(x) - 1$, where function f is defined in the accompanying table. What is the value of $g(k)$ when $h(k) = 5$?

(A) 1
(B) 2
(C) 3
(D) 4

x	0	1	4	5
$f(x)$	−2	5	0	2

x	0	2	3	−4
$g(x)$	2	−1	1	5

14. Some values of functions f and g are given by the tables above. What is the value of $g(f(5))$?

(A) −1
(B) 1
(C) 2
(D) 5

15. In 2012, a retail chain of fast food restaurants had 68 restaurants in California and started to expand nationally by adding 9 new restaurants each year thereafter. At this rate, which of the following functions f represent the number of restaurants there will be in this retail chain n years after 2012 assuming none of these restaurants close?

(A) $f(n) = 2{,}012 + 9n$
(B) $f(n) = 9 + 68n$
(C) $f(n) = 68 + 9(n - 2{,}012)$
(D) $f(n) = 68 + 9n$

16. According to market research, the number of magazine subscriptions that can be sold can be estimated using the function

$$n(p) = \frac{5{,}000}{4p - k},$$

where n is the number of thousands of subscriptions sold, p is the price in dollars for each individual subscription, and k is some constant. If 250,000 subscriptions were sold at \$15 for each subscription, how many subscriptions could be sold if the price were set at \$20 for each subscription?

(A) 50,000
(B) 75,000
(C) 100,000
(D) 125,000

Grid-In

1. Let h be the function defined by $h(x) = x + 4^x$. What is the value of $h\left(-\dfrac{1}{2}\right)$?

3. Let the function f be defined by $f(x) = x^2 + 12$. If n is a positive number such that $f(3n) = 3f(n)$, what is the value of n?

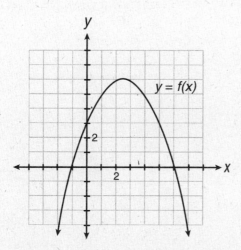

2. The above figure shows the graph of $y = f(x)$ where c is a nonzero constant. If $f(w + 1.7) = 0$ and $w > 0$, what is a possible value of w?

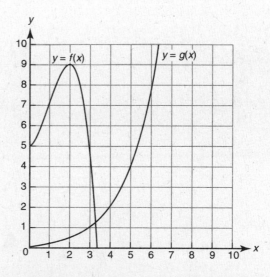

4. Let the functions f and g be defined by the graphs in the accompanying diagram. What is the value of $f(g(3))$?

Questions 5 and 6 Let the function f be defined by the graph below.

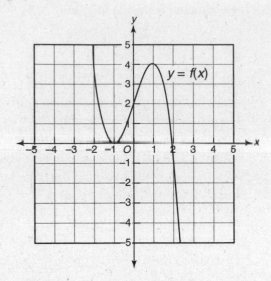

5. What is the integer value of $2f(-1) + 3f(1)$?

6. If n represents the number of different values of x for which $f(x) = 2$ and m represents the number of different values of x for which $f(x) = 4$, what is the value of mn?

7. Let g be the function defined by

   $g(x) = x - 1$. If $\dfrac{1}{2}g(c) = 4$, what is the value

   of $g(2c)$?

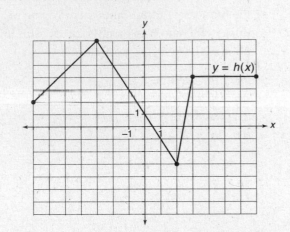

8. The figure above shows the graph of function h. If function f is defined by $f(x) = h(2x) + 1$, what is the value of $f(-1)$?

NOTE: See pages 391–421 for worked out solutions.

Lesson 3–1

1. B	6. A	11. A	16. B	21. A
2. A	7. B	12. C	17. B	**GRID-IN**
3. D	8. B	13. C	18. C	1. 2
4. D	9. C	14. C	19. A	2. 707
5. C	10. A	15. A	20. D	3. 32
				4. 4

Lesson 3–2

1. A	9. D	17. B	3. 7/10	11. 41
2. D	10. C	18. B	4. 42	12. 4
3. C	11. A	19. C	5. 36	13. 19/6
4. D	12. D	20. B	6. 7/2	14. 23.5
5. B	13. D	21. A	7. 15/7	15. 3.59
6. A	14. C	**GRID-IN**	8. 17.5	
7. A	15. B	1. 5/4	9. 6	
8. B	16. A	2. 4	10. 6,000	

Lesson 3–3

1. C	5. A	9. D	13. B	**GRID-IN**	4. 5
2. A	6. B	10. D	14. A	1. 8	5. 7/17
3. B	7. B	11. C	15. D	2. 5	
4. A	8. C	12. D	16. A	3. 18	

Lesson 3–4

1. D	5. B	9. C	13. D	17. A
2. A	6. B	10. D	14. B	**GRID-IN**
3. C	7. B	11. B	15. D	1. 19
4. D	8. C	12. D	16. D	2. 20
				3. 9/4

Lesson 3–5

1. B	3. A	5. B	7. A	9. D	11. D
2. A	4. C	6. A	8. C	10. A	12. C

Lesson 3–6

1. C	5. C	9. D	13. B	3. 13
2. D	6. B	10. B	GRID-IN	4. 16/3
3. A	7. C	11. A	1. 2	5. 4
4. D	8. D	12. C	2. 2	6. 4/5

Lesson 3–7

1. A	6. C	11. B	16. D	GRID-IN	5. 20
2. B	7. D	12. C	17. D	1. 3	6. 8
3. B	8. C	13. C	18. C	2. 26	7. .21
4. A	9. A	14. A	19. D	3. 1.45	8. 9.68
5. D	10. B	15. B	20. D	4. 206	

Lesson 3–8

1. A	5. A	9. D	13. C	2. 6	6. 40
2. D	6. A	10. D	14. D	3. .76	7. 8
3. B	7. B	11. B	GRID-IN	4. 6	8. 6.1
4. A	8. A	12. A	1. 4	5. 15	

Lesson 3–9

1. C	4. D	7. D	10. A	2. 1/2
2. A	5. A	8. A	GRID-IN	3. 16
3. B	6. D	9. A	1. 11	4. 652

Lesson 3-10

1. A	5. C	9. C	13. D	17. B	GRID-IN
2. D	6. B	10. D	14. A	18. B	1. 3/4
3. A	7. B	11. B	15. D	19. C	2. 3/2
4. D	8. C	12. A	16. B	20. C	3. 15/2
					4. 5
					5. 10

Lesson 3–11

1. C	4. D	7. B	10. C	2. 7/6
2. B	5. C	8. D	GRID-IN	3. 1/2
3. C	6. D	9. B	1. 15/2	4. 7

Lesson 3–12

1. A	6. C	11. D	16. D	4. 7
2. B	7. D	12. C	GRID-IN	5. 12
3. D	8. D	13. B	1. 0	6. 6
4. B	9. B	14. A	2. 4.3	7. 17
5. C	10. B	15. D	3. 2	8. 6

Problem Solving and Data Analysis

4

The beginning lessons of this chapter focus on mathematical reasoning involving algebraic representations, percent, and ratios. Because of their importance, linear and exponential functions receive special attention. The chapter also covers analyzing and summarizing data using graphs, scatter plots, tables, and basic statistical measures. "Problem Solving and Data Analysis" represents the second of the four major mathematics content groups tested by the redesigned SAT.

LESSONS IN THIS CHAPTER

Lesson 4-1   Working with Percent

Lesson 4-2   Ratio and Variation

Lesson 4-3   Rate Problems

Lesson 4-4   Converting Units of Measurement

Lesson 4-5   Linear and Exponential Functions

Lesson 4-6   Graphs and Tables

Lesson 4-7   Scatterplots and Sampling

Lesson 4-8   Summarizing Data Using Statistics

> ### OVERVIEW
>
> *Percent* means parts out of 100. Simple algebraic equations can be used to help solve a variety of problems involving percent.

THE THREE TYPES OF PERCENT PROBLEMS

You can solve each of the three basic types of percent problems by writing and solving an equation.

- **Type 1:** Finding a percent of a given number.

➥ **Example**

What is 15% of 80?

$$n = 0.15 \times 80$$
$$= 12$$

15% of 80 is 12.

- **Type 2:** Finding a number when a percent of it is given.

➥ **Example**

30% of what number is 12?

$$0.30 \times n = 12$$
$$0.30n = 12$$
$$10(0.3n) = 10(12)$$
$$3n = 120$$
$$n = \frac{120}{3}$$
$$= 40$$

30% of 40 is 12.

- **Type 3:** Finding what percent one number is of another.

➥ **Example**

What percent of 30 is 9?

$$\frac{p}{100} \times 30 = 9$$
$$\frac{p}{100} \times 30 = 9$$
$$\frac{\overset{3}{\cancel{30}}p}{\underset{10}{\cancel{100}}} = 9$$
$$p = \frac{10 \cdot 9}{3} = 30$$

9 is 30% of 30.

Finding an Amount After a Percent Change

If a 20% tip is left on a restaurant bill of $80, then to find the total amount of the bill including the tip, do the following:

(STEP 1) Find the amount of the tip: $80 × 0.20 = $16
(STEP 2) Add the tip to the bill: $80 + $16 = $96

TIME SAVER

Find the final result of increasing or decreasing a number by a percent in one step by multiplying the number by the *total* percentage:

- If 80 is *increased* by 20%, the total percentage is 100% + 20% = 120%, so the final amount is 80 × 1.20 = 96.
- If 50 is *decreased* by 30%, the total percentage is 100% − 30% = 70%, so the final amount is 50 × 0.70 = 35.

➡ Example

If the length of a rectangle is increased by 30% and its width is increased by 10%, by what percentage will the area of the rectangle be increased?

(A) 33%
(B) 37%
(C) 40%
(D) 43%

Solution

Pick easy numbers for the length and width of the rectangle. Assume the length and width of the rectangle are each 10 units, so the area of the original rectangle is 10 × 10 = 100 square units.

- The total percent increase of the length is 100% + 30% = 130%. The length of the new rectangle is 10 × 1.3 = 13.
- The total percent increase of the width is 100% + 10% = 110%. The width of the new rectangle is 10 × 1.1 = 11.
- The area of the new rectangle is 13 × 11 = 143 square units. Compared to the original area of 100 square units, this is a 43% increase.

The correct choice is **(D)**.

Finding an Original Amount after a Percent Change

If you know the number that results after a given number is increased by P%, you can find the original number by dividing the new amount by the total percentage:

$$\text{Original amount} = \frac{\text{New amount after an increase of } P\%}{100\% + P\%}$$

Similarly, if you know the number that results after a given number is decreased by $P\%$, you can find the original number by dividing the new amount by the total percentage:

$$\text{Original amount} = \frac{\text{New amount after a decrease of } P\%}{100\% - P\%}$$

➡ Example

A pair of tennis shoes cost $48.60 including sales tax. If the sales tax rate is 8%, what is the cost of the tennis shoes before the tax is added?

Solution

The total percentage is 100% + 8% = 108%.

$$\text{Cost of tennis shoes} = \frac{\text{Cost with tax included}}{\text{Total percentage}}$$

$$= \frac{48.60}{108\%}$$

$$= \frac{48.60}{1.08}$$

Use a calculator to divide: = 45

The tennis shoes cost **$45** without tax.

Finding the Percent of Increase or Decrease

When a quantity goes up or down in value, the percent of change can be calculated by comparing the amount of the change to the original amount:

$$\text{Percent of change} = \frac{\text{Amount of change}}{\text{Original amount}} \times 100\%$$

➡ Example

If the price of an item increases from $70 to $84, what is the percent of increase in price?

Solution

The original amount is $70, and the amount of increase is $84 − $70 = $14:

$$\text{Percent of change} = \frac{\text{Amount of increase}}{\text{Original amount}} \times 100\%$$

$$= \frac{14}{70} \times 100\%$$

$$= \frac{1}{5} \times 100\%$$

$$= 20\%$$

Multiple-Choice

1. By the end of the school year, Terry had passed 80% of his science tests. If Terry failed 4 science tests, how many science tests did Terry pass?

   (A) 12
   (B) 15
   (C) 16
   (D) 18

2. A soccer team has played 25 games and has won 60% of the games it has played. What is the minimum number of additional games the team must win in order to finish the season winning 80% of the games it has played?

   (A) 28
   (B) 25
   (C) 21
   (D) 18

3. In a movie theater, 480 of the 500 seats were occupied. What percent of the seats were NOT occupied?

   (A) 0.4%
   (B) 2%
   (C) 4%
   (D) 20%

4. In a certain mathematics class, the part of the class that are members of the math club is 50% of the rest of that class. The total number of math club members in this class is what percent of the entire class?

   (A) 20%
   (B) 25%
   (C) $33\frac{1}{3}\%$
   (D) 50%

5. After 2 months on a diet, John's weight dropped from 168 pounds to 147 pounds. By what percent did John's weight drop?

   (A) $12\frac{1}{2}\%$
   (B) $14\frac{2}{7}\%$
   (C) 21%
   (D) 25%

6. If 1 cup of milk is added to a 3-cup mixture that is $\frac{2}{5}$ flour and $\frac{3}{5}$ milk, what percent of the 4-cup mixture is milk?

   (A) 80%
   (B) 75%
   (C) 70%
   (D) 65%

7. If the result of increasing a by 300% of a is b, then a is what percent of b?

   (A) 20%
   (B) 25%
   (C) $33\frac{1}{3}\%$
   (D) 40%

8. After a 20% increase, the new price of a radio is $78.00. What was the original price of the radio?

   (A) $15.60
   (B) $60.00
   (C) $62.40
   (D) $65.00

9. After a discount of 15%, the price of a shirt is $51. What was the original price of the shirt?

   (A) $44.35
   (B) $58.65
   (C) $60.00
   (D) $64.00

10. Three students use a computer for a total of 3 hours. If the first student uses the computer 28% of the total time, and the second student uses the computer 52% of the total time, how many minutes does the third student use the computer?

(A) 24
(B) 30
(C) 36
(D) 42

11. In an opinion poll of 50 men and 40 women, 70% of the men and 25% of the women said that they preferred fiction to nonfiction books. What percent of the number of people polled preferred to read fiction?

(A) 40%
(B) 45%
(C) 50%
(D) 60%

12. In a factory that manufactures light bulbs, 0.04% of all light bulbs manufactured are defective. On the average, there will be three defective light bulbs out of how many manufactured?

(A) 2,500
(B) 5,000
(C) 7,500
(D) 10,000

13. A used-car lot has 4-door sedans, 2-door sedans, sports cars, vans, and jeeps. Of these vehicles, 40% are 4-door sedans, 25% are 2-door sedans, 20% are sports cars, 10% are vans, and 20 of the vehicles are jeeps. If this car lot has no other vehicles, how many vehicles are on the used-car lot?

(A) 300
(B) 400
(C) 480
(D) 600

14. Jack's weight first increased by 20% and then his new weight decreased by 25%. His final weight is what percent of his beginning weight?

(A) 95%
(B) 92.5%
(C) 90%
(D) 88.5%

15. The price of a stock falls 25%. By what percent of the new price must the stock price rise in order to reach its original value?

(A) 25%
(B) 30%
(C) $33\frac{1}{3}\%$
(D) 40%

VOTING POLL	
Candidate A	30%
Candidate B	50%
Undecided	20%

16. The table above summarizes the results of an election poll in which 4,000 voters participated. In the actual election, all 4,000 of these people voted, and those people who chose a candidate in the poll voted for that candidate. People who were undecided voted for candidate A in the same proportion as the people who cast votes for candidates in the poll. Of the people polled, how many voted for candidate A in the actual election?

(A) 1,420
(B) 1,500
(C) 1,640
(D) 1,680

17. A car starts a trip with 20 gallons of gas in its tank. The car traveled at an average speed of 65 miles per hour for 3 hours and consumed gas at a rate of 30 miles per gallon. What percent of the gas in the tank was used for the 3-hour trip?

(A) 32.5
(B) 33.0
(C) 33.5
(D) 34.0

Grid-In

1. A store offers a 4% discount if a consumer pays cash rather than paying by credit card. If the cash price of an item is $84.00, what is the credit-card purchase price of the same item?

2. During course registration, 28 students enroll in a certain college class. After three boys are dropped from the class, 44% of the class consists of boys. What percent of the original class did girls comprise?

3. A high school tennis team is scheduled to play 28 matches. If the team wins 60% of the first 15 matches, how many additional matches must the team win in order to finish the season winning 75% of its scheduled matches?

4. In a club of 35 boys and 28 girls, 80% of the boys and 25% of the girls have been members for more than 2 years. If n percent of the club have been members for more than 2 years, what is the value of n?

OVERVIEW

A **ratio** is a comparison by division of two quantities that are measured in the same units. For example, if Mary is 16 years old and her brother Gary is 8 years old, then Mary is 2 times as old as Gary. The ratio of Mary's age to Gary's age is 2:1 (read as "2 to 1") since

$$\frac{\text{Mary's age}}{\text{Gary's age}} = \frac{16 \text{ years}}{8 \text{ years}} = \frac{2}{1} \text{ or } 2:1$$

One quantity may be related to another quantity so that either the ratio or product of these quantities always remains the same.

RATIO OF *a* TO *b*

The ratio of a to b ($b \neq 0$) is the fraction $\dfrac{a}{b}$, which can be written as $a:b$ (read as "a to b").

➡ Example

The ratio of the number of girls to the number of boys in a certain class is $3:5$. If there is a total of 32 students in the class, how many girls are in the class?

Solution

Since the number of girls is a multiple of 3 and the number of boys is the same multiple of 5, let

$$3x = \text{the number of girls in the class}$$

and $5x = \text{the number of boys in the class}$

Then

$$3x + 5x = 32$$
$$8x = 32$$
$$x = \frac{32}{8} = 4$$

The number of girls $= 3x = 3(4) = 12$.

RATIO OF *A* TO *B* TO *C*

If $A:B$ represents the ratio of A to B and $B:C$ represents the ratio of B to C, then the ratio of A to C is $A:C$, provided that B stands for the same number in both ratios. For example, if the ratio of A to B is $3:5$ and the ratio of B to C is $5:7$, then the ratio of A to C is $3:7$. In this case, B represents the number 5 in both ratios.

→ Example

If the ratio of A to B is $3:5$ and the ratio of B to C is $2:7$, what is the ratio of A to C?

Solution

Change each ratio into an equivalent ratio in which the term that corresponds to B is the same number:

- The ratio of A to B is $3:5$, so the term corresponding to B in this ratio is 5. The ratio of B to C is $2:7$, so the term corresponding to B in this ratio is 2.
- The least common multiple of 5 and 2 is 10. You need to change each ratio into an equivalent ratio in which the term corresponding to B is 10.
- Multiplying each term of the ratio $3:5$ by 2 gives the equivalent ratio $6:10$. Multiplying each term of the ratio $2:7$ by 5 gives $10:35$.
- Since the ratio of A to B is equivalent to $6:10$ and the ratio of B to C is equivalent to $10:35$, the **ratio of A to C is $6:35$**.

Direct Variation

If two quantities change in value so that their ratio always remains the same, then one quantity is said to vary **directly** with the other. When one quantity varies directly with another quantity, a change in one causes a change in the other in the same direction—both increase or both decrease.

→ Example

If $y = kx$, where k is a constant and $y = 27$ when $x = 18$, what is the value of y when $x = 30$?

> **MATH REFERENCE FACT**
>
> If y varies *directly* as x, then $y = kx$ or, equivalently,
> $\frac{y}{x} = k$, where k is a nonzero constant.

Solution

The ordered pairs (18, 27) and (30, y) must satisfy the equation $y = kx$ or, equivalently, $\frac{y}{x} = k$. Since $\frac{27}{18} = k$ and $\frac{y}{30} = k$:

$$\frac{27}{18} = \frac{y}{30}$$
$$\frac{3}{2} = \frac{y}{30}$$
$$2y = 90$$
$$y = \frac{90}{2} = 45$$

The value of y is **45**.

➡ Example

If 28 pennies weigh 42 grams, what is the weight in grams of 50 pennies?

Solution

The number of pennies and their weight vary directly since multiplying one of the two quantities of pennies by a constant causes the other to be multiplied by the same constant. If x represents the weight in grams of 50 pennies, then

$$\frac{\text{Pennies}}{\text{Grams}} = \frac{28}{42} = \frac{50}{x}$$

Cross-multiply:
$$28x = 42(50)$$
$$x = \frac{2100}{28} = 75$$

The weight of 50 pennies is **75** grams.

Inverse Variation

If two quantities change in opposite directions, so that their product always remains the same, then one quantity is said to vary **inversely** with the other.

> **MATH REFERENCE FACT**
>
> If y varies *inversely* as x, then $xy = k$ where k is a nonzero constant.

➡ Example

If $xy = k$, where k is a constant and $y = 21$ when $x = 6$, what is the value of y when $x = 9$?

Solution

The ordered pairs (6, 21) and (9, y) must satisfy the equation $xy = k$. Since $6 \times 21 = k$ and $9 \times y = k$,

$$9y = (6)(21)$$
$$\frac{9y}{9} = \frac{126}{9}$$
$$y = 14$$

The value of y is **14**.

➡ Example

Four workers can build a house in 9 days. How many days would it take 3 workers to build the same house?

Solution

As the number of people working on the house *decreases*, the number of days needed to build the house *increases*. Since this is an inverse variation, the number of workers times the number of days needed to build the house stays constant.

If d represents the number of days that 3 workers take to build the house, then

$$3 \times d = 4 \times 9$$
$$3d = 36$$
$$d = \frac{36}{3} = 12$$

Three people working together would take **12** days to build the house.

DIVIDING A SEGMENT INTO TWO SEGMENTS WITH A GIVEN RATIO

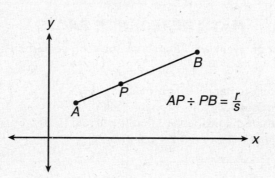

To determine the coordinates of the point P that divides the line segment drawn from point $A(x_A, y_A)$ to point $B(x_B, y_B)$ into two shorter segments such that the ratio of AP to PB is, say r to s:

- Calculate the difference in the x-coordinates ("run") and the difference in the y-coordinates ("rise") from point A to point B.

- Multiply both the run and the rise between the two given points by $\dfrac{r}{r+s}$ and then add the results to the corresponding coordinates of point A:

$$x\text{-coordinate of point } P: \quad x_A + \left(\text{run} \times \frac{r}{r+s} \right)$$
$$y\text{-coordinate of point } P: \quad y_A + \left(\text{rise} \times \frac{r}{r+s} \right)$$

➡ Example

What are the coordinates of the point P in the xy-plane that divides the line segment whose endpoints are $A(-1, 4)$ and $B(4, -6)$ into two segments such that the ratio of AP to PB is 2 to 3?

(A) (1, 0)
(B) (0, 2)
(C) (0.5, 1)
(D) (2.5, −3)

Solution

- Between points A and B, the run is $4 - (-1) = 5$ and the rise is $-6 - 4 = -10$.
- Multiply the run and the rise by $\dfrac{2}{2+3}\left(=\dfrac{2}{5}\right)$ and then add the products to the corresponding coordinates of point A:

$$P\left(-1+\frac{2}{5}\times 5, 4+\frac{2}{5}\times(-10)\right) = P(-1+2, 4-4) = P(1,0)$$

The correct choice is **(A)**.

MATH REFERENCE FACTS

The probability that an event will happen can be represented by a number from 0 to 1.

- If an event is certain to happen, its probability is 1.
- If an event is an impossibility, its probability is 0.
- $P(not\ E) = 1 - P(E)$. If the probability that it will rain is 20% or 0.2, then the probability it will *not* rain is $1 - 0.2 = 0.8$ or 80%.

SIMPLE PROBABILITY

To find the probability that some event E will happen, write the ratio of the number of favorable outcomes to the total number of different outcomes that are possible:

$$P(E) = \frac{\text{Number of ways } E \text{ can happen}}{\text{Total number of possible outcomes}}$$

The shorthand notation $P(E)$ is used to represent the probability that event E will happen. If 3 yellow marbles and 2 red marbles are placed in a hat and a marble is picked from the hat without looking, then

$$P(\text{yellow}) = \frac{3}{3+2} = \frac{3}{5}$$

Multiple-Choice

1. A recipe for 4 servings requires salt and pepper to be added in the ratio of $2:3$. If the recipe is adjusted from 4 to 8 servings, what is the ratio of the salt and pepper that must now be added?

   (A) $4:3$
   (B) $2:6$
   (C) $2:3$
   (D) $3:2$

2. On a certain map, $\frac{3}{8}$ of an inch represents 120 miles. How many miles does $1\frac{3}{4}$ inches represent?

   (A) 300
   (B) 360
   (C) 480
   (D) 560

3. The population of a bacteria culture doubles in number every 12 minutes. The ratio of the number of bacteria at the end of 1 hour to the number of bacteria at the beginning of that hour is

   (A) $64:1$
   (B) $60:1$
   (C) $32:1$
   (D) $16:1$

4. At the end of the season, the ratio of the number of games a team has won to the number of games it lost is $4:3$. If the team won 12 games and each game played ended in either a win or a loss, how many games did the team play during the season?

   (A) 9
   (B) 15
   (C) 18
   (D) 21

5. If s and t are integers, $8 < t < 40$, and $\frac{s}{t} = \frac{4}{7}$, how many possible values are there for t?

   (A) Two
   (B) Three
   (C) Four
   (D) Five

6. A school club includes only sophomores, juniors, and seniors, in the ratio of $1:3:2$. If the club has 42 members, how many seniors are in the club?

   (A) 6
   (B) 7
   (C) 12
   (D) 14

7. If $\frac{c-3d}{4} = \frac{d}{2}$, what is the ratio of c to d?

   (A) $5:1$
   (B) $3:2$
   (C) $4:3$
   (D) $3:4$

8. If 4 pairs of socks costs $10.00, how many pairs of socks can be purchased for $22.50?

   (A) 5
   (B) 7
   (C) 8
   (D) 9

9. Two boys can paint a fence in 5 hours. How many hours would it take 3 boys to paint the same fence?

   (A) $\frac{3}{2}$
   (B) 3
   (C) $3\frac{1}{3}$
   (D) $4\frac{2}{3}$

10. A car moving at a constant rate travels 96 miles in 2 hours. If the car maintains this rate, how many miles will the car travel in 5 hours?

(A) 480
(B) 240
(C) 210
(D) 192

11. The number of kilograms of corn needed to feed 5,000 chickens is 30 less than twice the number of kilograms needed to feed 2,800 chickens. How many kilograms of corn are needed to feed 2,800 chickens?

(A) 70
(B) 110
(C) 140
(D) 190

12. The number of calories burned while jogging varies directly with the number of minutes spent jogging. If George burns 180 calories by jogging for 25 minutes, how many calories does he burn by jogging for 40 minutes?

(A) 200
(B) 276
(C) 288
(D) 300

13. If y varies directly as x and $y = 12$ when $x = c$, what is y in terms of c when $x = 8$?

(A) $\dfrac{2c}{3}$

(B) $\dfrac{3}{2c}$

(C) $20c$

(D) $\dfrac{96}{c}$

$$\frac{x}{z} = \frac{1}{3}$$

14. If in the equation above x and z are integers, which are possible values of $\dfrac{x^2}{z}$?

I. $\dfrac{1}{9}$

II. $\dfrac{1}{3}$

III. 3

(A) II only
(B) III only
(C) I and III only
(D) II and III only

15. If $a - 3b = 9b - 7a$, then the ratio of a to b is

(A) $3:2$
(B) $2:3$
(C) $3:4$
(D) $4:3$

16. The ratio of A to B is $a:8$, and the ratio of B to C is $12:c$. If the ratio of A to C is $2:1$, what is the ratio of a to c?

(A) $2:3$
(B) $3:2$
(C) $4:3$
(D) $3:4$

17. If $8^r = 4^t$, what is the ratio of r to t?

(A) $2:3$
(B) $3:2$
(C) $4:3$
(D) $3:4$

18. If $\dfrac{a+b}{b} = 4$ and $\dfrac{a+c}{c} = 3$, what is the ratio of c to b?

(A) $2:3$
(B) $3:2$
(C) $2:1$
(D) $3:1$

19. In a certain college, the ratio of mathematics majors to English majors is $3:8$. If in the following school year the number of mathematics majors increases 20% and the number of English majors decreases 15%, what is the new ratio of mathematics majors to English majors?

   (A) $4:9$
   (B) $1:2$
   (C) $9:17$
   (D) $17:32$

20. At a college basketball game, the ratio of the number of freshmen who attended to the number of juniors who attended is $3:4$. The ratio of the number of juniors who attended to the number of seniors who attended is $7:6$. What is the ratio of the number of freshmen to the number of seniors who attended the basketball game?

   (A) $7:8$
   (B) $3:4$
   (C) $2:3$
   (D) $1:2$

21. It took 12 men 5 hours to build an airstrip. Working at the same rate, how many additional men could have been hired in order for the job to have taken 1 hour less?

   (A) Two
   (B) Three
   (C) Four
   (D) Six

22. If x represents a number picked at random from the set $\{-3, -2, -1, 0, 1, 2\}$, what is the probability that x will satisfy the inequality $4 - 3x < 6$?

   (A) $\dfrac{1}{6}$
   (B) $\dfrac{1}{3}$
   (C) $\dfrac{1}{2}$
   (D) $\dfrac{2}{3}$

23. What are the coordinates of the point P in the xy-plane that divides the segment whose endpoints are $A(-2, 9)$ and $B(7, 3)$ into two segments such that the ratio of AP to PB is 1 to 2?

   (A) $P(1, 5)$
   (B) $P(4, 1)$
   (C) $P(1, 7)$
   (D) $P(2, 6)$

Grid-In

1. A string is cut into 2 pieces that have lengths in the ratio of $2:9$. If the difference between the lengths of the 2 pieces of string is 42 inches, what is the length in inches of the shorter piece?

2. For integer values of a and b, $b^a = 8$. The ratio of a to b is equivalent to the ratio of c to d, where c and d are integers. What is the value of c when $d = 10$?

3. Jars A, B, and C each contain 8 marbles. What is the minimum number of marbles that must be transferred among the jars so that the ratio of the number of marbles in jar A to the number in jar B to the number in jar C is $1:2:3$?

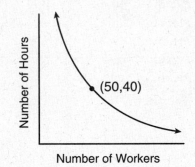

Number of Workers

4. A political campaign organizer has determined that the number of hours needed to get out a mailing for her candidate is inversely related to the number of campaign workers she has. If she uses the information in the accompanying graph, how many hours would it take to do the mailing if 125 workers are used?

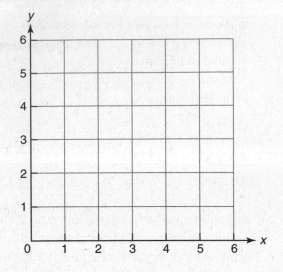

5. A square dartboard is placed in the first quadrant from $x = 0$ to 6 and $y = 0$ to 6, as shown in the accompanying figure. A triangular region on the dartboard is enclosed by the graphs of the equations $y = 2$, $x = 6$, and $y = x$ (not shown). Find the probability that a dart that randomly hits the dartboard will land in the triangular region formed by the three lines.

OVERVIEW

A ratio of two quantities that have different units of measurement is called a **rate**. For example, if a car travels a total distance of 150 miles in 3 hours, the average rate of speed is the distance traveled divided by the amount of time required to travel that distance:

$$\text{Rate} = \frac{\text{Distance}}{\text{Time}} = \frac{150 \text{ miles}}{3 \text{ hours}} = 50 \text{ miles per hour}$$

Rate problems are usually solved using the general relationship

$$\text{Rate (of } A \text{ per unit } B) \times B = A$$

You are using the correct rate relationship if the units check, as in

$$\underset{\downarrow}{\text{Rate}} \quad \times \quad \underset{\downarrow}{\text{Time}} \quad = \underset{\downarrow}{\text{Distance}}$$

$$\frac{\text{Miles}}{\cancel{\text{Hour}}} \times \cancel{\text{Hours}} = \text{Miles}$$

UNIT COST PROBLEMS

Unit cost problems require you to figure out a rate by calculating the cost per item. Multiplying this rate by a given number of items gives the total cost of those items.

➡ Example

If 5 cans of soup cost $1.95, how much do 3 cans of soup cost?

Solution 1

Since 5 cans of soup cost $1.95, the cost of 1 can is

$$\frac{\$1.95}{5 \text{ cans}} = 0.39 \text{ dollar per can}$$

To find the cost of 3 cans, multiply the rate of 0.39 dollar per can by 3:

$$\text{Cost of 3 cans} = 3 \text{ cans} \times 0.39 \frac{\$}{\text{can}} = \$1.17$$

Solution 2

The number of cans of soup varies directly with the cost of the cans. Form a proportion in which x represents the cost of 3 cans of soup:

$$\frac{\text{Cost}}{\text{Number of cans}} = \frac{x}{3} = \frac{1.95}{5}$$

$$5x = 3(1.95)$$

$$x = 5.85$$

$$x = \frac{5.85}{5} = 1.17$$

The cost of 3 cans of soup is **$1.17**.

MOTION PROBLEMS

The solution of motion problems depends on the relationship

$$\text{Rate} \times \text{Time} = \text{Distance}$$

➡ Example

John rode his bicycle to town at the rate of 15 miles per hour. He left the bicycle in town for minor repairs and walked home along the same route at the rate of 3 miles per hour. Excluding the time John spent in taking the bike into the repair shop, the trip took 3 hours. How many hours did John take to walk back?

Solution

Since the trip took a total of 3 hours,

let x = the number of hours John took to ride his bicycle to town,

and $3 - x$ = the number of hours John took to walk back from town.

	Rate	×	Time	=	Distance
To town	15 mph		x hours		$15x$
Return trip	3 mph		$3 - x$ hours		$3(3 - x)$

Since John traveled over the same route, the two distances must be equal. Hence,

$$15x = 3(3 - x)$$
$$15x = 9 - 3x$$
$$18x = 9$$
$$x = \frac{9}{18} = \frac{1}{2} \text{ hour}$$

John took $3 - \frac{1}{2} = 2\frac{1}{2}$ hours to walk back.

WORK PROBLEMS

When solving a problem involving two or more people or pieces of machinery working together to complete a job, use these relationships:

- The rate of work is the reciprocal of the time it takes to complete the entire job.
- Rate of work × Actual time worked = Fractional part of job completed

Suppose John can mow an entire lawn in 3 hours and Maria can mow the same lawn in 4 hours. The number of hours, x, it would take them to mow the lawn working together is given by the equation:

$$\underbrace{\left(\frac{1}{3}\right)x}_{\substack{\text{Part of job} \\ \text{completed by John}}} + \underbrace{\left(\frac{1}{4}\right)x}_{\substack{\text{Part of job} \\ \text{completed by Maria}}} = 1 \text{ whole job}$$

➥ Example

A new printing press can print 5,000 flyers in half the amount of time it takes for an older printing press to print the same 5,000 flyers. Working together, the two printing presses can complete the entire job in 3 hours. How long would it take the faster printing press working alone to complete the job?

Solution

If x represents the amount of time it takes the slower printing press to complete the job working alone, then the faster printing press can complete the job in $\dfrac{x}{2}$ hours. Hence,

$$\left(\frac{1}{x}\right)3 + \left(\frac{1}{x/2}\right)3 = 1$$
$$\left(\frac{3}{x}\right) + \left(\frac{6}{x}\right) = 1$$
$$3 + 6 = x$$

Since the slower printing press would take 9 hours working alone to complete the job, the faster press would take $\dfrac{9}{2} = 4.5$ hours working alone to complete the job.

If the original equation in the previous example was written in the equivalent form $\dfrac{1}{x} + \dfrac{2}{x} = \dfrac{1}{3}$, then

- $\dfrac{1}{x}$ represents the part of the whole job completed by the slower press in 1 hour.

- $\dfrac{2}{x}$ represents the part of the whole job completed by the faster press in 1 hour.

- $\dfrac{1}{3}$ represents the part of the whole job that is completed in 1 hour with both presses working together.

Multiple-Choice

1. If four pens cost $1.96, what is the greatest number of pens that can be purchased for $7.68?

   (A) 12
   (B) 14
   (C) 15
   (D) 16

2. Two pipes of different diameters may be used to fill a swimming pool. The pipe with the larger diameter working alone can fill the swimming pool 1.25 times faster than the other pipe when it works alone. One hour after the larger pipe is opened, the smaller pipe is opened, and the swimming pool is filled 5 hours later. Which equation could be used to find the number of hours, x, it would take for the larger pipe to fill the pool working alone?

   (A) $\left(\dfrac{1}{1.25x}\right)5+\left(\dfrac{1}{x}\right)6=1$

   (B) $\left(\dfrac{1}{x}\right)5+\left(\dfrac{1}{1.25x}\right)6=1$

   (C) $\left(\dfrac{x}{5}\right)1.25+\left(\dfrac{x}{6}\right)=1$

   (D) $\left(\dfrac{x}{5}\right)+\left(\dfrac{x}{6}\right)1.25=1$

3. On a certain map, 1.5 inches represent a distance of 120 miles. If two cities on this map are 1 foot apart, what is the distance, in miles, between the cities?

   (A) 180
   (B) 480
   (C) 960
   (D) 1,080

4. A freight train left a station at 12 noon, going north at a rate of 50 miles per hour. At 1:00 P.M. a passenger train left the same station, going south at a rate of 60 miles per hour. At what time were the trains 380 miles apart?

   (A) 3:00 P.M.
   (B) 4:00 P.M.
   (C) 4:30 P.M.
   (D) 5:00 P.M.

5. A man drove to work at an average rate of speed of 60 miles per hour and returned over the same route driving at an average rate of speed of 40 miles per hour. If his total driving time was 1 hour, what was the total number of miles in the round trip?

   (A) 12
   (B) 24
   (C) 30
   (D) 48

6. If x people working together at the same rate can complete a job in h hours, what part of the same job can one person working alone complete in k hours?

   (A) $\dfrac{k}{xh}$

   (B) $\dfrac{h}{xk}$

   (C) $\dfrac{k}{x+h}$

   (D) $\dfrac{kh}{x}$

7. An electrician can install 5 light fixtures in 3 hours. Working at that rate, how long will it take the electrician to install 8 light fixtures?

    (A) $3\dfrac{4}{5}$ hours

    (B) $4\dfrac{1}{5}$ hours

    (C) $4\dfrac{1}{2}$ hours

    (D) $4\dfrac{4}{5}$ hours

8. A freight train and a passenger train start toward each other at the same time from two towns that are 500 miles apart. After 3 hours, the trains are still 80 miles apart. If the average rate of speed of the passenger train is 20 miles per hour faster than the average rate of speed of the freight train, what is the average rate of speed, in miles per hour, of the freight train?

    (A) 40
    (B) 45
    (C) 50
    (D) 60

9. One machine can seal 360 packages per hour, and an older machine can seal 140 packages per hour. How many MINUTES will the two machines working together take to seal a total of 700 packages?

    (A) 48
    (B) 72
    (C) 84
    (D) 90

10. A motor boat traveling at 18 miles per hour traveled the length of a lake in one-quarter of an hour less time than it took when traveling at 12 miles per hour. What was the length in miles of the lake?

    (A)   6
    (B)   9
    (C) 12
    (D) 15

11. Carmen went on a trip of 120 miles, traveling at an average of x miles per hour. Several days later she returned over the same route at a rate that was 5 miles per hour faster than her previous rate. If the time for the return trip was one-third of an hour less than the time for the outgoing trip, which equation can be used to find the value of x?

    (A) $\dfrac{120}{x+5} = \dfrac{1}{3}$

    (B) $\dfrac{x}{120} = \dfrac{x+5}{120} - \dfrac{1}{3}$

    (C) $\dfrac{120}{x+(x+5)} = \dfrac{1}{3}$

    (D) $\dfrac{120}{x} = \dfrac{120}{x+5} + \dfrac{1}{3}$

12. Jonathan drove to the airport to pick up his friend. A rainstorm forced him to drive at an average speed of 45 miles per hour, reaching the airport in 3 hours. He drove back home at an average speed of 55 miles per hour. How long, to the *nearest tenth of an hour*, did the trip home take him?

    (A) 2.0 hours
    (B) 2.5 hours
    (C) 2.8 hours
    (D) 3.7 hours

13. A plumber works twice as fast as his apprentice. After the plumber has worked alone for 3 hours, his apprentice joins him and working together they complete the job 4 hours later. How many hours would it have taken the plumber to do the entire job by himself?

    (A)   9
    (B) 12
    (C) 14
    (D) 18

Grid-In

1. Fruit for a dessert costs $1.20 a pound. If 5 pounds of fruit are needed to make a dessert that serves 18 people, what is the cost of the fruit needed to make enough of the same dessert to serve 24 people?

2. A printing press produces 4,600 flyers per hour. At this rate, in how many *minutes* can the same printing press produce 920 flyers?

FOREIGN CURRENCY CONVERSIONS

U.S. Dollar to British Pound = 1.56 to 1

British Pound to Euro = 1 to 1.38

---

3. Foreign currency conversion rates for the British pound, U.S. dollar, and Euro are listed above. What would be the cost in U.S. dollars for a shirt that has a purchase price of 46 Euros, correct to the *nearest dollar*?

4. Joseph typed a 1,200-word essay in 25 minutes with an average of 240 words on a page. At this rate, how many 240-word pages can he type in 1 hour?

OVERVIEW

Changing from one unit of measurement to another requires using a fractional conversion factor that tells the relationship between the two units so that like units cancel as in

$$2 \text{ miles} = 2 \text{ miles} \times 5{,}280 \, \frac{\text{feet}}{\text{mile}} = 10{,}560 \text{ feet}$$

conversion factor
miles to feet

HOW TO CONVERT UNITS

To convert from one unit of measurement to another:

- Write the conversion factor as a fraction placing the old units in the denominator and the new units in the numerator. For example, to change from feet (old units) into an equivalent number of miles (new units), use $\dfrac{1 \text{ mile}}{5{,}280 \text{ feet}}$ as the conversion factor.

- Multiply the number of units you want to convert by the conversion factor.

- Cancel any units that are in both a numerator and a denominator. Then check that the answer is in the correct units.

➡ Example

If an object is moving at an average rate of speed of $18 \, \dfrac{\text{km}}{\text{min}}$, how many meters does it travel in 5 seconds?

Solution

Convert $18 \, \dfrac{\text{km}}{\text{min}}$ to an equivalent number of meters per second. Then multiply the result by 5 seconds:

- Since 1 kilometer = 1,000 meters, to convert from kilometers to meters use the conversion factor $\dfrac{1{,}000 \text{ m}}{1 \text{ km}}$. Thus,

$$18 \, \frac{\text{km}}{\text{min}} = 18 \, \frac{\text{km}}{\text{min}} \times \frac{1{,}000 \text{ meters}}{1 \text{ km}} = 18{,}000 \, \frac{\text{meters}}{\text{min}}$$

TIP

A conversion factor expresses the relationship between the two units as a fraction and is always numerically equal to 1. For example, because the conversion factor

of $\dfrac{5{,}280 \text{ feet}}{1 \text{ mile}}$

has the same distance in both its numerator and its denominator, it has a value of 1.

- The numerator for the conversion factor for time must contain minutes so that it cancels the minutes in the denominator of $18{,}000\,\frac{m}{min}$. Hence, use the conversion factor of $\frac{1\,min}{60\,sec}$:

$$18{,}000\,\frac{m}{min} = \frac{18{,}000\,m}{1\,\cancel{min}} \times \frac{1\,\cancel{min}}{60\,sec}$$

$$= \frac{18{,}000\,m}{60\,sec}$$

$$= 300\,\frac{m}{sec}$$

- Find the distance:

$$300\,\frac{m}{sec} \times 5\,sec = 1{,}500\,m$$

➡ Example

The average download speed for Max's computer's Internet connection is 30 megabits per second. Assuming no interruptions in Internet service, what is the best estimate for the maximum number of complete video files that Max can download to his computer in a six-hour period if each video file is 4.2 gigabytes in size? [1 megabyte = 8 megabits and 1 gigabyte = 1,024 megabytes]

(A) 15
(B) 18
(C) 19
(D) 21

Solution

- Convert from $\dfrac{megabits}{sec}$ to $\dfrac{megabits}{hour}$:

$$30\,\frac{megabits}{sec} = 30\,\frac{megabits}{\cancel{sec}} \times \frac{60\,\cancel{sec}}{1\,\cancel{min}} \times \frac{60\,\cancel{min}}{1\,hour} = 108{,}000\,\frac{megabits}{hour}$$

- Convert from $\dfrac{megabits}{hour}$ to $\dfrac{gigabytes}{hour}$:

$$108{,}000\,\frac{megabits}{hour} = 108{,}000\,\frac{\cancel{megabits}}{hour} \times \frac{1\,\cancel{megabyte}}{8\,\cancel{megabits}} \times \frac{1\,gigabyte}{1024\,\cancel{megabytes}} \approx 13.184\,\frac{gigabytes}{hour}$$

- In a 6-hour period, the total number of gigabytes that can be downloaded is

$$13.184\,\frac{gigabytes}{\cancel{hour}} \times 6\,\cancel{hours} = 79.104\,gigabytes$$

Since $79.104 \div 4.2 \approx 18.8$, a maximum of 18 video files can be download.

The correct choice is **(B)**.

SCIENTIFIC NOTATION

When a very large or very small number includes a sequence of trailing or leading zeros, particularly in science measurements, it is sometimes convenient to write that number as a decimal number between 1 and 10 times a power of 10.

- If the original number is greater than 10, the number of places the decimal point needs to be moved to the *left* becomes the power of 10. If necessary, insert the decimal point in the original number, as in

$$32{,}000{,}000. = 3.2 \times 10^7$$

7 places

- If the original number is between 0 and 1, make the power of 10 negative followed by the number of places the decimal point needs to be moved to the *right*, as in

$$0.00006\,08 = 6.08 \times 10^{-5}$$

5 places

Multiple-Choice

1. Which expression could be used to change 8 kilometers per hour to meters per minute?

   (A) $\dfrac{8\,km}{hr} \times \dfrac{1\,km}{1{,}000\,m} \times \dfrac{1\,hr}{60\,min}$

   (B) $\dfrac{8\,km}{hr} \times \dfrac{1{,}000\,m}{1\,km} \times \dfrac{60\,min}{1\,hr}$

   (C) $\dfrac{8\,km}{hr} \times \dfrac{1{,}000\,m}{1\,km} \times \dfrac{1\,hr}{60\,min}$

   (D) $\dfrac{8\,km}{hr} \times \dfrac{1\,km}{1{,}000\,m} \times \dfrac{60\,min}{1\,hr}$

2. Which expression represents 72 kilometers per hour expressed as meters per hour?

   (A) 7.2×10^{-2}

   (B) 7.2×10^{2}

   (C) 7.2×10^{-3}

   (D) 7.2×10^{4}

3. If the mass of a proton is 1.67×10^{-24} gram, what is the number of grams in the mass of 1,000 protons?

   (A) 1.67×10^{-27}

   (B) 1.67×10^{-23}

   (C) 1.67×10^{-22}

   (D) 1.67×10^{-21}

4. There are 12 players on a basketball team. Before a game, both ankles of each player are taped. Each roll of tape will tape three ankles. Which product can be used to determine the number of rolls of tape needed to tape the players' ankles?

   (A) $12\,players \cdot \dfrac{1\,player}{2\,ankles} \cdot \dfrac{3\,ankles}{1\,roll}$

   (B) $12\,players \cdot 2\,ankles \cdot \dfrac{3\,rolls}{1\,ankle}$

   (C) $12\,players \cdot \dfrac{2\,ankles}{1\,player} \cdot \dfrac{1\,roll}{3\,ankles}$

   (D) $12\,players \cdot \dfrac{1\,roll}{3\,players} \cdot \dfrac{3\,ankles}{roll}$

$$\dfrac{40\,yd}{4.5\,sec} \cdot \dfrac{3\,ft}{1\,yd} \cdot \dfrac{5{,}280\,ft}{1\,mi} \cdot \dfrac{60\,sec}{1\,min} \cdot \dfrac{60\,min}{1\,hr}$$

5. A sprinter who can run the 40-yard dash in 4.5 seconds converts his speed into miles per hour, using the product above. Which fraction in the product is *incorrectly* written to convert his speed?

   (A) $\dfrac{3\,ft}{1\,yd}$

   (B) $\dfrac{5{,}280\,ft}{1\,mi}$

   (C) $\dfrac{60\,sec}{1\,min}$

   (D) $\dfrac{60\,min}{1\,hr}$

6. A star constellation is approximately 3.1×10^{4} light years from Earth. One light year is about 5.9×10^{12} miles. What is the approximate distance, in miles, between Earth and the constellation?

   (A) 1.83×10^{17}

   (B) 9.0×10^{49}

   (C) 1.9×10^{8}

   (D) 9.0×10^{16}

7. An eye medication that is used to treat increased pressure inside the eye is packaged in 2.5 milliliter bottles. During the manufacturing process, a 10 decaliter capacity bin is used to fill the bottles. If 1 decaliter is equivalent to 10 liters and 1 liter is equivalent to 1,000 milliliters, what is the maximum number of bottles that can be filled?

   (A) 4×10^{5}

   (B) 4×10^{4}

   (C) 2.5×10^{3}

   (D) 2.5×10^{2}

Grid-In

1. At a party, six 1-liter bottles of soda are completely emptied into 8-ounce cups. What is the *least* number of cups that are needed? [There are approximately 1.1 quarts in 1 liter.]

2. On a certain map, 1 inch represents 2 kilometers. A region is located on the map that is 1.5 inches by 4.0 inches. What is the actual area of the region in square *miles* if 1 kilometer is equal to 0.6 mile?

3. The distance from Earth to Mars is 136,000,000 miles. A spacecraft travels at an average speed of 28,500 kilometers per hour. Determine, to the *nearest day*, how long it will take the spacecraft to reach Mars. [1 kilometer = 0.6 miles]

4. A certain generator will run for 1.5 hours on one liter of gas. If the gas tank has the shape of a rectangular box that is 25 cm by 20 cm by 16 cm, how long will the generator run on a full tank of gas? [1 liter = 1,000 cubic centimeters]

5. One knot is one nautical mile per hour, and one nautical mile is 6,080 feet. If a cruiser ship has an average speed of 3.5 knots, how many feet does the ship travel in 24 minutes?

OVERVIEW

The SAT assumes that you are familiar with the characteristics of two important classes of functions: the linear function and the exponential function.

- In a linear function, the greatest power of the variable is 1, as in $y = 3x + 2$. A key distinguishing feature of a linear function is that the rate of change between the variables is *constant*.
- In an exponential function, the variable is in an exponent, as in $y = 2^x$. The rate of change between the variables is *not* constant but changes by a fixed percent of its previous value as when interest on a savings account is compounded year after year.

LINEAR FUNCTIONS

The graph of a linear function is a straight line. When a linear function is written in the form $f(x) = mx + b$, m is the slope of the line and b is its y-intercept. If $y = f(x) = 3x + 2$, then each time x increases by 1 unit, y increases by 3 units, so the rate of change is fixed at $\frac{3}{1}$ or 3. Thus, the slope of a line represents the rate at which a linear function changes in value with respect to x. Linear functions can be used to represent situations that involve *constant* rates of change.

➡ Example

A certain 4 inch spring stretches 1.5 inches for each ounce of weight attached to it.

Solution

The length of the spring, L, when an x-ounce weight is attached to it, can be represented by the function $L(x) = 4 + 1.5x$.

➡ Example

A car starts a trip with 20 gallons of gasoline in its tank and consumes gas at a rate of 1 gallon for each 25 miles traveled.

Solution

After traveling x miles, the amount of gas, g, remaining in the tank is $g(x) = 20 - \frac{1}{25}x$.

INTERPRETING THE PARTS OF A LINEAR MODEL

When a linear equation of the form $y = a + bx$ is used to represent or model a real-world situation, the values of the constants a and b have specific meanings within the particular setting described. Typically,

- The constant a represents some starting value for y or some initial condition.
- The constant b tells for each unit change in x how much y increases or decreases.

If the dollar value, L, of a certain laptop computer decreases m months after its purchase according to the equation $L = 2,100 - 28m$, then -28 represents the number of dollars the value of the laptop *decreases* each month after its purchase, whereas 2,100 represents the starting or purchase price of the laptop.

➥ Example

When there are x milligrams of a certain drug in a patient's bloodstream, a patient's heart rate, h, in beats per minute, can be modeled by the equation $h = 70 + 0.5x$. Which statements are true?

  I. 10 minutes after taking the drug, the patient's heart rate increases to 75 beats per minute.

 II. When the drug is not in the bloodstream, the patient's heart rate is 70 beats per minute.

III. For each 10 milligram increase of the drug in the patient's bloodstream, the patient's heartbeat increases 5 beats per minute.

    (A) I and II

    (B) I and III

    (C) II and III

    (D) I, II, and III

Solution

- I. The model does not explicitly consider time so Statement I is *not* correct.
- II. When $x = 0$, $h = 70 + 0.5(0) = 70$ so Statement II is true.
- III. Statement III is true since each time x increases by 10, h increases by $0.5 \times 10 = 5$ beats per minute.

The correct choice is (**C**).

GRAPHS AS MODELS

A graph of a function can model a situation by giving a visual picture of how one real-world quantity measured along the vertical y-axis depends on another quantity measured along the horizontal x-axis. For example, the graph in Figure 4.1 shows how the cost C, in dollars, of manufacturing a certain product depends on the number of units manufactured.

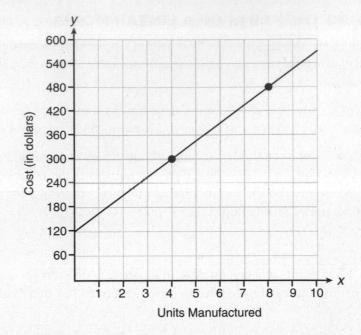

Figure 4.1 Cost as a function of the number of units manufactured

Since the graph is a straight line, the rate at which the cost changes for each additional unit manufactured is constant and equal to the slope of the line.

TIP

If the relationship between two quantities is represented by a line, then you can determine the rate of change between these quantities by finding the slope of the line.

■ To find the slope of the line, choose any two convenient points with integer coordinates such as (4, 300) and (8, 480):

$$\text{slope} = \frac{\Delta y}{\Delta x} = \frac{480-300}{8-4} = \frac{180}{4} = 45\,\frac{\text{dollars}}{\text{unit}}$$

Thus, the cost of manufacturing each additional unit is **$45**.

■ To find an equation of the line from the graph, write the general form of the equation as $C = ax + b$ where a is the slope of the line and b is the y-intercept. Because $a = 45$ and $b = 120$, the equation of the line is $C = 45x + 120$.

➡ **Example**

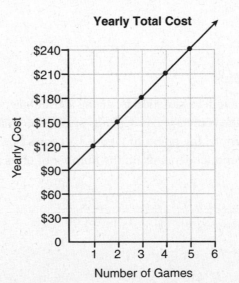

The graph on page 222 represents the yearly cost of playing 0 to 5 games of golf at the Sunnybrook Golf Course, which includes the yearly membership fee.

a. What is the cost of playing one game of golf?

b. Write a linear function that expresses the total cost in dollars, C, of joining the club and playing n games during the year. What is the cost of playing 10 games of golf?

Solution

a. The slope of the line represents the cost per game of golf. Pick any two points on the line with integer coordinates such as (1, 120) and (2, 150) and find the slope of the line:

$$\text{slope} = \frac{\Delta y}{\Delta x} = \frac{150 - 120}{2 - 1} = 30 \frac{\text{dollars}}{\text{game}}$$

The cost of one game of golf is **$30**.

b. When the number of games is 0, the cost is $90. Hence, the y-intercept of the line represents the yearly cost of membership. Because the slope of the line is 30 and its y-intercept is 90, an equation of the line is $C = 30n + 90$. To find the cost of playing 10 games, substitute 10 for n in $C = 30n + 90$, which gives $C = 30 \cdot 10 + 90 = 390$. Hence, the total cost of joining the club and playing 10 games during the year is **$390**.

THE EXPONENTIAL FUNCTION

An **exponential function** is an equation of the form $y = Ab^x$, where $A \neq 0$ and b is a positive constant other than 1. The constant A represents the initial amount of y which for successive integer values of x (= 1, 2, 3, 4, ...) is repeatedly multiplied by the base b. Figure 4.2 shows the effect of b on the graph of $y = b^x$.

- If $b > 1$, then as x increases, y increases and the graph rises at an increasingly faster rate.
- If $0 < b < 1$, then as x increases, y decreases and the graph falls but at a slower and slower rate.

> **MATH REFERENCE FACT**
>
> In a linear function, y changes at a constant rate—each time x changes by 1 unit, the amount y changes stays the same. In an exponential function of the form $y = Ab^x$, the rate at which y changes is *not* constant—each time x increases by 1 unit, y increases to b times its previous value.

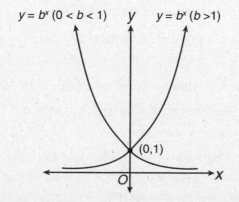

Figure 4.2 The exponential function $y = b^x$

Illustrating Exponential Change Graphically

The slope of a line is constant across the line. The slope or the *rate* at which y changes along an exponential curve. The accompanying table of values represent points from the graph of $y = 5 \cdot 2^x$ shown in Figure 4.3.

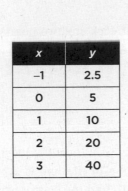

x	y
−1	2.5
0	5
1	10
2	20
3	40

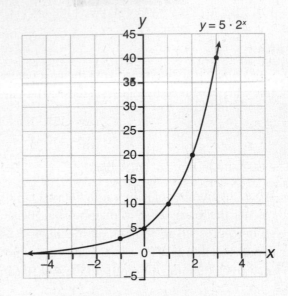

Figure 4.3 Graph of $y = 5 \cdot 2^x$

You should verify that each time x increases by 1 in Figure 4.3, y increases to twice its previous value. In terms of percent, y increases by 100% of its previous value each time x increases by 1. This type of behavior represents exponential *growth*.

Illustrating Exponential Change Algebraically

Suppose $320 is invested in an account that earns 7% interest compounded annually.

- After one year, the balance in the account is

$$\underbrace{320}_{\text{initial amount}} + \underbrace{320 \times 0.07}_{\text{interest}} = 320(1+0.07) = 320\underbrace{(1.07)}_{\text{multiplying factor}}$$

- After the second year, the balance in the account is

$$\underbrace{320(1.07)}_{\text{old amount}} + \underbrace{[320(1.07)] \times (0.07)}_{\text{interest}} = 320(1.07)(1+0.07) = 320\underbrace{(1.07)^2}_{x}$$

- After n years, the balance, y, has increased by a factor of 1.07 for a total of n times. Thus,

$$y = 320\underbrace{(1.07)(1.07)\cdots(1.07)}_{n\,\text{factors}} = 320\underbrace{(1.07)^n}_{x}$$

If function f represents the balance after n years, then $f(n) = 320(1.07)^n$. This exponential function has a starting amount of 320 and a growth, or multiplying, factor of 1.07. Many real-life processes involve either exponential growth or exponential decay. Exponential *growth* occurs when the multiplying factor is greater than 1. Exponential *decay* happens if the multiplying factor is a positive number less than 1. Here are more examples of exponential change:

- Assume a certain population of 1,000 insects triples every 6 days. To represent this process as an exponential function, make a table with the first few terms and then generalize:

Number of Days	Population
6	1000×3
12	$(1000 \times 3) \times 3 = 1000 \times 3^2$
18	$(1000 \times 3 \times 3) \times 3 = 1000 \times 3^3$
...	...
n	$1000 \times 3^{\frac{n}{6}}$

Thus, the exponential function $f(n) = 1000 \times 3^{\frac{n}{6}}$ describes this growth process, where n is the number of days that have elapsed.

- In the geometric sequence

$$400, 200, 100, 50, 25, \ldots$$

each term after the first is obtained by multiplying the term that comes before it by $\frac{1}{2}$.

To represent this process as an exponential function, make a table with the first few terms and then generalize:

Term Number	Term
1	400
2	$400 \times \frac{1}{2}$
3	$\left(400 \times \frac{1}{2}\right) \times \frac{1}{2} = 400 \times \left(\frac{1}{2}\right)^2$
4	$\left(400 \times \frac{1}{2} \times \frac{1}{2}\right) \times \frac{1}{2} = 400 \times \left(\frac{1}{2}\right)^3$
...	...
n	$400 \times \left(\frac{1}{2}\right)^{n-1}$

Thus, the exponential function $a(n) = 400 \times \left(\frac{1}{2}\right)^{n-1}$ describes this decay process, where $a(n)$ represents the nth term of this number sequence.

GENERAL FORMULAS FOR EXPONENTIAL GROWTH AND DECAY

If an initial quantity A changes exponentially at a rate of $r\%$ per time period, then after x successive time periods, the amount present, y, is given by the formulas in the accompanying table.

Process	Exponential Function
Growth	$y = A(1 + r)^x$
Decay	$y = A(1 - r)^x$

In an exponential function of the form $y = Ab^x$ where $b = 1 \pm r$.

- when $b > 1$, the initial amount A is *increasing* exponentially at a rate of $r\%$. If a colony of insects begins with 20 insects and increases exponentially in number at a rate of 35% every 5 days, then after 30 days the number of insects, y, is

$$y = 20(1 + 0.35)^{\frac{30}{5}} = 20(1.35)^6$$

- when $0 < b < 1$, the initial amount A is *decreasing* exponentially at a rate of $r\%$. If the population of a town with a current population of 5,400 is decreasing exponentially at a rate of 13% per year, then after 8 years the population, y, is

$$y = 5,400(1 - 0.13)^8 = 5,400(0.87)^8$$

- The quantity b that is being raised to a power is sometimes referred to as the **growth factor**.

➡ Example

The current population of a town is 10,000. If the population, P, increases by 3.5% every six months, which equation could be used to find the population after t years?

(A) $P = 10,000(1.035)^{\frac{t}{2}}$
(B) $P = 10,000\,(0.965)^{2t}$
(C) $P = 10,000\,(1.035)^{2t}$
(D) $P = 10,000(0.965)^{\frac{t}{2}}$

Solution

Since the growth factor is $1 + 3.5\% = 1 + 0.035 = 1.035$, eliminate choices (B) and (D). Because the growth cycle is every 6 months, there are 2 growth cycles or compounding periods each year in which the population increases by 3.5% over its current amount, $P = 10,000\,(1.035)^{2t}$.

The correct choice is **(C)**.

➡ Example

A car loses its value at a rate of 4.5% annually. If a car is purchased for $24,500, which equation can be used to determine the value of the car, V, after 5 years?

(A) $V = 24,500\,(0.045)^5$
(B) $V = 24,500\,(0.55)^5$
(C) $V = 24,500\,(1.045)^5$
(D) $V = 24,500\,(0.955)^5$

Solution

Since the growth factor is $1 - 4.5\% = 1 - 0.045 = 0.955$, $V = 24,500(0.955)^5$.

The correct choice is **(D)**.

Multiple-Choice

1. On January 1, a share of a certain stock costs
   $180. Each month thereafter, the cost of a
   share of this stock decreased by one-third. If
   x represents the time, in months, and *y*
   represents the cost of the stock, in dollars,
   which graph best represents the cost of a
   share over the following 5 months?

   (A) Graph A
   (B) Graph B
   (C) Graph C
   (D) Graph D

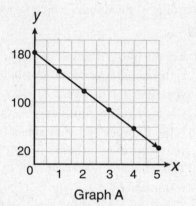

Graph A

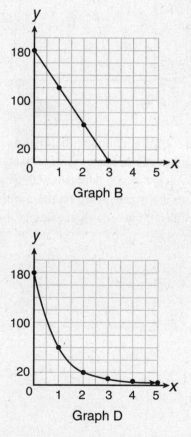

Graph B

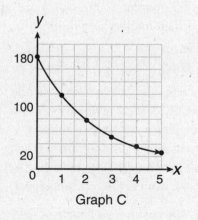

Graph C

Graph D

2. A certain population of insects starts at 16 and doubles every 6 days. What is the population after 60 days?

(A) 2^6
(B) 2^{10}
(C) 2^{14}
(D) 2^{32}

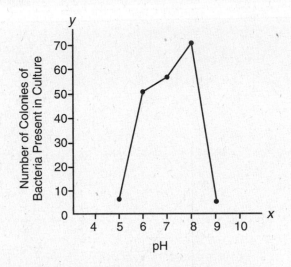

3. The accompanying graph illustrates the presence of a certain strain of bacteria at various pH levels. Between which two pH values is the rate at which the number of colonies of bacteria increasing at the lowest rate?

(A) 5 to 6
(B) 6 to 7
(C) 7 to 8
(D) 8 to 9

4. After a single sheet of paper is folded in half, there are two layers of paper. The same sheet of paper is repeatedly folded in half. If function f represents the number of layers of paper that results when the original sheet of paper is folded a total of x times, then which equation could represent this function?

(A) $f(x) = 2x$
(B) $f(x) = x^2$
(C) $f(x) = 2^x$
(D) $f(x) = 4^{\frac{x}{2}}$

5. The yearly growth in the number of fast food restaurants in a certain retail chain can be modeled by the function $f(n) = 5 + 8n$. According to this model which statement is true?

(A) 8 is the initial number of restaurants; 5 is the number of restaurants added each year after the first year.
(B) 5 is the starting number of restaurants; 8 is the number of restaurants added each year after the first year.
(C) The retail chain opened with 13 restaurants.
(D) The y-intercept of the graph of function f shows the year in which the retail chain made a zero profit.

$$C = 60 + 0.05d$$

6. The equation above represents the total monthly cost, C, in dollars of a data plan offered by a cell phone company when the data usage in a month exceeds a 1 gigabyte limit by d megabytes. According to the model, what is the meaning of 0.05?

(A) The cost per megabyte of data used
(B) The cost per gigabyte of data used
(C) The cost per megabyte of data after one gigabyte of data is used
(D) The cost of each additional gigabyte of data after the first megabyte of data is used

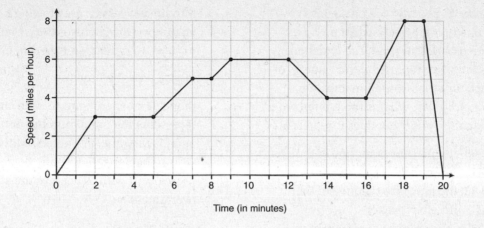

Time (in minutes)

7. The graph above represents a jogger's speed during her 20-minute jog around her neighborhood. Which statement best describes what the jogger was doing during the 9 to 12 minute interval of her jog?

(A) She was standing still.
(B) She was increasing her speed.
(C) She was decreasing her speed.
(D) She was jogging at a constant rate.

8. If a_n represents the nth term of the sequence 54, 18, 6, ..., and a_1 represents the first term, then $a_n =$

(A) $6\left(\dfrac{1}{3}\right)^n$

(B) $6\left(\dfrac{1}{3}\right)^{n-1}$

(C) $54\left(\dfrac{1}{3}\right)^n$

(D) $54\left(\dfrac{1}{3}\right)^{n-1}$

9. The owner of a small computer repair business has one employee, who is paid an hourly rate of $22. The owner estimates his weekly profit using the function $P(x) = 8,600 - 22x$. In this function, x represents the number of

(A) computers repaired per week.
(B) employee's hours worked per week.
(C) customers served per week.
(D) days worked per week.

10. The breakdown of a sample of a chemical compound is represented by the function $p(n) = 300(0.5)^n$, where $p(n)$ represents the number of milligrams of the substance that remains at the end of n years. Which of the following is true?

I. 300 represents the number of milligrams of the substance that remains after the first year.
II. 0.5 represents the fraction of the starting amount by which the substance gets reduced by the end of each year.
III. Each year the substance gets reduced by one-half of 300.

(A) I only
(B) II only
(C) I and III only
(D) II and III only

11. Some banks charge a fee on savings accounts that are left inactive for an extended period of time. The equation $y = 5,000(0.98)^x$ represents the value, y, of one account that was left inactive for a period of x years. What is the y-intercept of this equation and what does it represent?

(A) 0.98, the percent of money in the account initially
(B) 0.98, the percent of money in the account after x years
(C) 5,000, the amount of money in the account initially
(D) 5,000, the amount of money in the account after x years

12. Chris plans to purchase a car that loses its value at a rate of 14% per year. If the intial cost of the car is $27,000, which of the following equations should Chris use to determine the value, v, of the car after 4 years?

(A) $v = 27,000 (1.14)^4$
(B) $v = 27,000 (0.14)^4$
(C) $v = 27,000 (0.86)^4$
(D) $v = 27,000 (0.86 \times 4)$

13. Vanessa plans to invest $10,000 for 5 years at an annual interest rate of 6% compounded annually. Which equation could be used to determine the profit, P, Vanessa earns from her initial investment?

(A) $P = 10,000 \times (1.06)^5$
(B) $P = 10,000 \times [(1.06)^5 - 1]$
(C) $P = 10,000 \times [(1.06)^5 + 1]$
(D) $P = 10,000 \times [5(1.06) - 1]$

14. Miriam and Jessica are growing bacteria in a laboratory. Miriam uses the growth function $f(t) = n^{2t}$ to model her experiment while Jessica uses the function $g(t) = \left(\dfrac{n}{2}\right)^{4t}$ to model her experiment. In each case, n represents the initial number of bacteria, and t is the time, in hours. If Miriam starts with 16 bacteria, how many bacteria should Jessica start with to achieve the same growth over time?

(A) 32
(B) 16
(C) 8
(D) 4

15. The number of square units, A, in the area covered by a bacteria culture is increasing at a rate of 20% every 7 days. If the bacteria culture initially covers an area of 10 square centimeters, which equation can be used to find the number of square units in the area covered by the bacteria culture after d days?

(A) $A = 10(1.20)^{\frac{d}{7}}$
(B) $A = 10(1.20)^{7d}$
(C) $A = 10(0.80)^{\frac{d}{7}}$
(D) $A = (12)^{7d}$

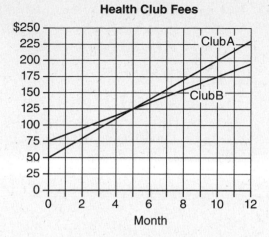

Health Club Fees

16. Two health clubs offer different membership plans. The accompanying graph represents the total yearly cost of belonging to Club A and Club B for one year. The yearly cost includes a membership fee plus a fixed monthly charge. By what amount does the monthly charge for Club A exceed the monthly charge for Club B?

(A) $5.00
(B) $7.50
(C) $10.00
(D) $12.50

x	y
0.5	9.0
1	8.75
1.5	8.5
2	8.25
2.5	8.0

17. Based on the data in the table above, which statement is true about the rate of change of y with respect to x?

(A) It is constant and equal to $-\frac{1}{8}$.

(B) It is constant and equal to 2.

(C) It is constant and equal to $-\frac{1}{2}$.

(D) It is not constant.

18. The City Tunnel and Bridge Authority in a certain city estimates that 40,000 vehicles currently travel over a certain bridge per year but, due to highway construction and changing traffic patterns, vehicle traffic over the bridge will begin to decline by 12% every 5 years. Which of the following expressions best represents the vehicle traffic projections for this bridge n years from now?

(A) $40,000(0.12)^{\frac{n}{5}}$
(B) $40,000(0.88)^{5n}$
(C) $40,000(0.12)^{5n}$
(D) $40,000(0.88)^{\frac{n}{5}}$

19. Which of the accompanying tables that show how population is changing over time illustrate exponential decay?

Time (months)	Population
0	10,000
6	7,000
12	4,000
18	1,000

Table I

Time (months)	Population
0	10,000
6	5,000
12	2,500
18	1,250

Table II

Time (months)	Population
0	10,000
6	20,000
12	40,000
18	80,000

Table III

Time (months)	Population
0	10,000
6	15,000
12	20,000
18	25,000

Table IV

(A) Table I
(B) Table II
(C) Table III
(D) Table IV

20. A radioactive substance has an initial mass of 100 grams, and its mass is reduced by 40% every 5 years. Which equation could be used to find the number of grams in the mass, y, that remains after x years?

(A) $y = 100(0.4)^{5x}$
(B) $y = 100(0.6)^{5x}$
(C) $y = 100(0.4)^{\frac{x}{5}}$
(D) $y = 100(0.6)^{\frac{x}{5}}$

21. The gas tank in a car holds a total of 16 gallons of gas. The car travels 75 miles on 4 gallons of gas. If the gas tank is full at the beginning of a trip, which graph represents the rate of change in the amount of gas in the tank?

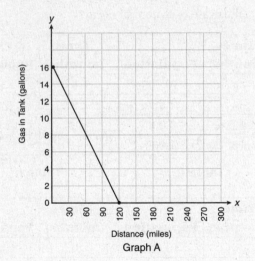

Graph A

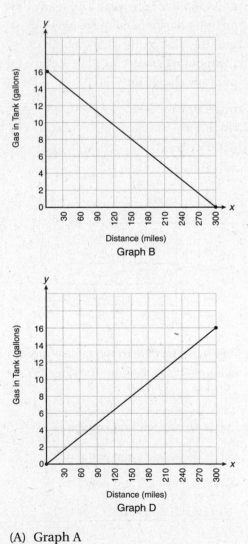

Graph B

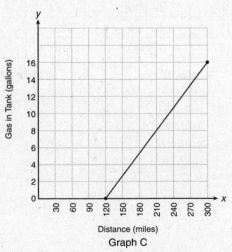

Graph C

Graph D

(A) Graph A
(B) Graph B
(C) Graph C
(D) Graph D

Grid-In

1. Jacob begins painting at 12:00 noon. At 12:30 P.M. he estimates that 13 gallons of paint are left, and at 3:30 he estimates that 4 gallons of paint remain. If the paint is being used at a constant rate, how many gallons of paint did Jacob have when he started the job?

2. The number of hours, H, needed to manufacture X computer monitors is given by the function $H = kX + q$, where k and q are constants. If it takes 270 hours to manufacture 100 computer monitors and 410 hours to manufacture 160 computer monitors, how many *minutes* are required to manufacture each additonal computer monitor?

3. Given a starting population of 100 bacteria, the formula $b(t) = 100(2^t)$ can be used to determine the number of bacteria, b, after t periods of time. If each time period is 15 minutes long, how many minutes will it take for the population of bacteria to reach 51,200?

4. A certain drug raises a patient's heart rate, h, in beats per minute, according to the equation $h(x) = 65 + 0.2x$, where x is the number of milligrams of the drug in the patient's bloodstream. After t hours, the level of the drug in the patient's bloodstream decreases exponentially according to the equation $x(t) = 512(0.7)^t$. After 5 hours, what is the number of beats per minute in the patient's heart rate, correct to the *nearest whole number*?

5. The breakdown of a sample of a chemical compound is represented by the function $p(t) = 300\left(\dfrac{1}{2}\right)^t$, where $p(t)$ represents the number of milligrams of the substance, and t represents the time, in years. If $t = 0$ represents the year 2015, what will be the first year in which the amount of the substance remaining falls to less than 5 milligrams?

6. Sasha invested $1,200 in a savings account at an annual interest rate of 1.6% compounded annually. She made no further deposits or withdrawals. To the *nearest dollar*, how much more money did she have in the account after 3 years than after 2 years?

> ### OVERVIEW
>
> Graphs and tables are used to visually summarize data in a way that is concise and easy to read. Before attempting to answer questions based on a graph, you should quickly scan the graph to get an overview of what data and information are being presented. Pay close attention to the type of graph and its title. Take note of any descriptive labels and units of measurement along horizontal and vertical sides of a bar or line graph. If a circle graph is given, look for the whole amount that the circle represents.

COMMON TYPES OF GRAPHS

Recognizing the type of graph may help you answer questions about the graph. In general,

- **A circle graph** or pie chart shows how the parts that comprise a whole compare with the whole and to each other. Each "slice" of a pie chart is typically labeled with the percent that part is of the whole. The sum of the "slices" of a pie chart add up to 100% as shown in Figure 4.4.

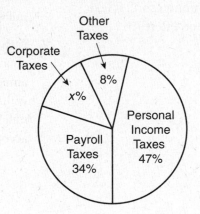

Figure 4.4 Circle graph (pie chart)

Since $x + 47 + 34 + 8 = 100$, you can conclude that the amount of tax revenue from corporate tax sources is 11% of 3.2 trillion dollars, which is 0.352 trillion dollars or, equivalently, 352 billion dollars.

- **A bar graph** compares similar categories of items using rectangular bars. The base of each bar has a nonnumerical label that describes a category. The height or length of each bar represents a numerical amount associated with that category. Figure 4.5 summarizes the number of different colleges to which five high school seniors applied for admission.

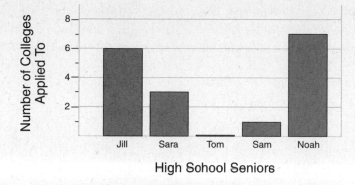

High School Seniors

Figure 4.5 High School Seniors

The average number of colleges applied to per student is

$$\frac{6+3+0+1+5}{5} = \frac{15}{5} = 3 \text{ colleges per student}$$

- A **histogram** is similar to a bar graph except that the base of each bar of a histogram is a single *numerical* value or an interval of values. The length of the bar is the corresponding amount of data items that have that numerical value or that fall into that interval of values. The descriptive labels along the axes, as well as the title of the graph, explain what the numerical quantities represent. Figure 4.6 shows a histogram that summarizes the number of colleges applied to for 18 high school seniors selected at random.

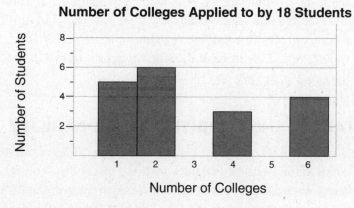

Figure 4.6 Number of Colleges

Notice that there are 5 students who applied to one college, 6 students who applied to two colleges, and so forth. The average number of colleges applied to per student is

$$\frac{(5 \times 1) + (6 \times 2) + (0 \times 3) + (3 \times 4) + (0 \times 5) + (4 \times 6)}{14} = \frac{53}{18} \approx 3 \text{ colleges per student.}$$

■ **A line graph** is used to represent how one data variable changes with another particularly when one of the variables is time. For example, a line graph could be used to show how the price of a stock rises and falls over time, as shown in Figure 4.7.

Figure 4.7 Line graph of the price of a stock from January through May

If a line segment slants up in a time interval, as in January in Figure 4.7, the amount is increasing for that time interval. If a line segment falls in a time interval, as in April, the amount is decreasing in that interval. A horizontal line segment indicates no change, as in February. Since over the 5-month period the price of the stock increases from $15 to $75, the average increase per month of the stock is $\frac{\$75 - \$15}{5}$ = $12 per month.

TWO-WAY TABLES AND CONDITIONAL PROBABILITY

A **two-way table** displays the set of all possible pairings of the values of *two* different categories of data. The values for one of these variables are listed in a vertical column and the values of the other are placed along a horizontal row as illustrated in the accompanying table, which shows the results of a survey of the types of books downloaded by 140 individuals according to age group. Since the horizontal row for the 22–50 age group intersects the vertical column for action novels in the cell that contains the number 10, you know that the age group 22–50 downloaded 10 action novels to their tablets. The totals for each vertical column and each horizontal row are given in the last cells for each column and row. The sum of the row totals equals the sum of the column totals.

Age Group	Type of Book Downloaded to a Computer Tablet				
	Non-Fiction	Action Novel	Romance Novel	Science Fiction Novel	Total
Under 21	2	14	15	7	38
22–50	12	10	19	18	59
Over 50	14	13	11	5	43
Total	28	37	45	30	140

Based on the table, you should be able to answer questions such as these:

- What percent of the individuals surveyed downloaded a novel? Since 28 people downloaded a non-fiction book, $140 - 28 = 112$ downloaded some type of novel. Thus, $\frac{112}{140} = 0.8 = 80\%$ of those surveyed, downloaded a novel.

- If a person over 21 is selected at random, what is the probability the person downloaded either a non-fiction book or a romance novel? Of the 140 persons surveyed, $12 + 14 = 26$ individuals over 21 years of age downloaded a non-fiction book and $19 + 11 = 30$ individuals over 21 years of age downloaded a romance novel. Hence, the probability asked for is $\frac{26 + 30}{140} = \frac{56}{140} = \frac{2}{5}$.

Multiple-Choice

SAT Math Scores at Cedar Lane High School

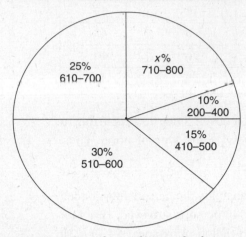

Questions 1 and 2 refer to the graph above.

1. If there are 72 SAT Math scores between 510 and 600, how many SAT Math scores are above 700?

   (A) 40
   (B) 48
   (C) 56
   (D) 64

2. If 20% of the students with SAT Math scores from 610 to 700 received college scholarships, how many students with SAT Math scores from 610 to 700 received college scholarships?

   (A) 12
   (B) 18
   (C) 30
   (D) 48

Minimum Age Requirement (years)	Number of States
14	7
15	12
16	27
17	2
18	2

Questions 3 and 4 refer to the table above.

3. The table above shows the minimum age requirement for obtaining a driver's license. In what percent of the states can a person obtain a driver's license before the age of 16?

   (A) 94%
   (B) 47%
   (C) 38%
   (D) 19%

4. If a state is chosen at random, what is the probability that the minimum age for obtaining a driver's license in that state will be at least 16?

   (A) $\dfrac{1}{25}$

   (B) $\dfrac{2}{25}$

   (C) $\dfrac{19}{50}$

   (D) $\dfrac{31}{50}$

Growth of Laptop Computers in U.S. Households

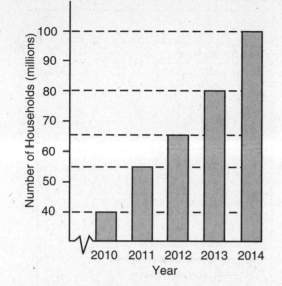

Investment Portfolio Valued at $250,000

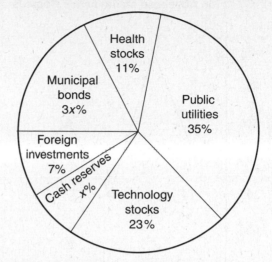

Questions 5 and 6 refer to the graph above.

The graph above shows the number of U.S. households with laptop computers for the years 2010 to 2014.

5. What was the percent of increase in the number of households with laptops from 2010 to 2014?

   (A) 40%
   (B) 60%
   (C) 120%
   (D) 150%

6. The greatest percent of increase in the number of households with laptops occurred in which two consecutive years?

   (A) 2010 to 2011
   (B) 2011 to 2012
   (C) 2012 to 2013
   (D) 2013 to 2014

Questions 7 and 8 refer to the graph above.

The graph above shows how $250,000 is invested.

7. How much money is invested in municipal bonds?

   (A) $45,000
   (B) $37,500
   (C) $35,000
   (D) $30,000

8. After 20% of the amount that is invested in technology stocks is reinvested in health stocks, how much money is invested in health stocks?

   (A) $77,500
   (B) $65,000
   (C) $45,000
   (D) $39,000

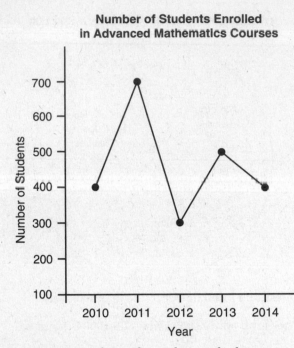

Number of Students Enrolled in Advanced Mathematics Courses

Questions 9 and 10 refer to the graph above.

9. The percent increase in the number of students enrolled in advanced mathematics courses from 2010 to 2011 exceeded the percent increase from 2012 to 2013 by approximately what percent?

   (A) 133
   (B) 75
   (C) 67
   (D) 8

10. From 2014 to 2015 the number of students enrolled in advanced mathematics courses increased by the same percent that student enrollment in advanced mathematics courses dropped from 2013 to 2014. What was the approximate number of students enrolled in advanced mathematics courses in 2015?

   (A) 420
   (B) 440
   (C) 450
   (D) 480

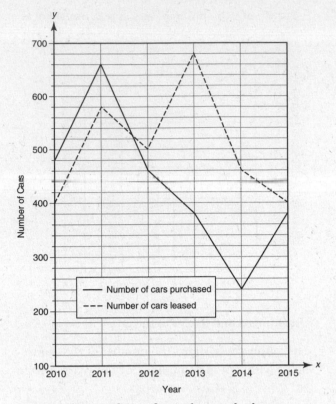

Questions 11 and 12 refer to the graph above.

11. In 2012, the number of cars purchased was x percent of the number of cars leased. What is the best approximation for x?

   (A) 75
   (B) 80
   (C) 85
   (D) 90

12. Which of the following is the best approximation for the decrease in the number of cars purchased per year between 2011 and 2014?

   (A) 105
   (B) 140
   (C) 300
   (D) 420

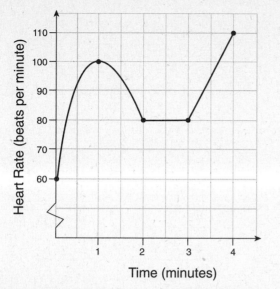

Time (minutes)

Questions 13 and 14 refer to the graph above which shows the heart rate, in beats per minute, of a jogger during a 4-minute interval.

13. Over what interval of time, in minutes, was the jogger's heart rate changing at a constant rate?

    (A) 0 to 1
    (B) 1 to 2
    (C) 2 to 3
    (D) 3 to 4

14. The greatest percent of increase in the jogger's heart rate occurred over what interval of time, in minutes?

    (A) 0 to 1
    (B) 1 to 2
    (C) 2 to 3
    (D) 3 to 4

HEAD CIRCUMFERENCE GROWTH SPEED	
Age	**Growth of Head Circumference (in centimeters)**
First year	Circumference $= \dfrac{\text{Height} + 12}{2}$
1 to 3 years	1 centimeter every 6 months
3 to 5 years	1 centimeter every year

15. The table above can be used to approximate the circumference of the head, in centimeters, during the first 5 years after birth. At 5 years of age, Jacob's head circumference was 81 cm. Based on the table, what was his approximate height, in centimeters, at 1 years old?

    (A) 138
    (B) 145
    (C) 152
    (D) 157

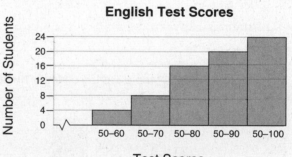

Test Scores

16. The cumulative histogram above shows the distribution of scores that 24 students received on an English test. If a student is selected at random, what is the probability that the student will have a score between 71 and 80?

    (A) $\dfrac{1}{6}$

    (B) $\dfrac{1}{3}$

    (C) $\dfrac{1}{2}$

    (D) $\dfrac{2}{3}$

Gender	Tennis Team Juniors	Seniors	Total
Male	14	11	25
Female	5	10	15
Total	19	21	40

17. The table above shows the composition of a coed high school tennis team with a total of 40 members. A player who will be selected at random from the team will be given two free tickets to attend the U.S. Open Tennis Tournament. What is the probability that the tickets will be given to either a female junior player or a male senior player?

(A) $\dfrac{1}{8}$

(B) $\dfrac{1}{4}$

(C) $\dfrac{2}{5}$

(D) $\dfrac{1}{2}$

Grid-In

Survey of 250 People Exiting a Movie Theater

Movie Category

Questions 1 and 2 refer to the graph above that summarizes a survey of a group of 250 people who were randomly selected when leaving a multiplex movie theater and asked what type of movie they had seen.

1. What percent of the people surveyed saw either an action or a science fiction movie?

2. If a total of 1,700 tickets were sold for the five types of movies represented in the histogram, what is the best estimate for the number of tickets sold to the Romance movie?

Number of Text Messages Sent in a Day

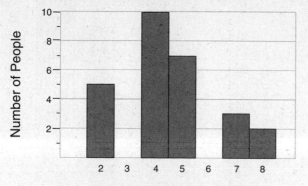

Number of Text Messages

3. The histogram above shows the number of mobile text messages sent by a randomly selected group of 27 people on a given day. The average number of text messages sent per person is closest to what whole number?

Grade	1 Club	2 Clubs	3 or More Clubs	Total
9th	37	43	18	98
10th	48	38	22	108
11th	52	27	31	110
12th	75	30	29	134
Total	212	138	100	450

Questions 4–6 refer to the above table, which summarizes the results of a survey of the student body of a high school about club membership in which each student enrolled in the high school enumerated the clubs in which they were members.

4. If a student is selected at random, what is the probability that the student does not belong to 3 or more clubs?

5. If 28% of the students who are enrolled in this school belong to at least two clubs, how many students who are enrolled in this school did *not* participate in the survey?

6. If a student is selected at random, what is the probability that the student is either a 9th grade student or belongs to 3 or more clubs?

Gender	Eye Color			Total
	Brown	**Hazel**	**Blue**	
Male		13		
Female		32		
Total	133			200

7. The partially completed table above describes the distribution of 200 subjects in a study involving eye color in which there were 4 times as many males with brown eyes as with blue eyes and 7 times as many females with brown eyes as with blue eyes. What percent of the subjects were either male with blue eyes or female with brown eyes?

Swimming Distances

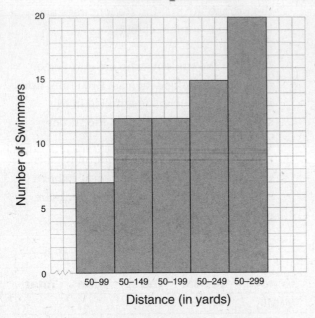

Errors on French Test

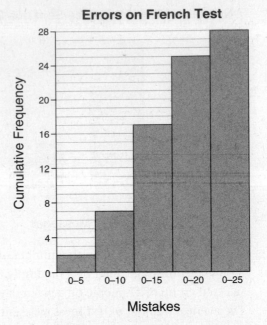

8. Based on the cumulative histogram above, what percent of the total number of swimmers swam between 200 and 249 yards?

9. The cumulative histogram above shows the distribution of mistakes 28 students in a French language class made on a test. What is the probability that a student selected at random made more than 10 mistakes?

OVERVIEW

A **scatterplot** is a graph that represents two sets of data as ordered pairs and shows them as points in the first quadrant of the *xy*-plane. The SAT may also include questions about sampling and errors associated with it. *Sampling* involves the selection of a smaller subset of individuals from within a larger statistical population for the purpose of making a generalization about the entire statistical population.

LINE OF BEST FIT

A scatterplot can indicate visually the type of relationship, if any, that exists between two sets of data. If the dots in a scatterplot are closely clustered about a line, as in Figure 4.8, then the relationship between the two sets of data can be approximated by a line. The **line of best fit** (or trend line) is the line that best represents the relationship between the two sets of data graphed in a scatterplot. Although there is a precise statistical method for determining the equation of the line of best fit, visually it is the line that can be drawn closest to all of the data points. Typically, the line of best fit passes through some, but not all of the plotted points with approximately the same number of data points falling on either side of it. A line of best fit is useful for predicting the *y*-value for any particular *x*-value that was not included in the original data.

Referring to Figure 4.8,

- Since the point (500, 560) lies on the line of best fit, a student with a verbal SAT score of 500 has a *predicted* math SAT score of 560.
- For only one data point, (400, **620**), does the math score differ by more than 100 points from the math score predicted by the line of best fit of (400, **480**).

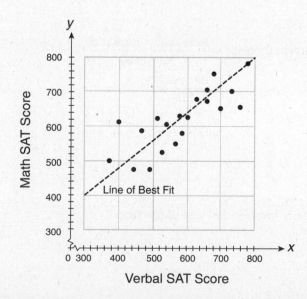

Figure 4.8 Scatterplot showing an approximately linear relationship between Verbal and Math SAT scores for students in the same SAT preparation class

MATH REFERENCE FACT

A scatterplot can show, at a glance, the relationship, if any, between two sets of data. It can show the type of relationship (linear, non-linear, or no relationship), the strength of the relationship (as indicated by how closely the points fit a line or curve), and whether the data variables move in the same direction (if $x \uparrow$, $y \uparrow$) or in the opposite direction (if $x \uparrow$, $y \downarrow$).

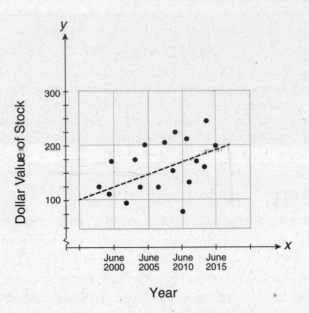

The graph above shows how the value of a stock has increased over time. The line of best fit is shown.

a. The value of the stock increased from June 2000 to June 2005 by approximately what percent?

(A) 16%
(B) 20%
(C) 25%
(D) 28%

Solution

$$\text{Percent increase} = \frac{\text{June 2005} - \text{June 2000}}{\text{June 2000}}$$
$$= \frac{150 - 125}{125}$$
$$= \frac{25}{125}$$
$$= 20\%$$

The correct choice is **(B)**.

b. What is the average yearly increase in value of the stock?

(A) $1
(B) $5
(C) $10
(D) $25

Solution

Find the slope of the line using the point (June 2000, 125) and (June 2005, 150):

$$\text{slope} = \frac{\$150 - \$125}{5 \text{ years}}$$
$$= \$5 \text{ per year}$$

The correct choice is **(B)**.

  c. What is the greatest difference between the actual value of the stock and the value of the stock predicted by the line of best fit?

    (A) $25
    (B) $55
    (C) $75
    (D) $95

Solution

Greatest difference occurred in June 2010:

        Predicted value − Actual value = $170 − $75 = $95

The correct choice is **(D)**.

RECOGNIZING THE TYPE OF ASSOCIATION FROM A SCATTERPLOT

A scatterplot may suggest different types of relationships, or no relationship, between two sets of measurements.

  ☐ Scatterplots may show a linear association between variables as in Figure 4.9.

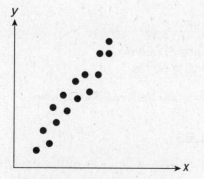
Positive Association: when
x increases, *y* increases

Negative Association: when
x increases, *y* decreases

Figure 4.9 Comparing scatterplots with positive and negative linear associations

☐ Scatterplots may show a nonlinear relationship between the data variables, as in Figure 4.10.

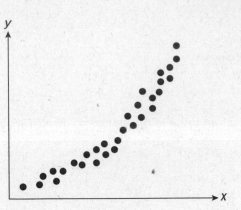
Exponential Association: an exponential curve can be fitted to the data points

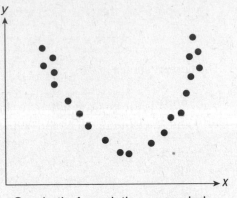
Quadratic Association: a parabola can be fitted to the data points

Figure 4.10 Comparing scatterplots with exponential and quadratic associations

☐ A scatterplot having no recognizable pattern indicates there is no relationship between the data variables, as illustrated in Figure 4.11.

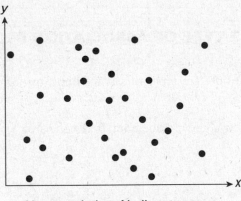
No association: No line or curve can be fitted to the data points

Figure 4.11 Scatterplot with no association between the variables

TIP

Although a scatterplot may indicate two variables are related, it does NOT provide any information about cause and effect. For example, a scatterplot of children's shoe sizes plotted against their reading levels would show a positive association (since both are related to age), although a larger shoe size does not cause a higher reading level.

UNBIASED VERSUS BIASED SAMPLES

Suppose you wanted to conduct a survey to find out how students in your school feel about an issue that affects the entire student body. It may not be practical or even possible to interview every student. In such cases, a smaller but representative group of students can be selected to be interviewed using some random selection process such as choosing one out of every five students entering the school cafeteria on a regular school day. The entire school population represents the target **population** and the smaller subset of students selected to be interviewed is the **sample**. A sample may be unbiased or biased.

- A sample is **unbiased** if each member of the target population has an equal chance of being selected for the sample.
- If the selection process is not random and favors a particular group within the target population, then the sample is **biased**.

Because the students selected to participate in this survey are chosen at random from the entire student body, the sample is *unbiased*. If only senior boys who are members of an athletic team were selected to participate in the survey, then the sample would be biased. Any general conclusions drawn from a biased survey may not be valid.

MARGIN OF ERROR

The **margin of error** for a survey based on a sample is a statistic that estimates the extent to which the conclusions drawn may differ from the results that would have been obtained from surveying the entire statistical population. In general, the larger and more random the sample, the smaller the margin of error.

CONFIDENCE INTERVALS

Anytime a sample is collected and a statistic such as the mean of the sample is calculated from it, the question arises regarding how accurately the sample mean represents the true population mean. A **confidence interval** for a sample statistic such as the mean is a range of values for which we expect the true population mean to lie within. Associated with a confidence interval is a probability value that reflects the degree of certainty that confidence intervals drawn from the same population will contain the actual population statistic. A "95% confidence interval for the mean" expresses the idea that if repeated samples were drawn from the same target population and a 95% confidence interval calculated for each sample, then 95% of those samples will contain the population mean. To illustrate this idea, suppose at the service center of a car dealership 65% of the people surveyed reported that they expect to buy or lease a new car within the next 3 years. A statistician calculates the confidence level to be 95% for an interval of 5% below and above the 65% mark. This means that if the same survey were to be repeated 100 times, we could assume that the percentage of people who would report that they expect to buy or lease a new car within the next 3 years would range between 60% and 70% in 95 of the 100 surveys.

When discussing confidence intervals, it is important to remember that:

- Probability values other than "95%" can be used.
- Other sample statistics such as standard deviation can be used.
- Reducing the width of the confidence interval, produces a more accurate estimate of the actual population statistic. This can often be accomplished by increasing the sample size and by reducing the variability of the data. Reducing the variability of the data and, as a result, reducing the standard deviation, may require improving the accuracy of numerical data measurements or modifying the design of the survey.

> **TIP**
>
> Experimental or laboratory data may also be subject to a margin of error due to inaccuracies in observation and measurement. The SAT may ask questions about situations that involve a margin of error, but it will *not* ask you to calculate it.

Multiple-Choice

1. Which survey is most likely to have the *least* bias?

   (A) surveying a sample of people leaving a movie theater to determine which flavor of ice cream is the most popular
   (B) surveying the members of a football team to determine the most watched TV sport
   (C) surveying a sample of people leaving a library to determine the average number of books a person reads in a year
   (D) surveying a sample of people leaving a gym to determine the average number of hours a person exercises per week

2. Erica is conducting a survey about the proposed increase in the sports budget in the Hometown School District. Which survey method would likely contain the *most* bias?

   (A) Erica asks every third person entering the Hometown Grocery Store
   (B) Erica asks every third person leaving the Hometown Shopping Mall this weekend
   (C) Erica asks every fifth student entering Hometown High School on Monday morning
   (D) Erica asks every fifth person leaving Saturday's Hometown High School football game

Ages of People in Survey on Driving Habits

Age Group	Number of Drivers
16–25	150
26–35	129
36–45	33
46–55	57
56–65	31

3. The table above summarizes the number of people by age group who were included in a survey of driving habits. Which of the following statements is true?

   (A) The survey was not biased since different age groups were included.
   (B) The survey was biased because individuals 36 and older were underrepresented.
   (C) The survey was biased because it did not differentiate between males and females.
   (D) The survey was not biased since a large number of drivers were polled.

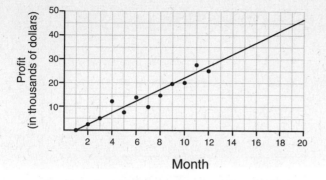

Month

4. The scatterplot above shows the profit, by month, for a new company for the first year of operation. A line of best fit is also shown. Using this line, by what dollar amount did the profit in the 18th month exceed the profit in the 13th month?

(A) $5,000
(B) $7,750
(C) $12,500
(D) $15,000

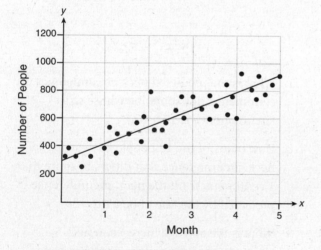

Month

Questions 5–7 refer to the scatterplot above.

A new fitness class was started at several fitness clubs owned by the same company. The scatterplot shows the total number of people attending the class during the first five months in which the class was offered. The line of best fit is drawn.

5. For month 4, the predicted number of people attending the class was approximately what percent greater than the actual number of people attending the class?

(A) 15%
(B) 20%
(C) 30%
(D) 36%

6. During the five-month period, the average increase in the number of people attending the class per month is closest to which of the following?

(A) 80
(B) 100
(C) 120
(D) 140

7. At the beginning of which month did the actual number of people attending the class differ by the greatest amount from the number predicted by the line of best fit?

(A) month 2
(B) month 3
(C) month 4
(D) month 5

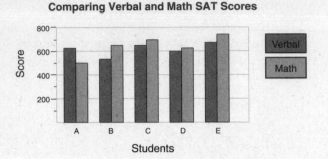

8. The bar graph above shows the verbal and math SAT scores for five students labeled A through E. If a scatterplot of the data in the bar graph is made such that the math SAT score for each student is plotted along the *x*-axis and their verbal SAT score is plotted along the *y*-axis, how many of the data points would lie above the line $y = x$?

(A) 1
(B) 2
(C) 3
(D) 4

Grid-In

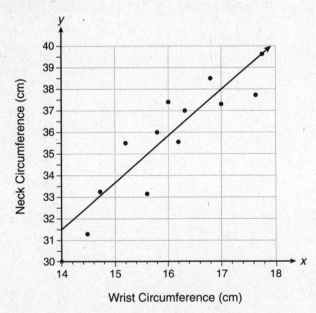

Wrist Circumference (cm)

1. What is the predicted neck circumference, in centimeters, for someone whose wrist circumference is 17.0 cm?

2. How many of the 12 people have an actual neck circumference that differs by more than 1 centimeter from the neck circumference predicted by the line of best fit?

3. What is the average increase in neck circumference per centimeter increase in wrist circumference, correct to the *nearest tenth*?

Questions 1–3 refer to the above scatterplot, which shows wrist and neck circumference measurements, in centimeters, for 12 people. The line of best fit is drawn.

> ### OVERVIEW
>
> A **statistic** is a single number used to describe a set of data values. The *mean*, *median*, and *mode* are statistics that measure *central tendency* by describing the central location of the data within the set. The *range* and *standard deviation* are statistics that describe the extent to which the data are spread out. The SAT may ask you to calculate the mean, median, mode, and range, but not the standard deviation. You may be asked a question, however, that asks you to compare the standard deviations of two sets of data based on how spread out the individual data values in each set appear to be from the means.

FINDING THE AVERAGE (ARITHMETIC MEAN)

To find the average of a set of n numbers, add all the numbers together and then divide the sum by n. Thus,

$$\text{Average} = \frac{\text{Sum of the } n \text{ values}}{n}$$

For example, the average of three exam grades of 70, 80, and 72 is 74 since

$$\text{Average} = \frac{70 + 80 + 72}{3} = \frac{222}{3} = 74$$

FINDING AN UNKNOWN NUMBER WHEN AN AVERAGE IS GIVEN

If you know the average of a set of n numbers, you can find the sum of the n values by using this relationship:

$$\text{Sum of the } n \text{ values} = \text{Average} \times n$$

➡ Example

The average of a set of four numbers is 78. If three of the numbers in the set are 71, 74, and 83, what is the fourth number?

Solution

If the average of four numbers is 78, the sum of the four numbers is $78 \times 4 = 312$. The sum of the three given numbers is $71 + 74 + 83 = 228$. Since $312 - 228 = 84$, the fourth number is **84**.

➡ Example

The average of w, x, y, and z is 31. If the average of w and y is 24, what is the average of x and z?

Solution 1

- Since $\dfrac{w + x + y + z}{4} = 31$, $w + x + y + z = 4 \times 31 = 124$.

- Since $\frac{w + y}{2} = 24$, $w + y = 2 \times 24 = 48$.

- Thus, $x + z + 48 = 124$, so $x + z = 76$.

- Since $x + z = 76$, $\frac{x + z}{2} = \frac{76}{2} = 38$.

Hence, the average of x and z is **38**.

Solution 2

Since the average of two of the four numbers is 7 (= 31 − 24) less than the average of the four numbers, the average of the other two numbers must be 7 more than the average of the four numbers. Hence, the average of x and z is 31 + 7 = 38.

FINDING THE WEIGHTED AVERAGE

A **weighted average** is the average of two or more sets of numbers that do not contain the same number of values. To find the weighted average of two or more sets of numbers, proceed as follows:

- Multiply the average of each set by the number of values in that set, and then add the products.
- Divide the sum of the products by the total number of values in all of the sets.

➡ Example

In a class, 18 students had an average midterm exam grade of 85 and the 12 remaining students had an average midterm exam grade of 90. What is the average midterm exam grade of the entire class?

Solution

$$\text{Weighted average} = \frac{\overbrace{(\text{Average} \times \text{Number})}^{\text{Group 1}} + \overbrace{(\text{Average} \times \text{Number})}^{\text{Group 2}}}{\text{Total number of students}}$$

$$= \frac{(85 \times 18) + (90 \times 12)}{30}$$

$$= \frac{1{,}530 \quad + 1{,}080}{30}$$

$$= \frac{2{,}610}{30}$$

$$= 87$$

FINDING THE MEDIAN

To find the median of a set of numbers, first arrange the numbers in size order.
- If a set contains an odd number of values, the median is the middle value. For example, the median of the set of numbers

$$8, 12, 15, \underset{\text{Median}}{17}, 19, 20, 25$$

is 17 since the number of values in the set that are less than 17 is the same as the number of values in the set that are greater than 17.

- If a set contains an even number of values, the median is the average (arithmetic mean) of the two middle values. For example, the set of numbers

$$10, 20, 24, 30, 40, 50$$

contains six values. Since the two middle values in the set are 24 and 30, the median is their average:

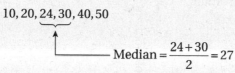

$$\text{Median} = \frac{24+30}{2} = 27$$

FINDING THE MODE

The mode of the set

$$7, 2, 6, 3, 2, 6, 7, 3, 9, 6$$

is 6, since 6 appears more times than any other number in the set.

RANGE AND STANDARD DEVIATION

The *range* and *standard deviation* provide information about how spread out the data are.

- The **range** is the difference between the greatest and smallest data values.
- The **standard deviation** is a statistic that measures how far apart the individual data scores are from the mean. A set of data scores in which the data values are clustered around the mean will have a smaller standard deviation than a data set in which the individual data values are more spread out and further from the mean.

TRANSFORMING AN ENTIRE SET OF DATA

If each score in a set of data values is changed in the same way, you can predict how the different statistical measures are affected without recalculating these statistics.

- When each value in a set of data is increased (or decreased) by the same nonzero number, then the mean, median, and mode are each increased (or decreased) by that number. The *range*, which is the difference between the highest and lowest values in the data set, is *not* affected. If the mean of a set of test scores is 78 and *each* test score is increased by 6, then the mean of the revised set of test scores is 78 + 6 = 84.
- When each value in a set of data is multiplied (or divided) by the same nonzero number, then the mean, median, mode, and range are each multiplied (or divided) by that number. Suppose the mean number of music CDs that five friends own is 13. Mary predicts that six months from now each of the five friends will own twice as many music CDs as they do now. Based on Mary's prediction, the mean number of music CDs they will own is 13 × 2 = 26.

Multiple-Choice

1. The average (arithmetic mean) of a set of seven numbers is 81. If one of the numbers is discarded, the average of the remaining numbers is 78. What is the value of the number that was discarded?

   (A) 98
   (B) 99
   (C) 100
   (D) 101

2. The arithmetic mean of a set of 20 test scores is represented by x. If each score is increased by y points, which expression represents the arithmetic mean of the revised set of test scores?

   (A) $x + y$

   (B) $x + 20y$

   (C) $x + \dfrac{y}{20}$

   (D) $\dfrac{x + y}{20}$

3. What is the area of the circle whose radius is the average of the radii of two circles with areas of 16π and 100π?

   (A) 25π
   (B) 36π
   (C) 49π
   (D) 64π

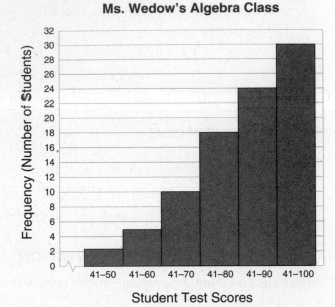

Ms. Wedow's Algebra Class

Frequency (Number of Students) vs. Student Test Scores

4. The diagram above shows a graph of the students' test scores in Ms. Wedow's algebra class. Which *ten-point* interval contains the median?

   (A) 61–70
   (B) 71–80
   (C) 81–90
   (D) 91–100

5. If k is a positive integer, which of the following represents the average of 3^k and 3^{k+2}?

   (A) $\dfrac{1}{2} \cdot 3^{k+1}$

   (B) $5 \cdot 3^k$

   (C) $6^{\frac{3}{2}k}$

   (D) $\dfrac{1}{2} \cdot 3^{3k}$

6. When x is subtracted from $2y$, the difference is equal to the average of x and y. What is the value of $\dfrac{x}{y}$?

    (A) $\dfrac{1}{2}$

    (B) $\dfrac{2}{3}$

    (C) 1

    (D) $\dfrac{3}{2}$

7. If the average of x, y, and z is 32 and the average of y and z is 27, what is the average of x and $2x$?

    (A) 42
    (B) 45
    (C) 48
    (D) 63

Company 1

Worker's Age in Years	Salary in Dollars
25	30,000
27	32,000
28	35,000
33	38,000

Company 2

Worker's Age in Years	Salary in Dollars
25	29,000
28	35,500
29	37,000
31	65,000

8. Which of the following statements is true about the data in the tables above?

    I. The mean salaries for both companies are greater than $35,000.
    II. The mean age of workers in Company 1 is greater than the mean age of workers in Company 2.
    III. The salary range in Company 2 is greater than the salary range in Company 1.

    (A) I only
    (B) III only
    (C) I and II only
    (D) II and III only

9. A man drove a car at an average rate of speed of 45 miles per hour for the first 3 hours of a 7-hour car trip. If the average rate of speed for the entire trip was 53 miles per hour, what was the average rate of speed in miles per hour for the remaining part of the trip?

    (A) 50
    (B) 55
    (C) 57
    (D) 59

10. In a set of n data values, m represents the median. If each number in the set is decreased by 3, which expression represents the median of the revised set of data values?

    (A) m
    (B) $m - 3$
    (C) $m - \dfrac{3}{n}$
    (D) $\dfrac{m - 3}{n}$

11. Susan received grades of 78, 93, 82, and 76 on four math exams. What is the lowest score she can receive on her next math exam and have an average of at least 85 on the five exams?

    (A) 96
    (B) 94
    (C) 92
    (D) 90

12. What is the average of $(x + y)^2$ and $(x - y)^2$?

    (A) $\dfrac{x + y}{2}$

    (B) xy

    (C) $x^2 - y^2$

    (D) $x^2 + y^2$

13. The average of the test scores of a group of x students is 76 and the average of the test scores of a group of y students is 90. When the scores of the two groups of students are combined, the average test score is 85. What is the value of $\frac{x}{y}$?

(A) $\frac{4}{7}$

(B) $\frac{5}{9}$

(C) $\frac{2}{3}$

(D) $\frac{7}{4}$

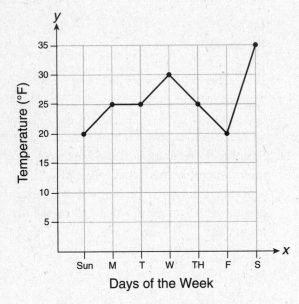

Days of the Week

14. The graph above shows the average daily temperature during a particular week in January in a certain city. Which statement best describes the temperature data in the graph above?

(A) median = mean
(B) mean < mode
(C) median = mode
(D) mean = mode

15. The average of a, b, c, d, and e is 28. If the average of a, c, and e is 24, what is the average of b and d?

(A) 31
(B) 32
(C) 33
(D) 34

16. If $2a + b = 7$ and $b + 2c = 23$, what is the average of a, b, and c?

(A) 5
(B) 7.5
(C) 15
(D) 12.25

Minutes	14	15	16	17	18	19	20
Number of Students	5	3	x	5	2	10	1

17. The number of minutes students took to complete a quiz is summarized in the table above. If the mean number of minutes was 17, which equation could be used to calculate x?

(A) $17 = \dfrac{119 + x}{x}$

(B) $17 = \dfrac{119 + 16x}{x}$

(C) $17 = \dfrac{446 + x}{26 + x}$

(D) $17 = \dfrac{446 + 16x}{26 + x}$

18. The average of a, b, c, and d is p. If the average of a and c is q, what is the average of b and d in terms of p and q?

(A) $2p + q$
(B) $2p - q$
(C) $2q + p$
(D) $2q - p$

19. The lowest value in a set of ordered scores is x and the highest value is y. If each score is increased by k, then which of the following must be true of the revised set of scores?

I.   The mean is increased by k.
II. The range is k.
III. The median remains unchanged.

(A) I only
(B) II only
(C) I and III
(D) II and III

Player's Annual Salaries (millions of dollars)					
0.5	0.5	0.6	0.7	0.75	0.8
1.0	1.0	1.1	1.25	1.3	1.4
1.4	1.8	2.5	3.7	3.8	4
4.2	4.6	5.1	6	6.3	7.2

21–22 The table above shows the annual salaries for the 24 members of a professional sports team in terms of millions of dollars.

21. If each player's salary is increased by 10%, which of the following statistics does *not* increase by 10%?

(A) median
(B) mean
(C) mode
(D) all increase by 10%

22. The team signs an additional player to a contract with an annual salary of 7.5 million dollars per year, which brings the sum of the salaries of the 25 players to 69 million dollars. By what amount, in dollars, does the mean increase?

(A) 197,500
(B) 256,250
(C) 300,000
(D) It cannot be determined.

Class X

Grade	Frequency
A	4
B	11
C	3
D	2
F	1

Class Y

Grade	Frequency
A	6
B	4
C	2
D	6
F	3

20. The tables above give the distribution of grades for 21 students in two different college mathematics classes. For purposes of making statistical calculations, A = 4, B = 3, C = 2, D = 1, and F = 0. Which of the following statements is true about the data shown for these two classes?

I.   The standard deviation of grades is greater for class X.
II. The standard deviation of grades is greater for class Y.
III. The median letter grade is the same for classes X and Y.

(A) I only
(B) II only
(C) I and III only
(D) II and III only

Grid-In

1. The average of r and s is 7.5, and the average of r, s and t is 11. What is the value of t?

2. If the average of x, y, and z is 12, what is the average of $3x$, $3y$, and $3z$?

3. In order to compensate for a difficult midterm exam, Danielle's mathematics teacher adjusted each of the 25 students' midterm exam scores by replacing it by one-half of the original score increased by 50. If the mean of the revised set of midterm scores is 82, what was the mean of the original set of scores?

4. On a test that has a normal distribution of scores, a score of 58 falls two standard deviations below the mean, and a score of 85 is one standard deviation above the mean. What is the mean score of this test?

NOTE: See pages 422–441 for worked out solutions.

Lesson 4–1

1. C	5. A	9. C	13. B	17. A	3. 12
2. B	6. C	10. C	14. C	GRID-IN	4. 5/9
3. C	7. B	11. C	15. C	1. 87.5	
4. C	8. D	12. C	16. B	2. 50	

Lesson 4–2

1. C	6. D	11. C	16. C	21. B	2. 15
2. D	7. A	12. C	17. A	22. C	3. 4
3. C	8. D	13. D	18. B	23. C	4. 16
4. D	9. C	14. D	19. C	GRID-IN	5. 2/9
5. C	10. B	15. A	20. A	1. 12	

Lesson 4–3

1. C	4. B	7. D	10. B	13. A	2. 12
2. A	5. D	8. D	11. D	GRID-IN	3. 52
3. C	6. A	9. C	12. B	1. 8	4. 12

Lesson 4–4

1. C	4. C	7. B	2. 8.64	5. 8,512
2. D	5. B	GRID-IN	3. 332	
3. D	6. A	1. 27	4. 12	

Lesson 4–5

1. C	6. C	11. C	16. A	21. B	4. 82
2. C	7. D	12. C	17. C	GRID-IN	5. 2021
3. B	8. D	13. B	18. D	1. 14.5	6. 20
4. C	9. B	14. C	19. B	2. 140	
5. B	10. B	15. A	20. D	3. 135	

Lesson 4-6

1. B	6. A	11. D	16. B	3. 5	8. 15
2. A	7. A	12. B	17. C	4. 7/9	9. 3/4
3. C	8. D	13. D	GRID-IN	5. 400	
4. D	9. D	14. A	1. 44	6. .4	
5. D	10. D	15. A	2. 204	7. 56	

Lesson 4-7

1. A	3. B	5. C	7. A	GRID-IN	2. 5
2. D	4. C	6. C	8. A	1. 38	3. 2.2

Lesson 4-8

1. B	6. C	11. A	16. A	21. B	3. 64
2. B	7. D	12. D	17. D	22. A	4. 76
3. C	8. B	13. B	18. B	GRID-IN	
4. B	9. D	14. C	19. A	1. 18	
5. B	10. B	15. D	20. B	2. 36	

Passport to Advanced Math

5

This chapter extends the facts, concepts, and skills covered in the Heart of Algebra chapter. Particular emphasis is placed on analyzing and manipulating more complicated algebraic expressions, equations, and functions. "Passport to Advanced Math" represents the third of the four major mathematics content groups tested by the redesigned SAT.

OVERVIEW

In order to give meaning to expressions such as 4^{-1} or 4^0 or $4^{\frac{1}{2}}$, the rules for exponents can be extended to include exponents that are rational numbers. Equations may contain variables with rational exponents or variables underneath a radical sign.

ZERO, NEGATIVE, AND FRACTIONAL EXPONENTS

The rules for working with zero, negative, and fractional exponents are summarized in Table 5.1.

Table 5.1 Rules for Rational Exponents

Zero Exponent Rule $(x \neq 0)$	Negarive Exponent Rule $(x \neq 0)$	Fractional Exponent Rule $(x \geq 0$ when n is even$)$
$x^0 = 1$	• $x^{-a} = \dfrac{1}{x^a}$ • $\dfrac{1}{x^{-a}} = x^a$	• $\dfrac{1}{x^n} = \sqrt[n]{x}$ • $x^{\frac{a}{n}} = \sqrt[n]{x^a} = \left(\sqrt[n]{x}\right)^a$
EXAMPLE: $(2x)^0 = 1$	EXAMPLE: $\dfrac{a^{-4}}{b^{-2}c^5} = \dfrac{b^2}{a^4c^5}$	EXAMPLE: $(-27)^{\frac{2}{3}} = \left(\sqrt[3]{-27}\right)^2 = (-3)^2 = 9$

MATH REFERENCE FACT

$x^{\frac{1}{2}} = \sqrt{x}$, provided $x \geq 0$

Here are some examples:

- $3x^{-2} = \dfrac{3}{x^2}$

- $\dfrac{x^5 y^2}{\left(xy^3\right)^2} = \dfrac{x^5 y^2}{x^2 y^6} = \dfrac{x^3}{y^4}$ or $x^3 y^{-4}$

- $\dfrac{\sqrt[3]{x}}{\sqrt{x}} = \dfrac{x^{\frac{1}{3}}}{x^{\frac{1}{2}}} = x^{-\frac{1}{6}}$ or $\dfrac{1}{x^{\frac{1}{6}}}$

- $8^{\frac{4}{3}} = \left(\sqrt[3]{8}\right)^4 = (2)^4 = 16$

- $\dfrac{bc^{-5}}{2b^{-3}} = \dfrac{b \cdot b^3}{2c^5} = \dfrac{b^4}{2c^5}$

- $\dfrac{x^2}{\sqrt[3]{x}} = \dfrac{x^2}{x^{\frac{1}{3}}} = x^{2-\frac{1}{3}} = x^{\frac{5}{3}}$ or $\sqrt[3]{x^5}$ or $\left(\sqrt[3]{x}\right)^5$

➡ **Example**

If $x > 1$ and $\dfrac{\sqrt{x^3}}{x^2} = x^n$, what is the value of n?

(A) $-\dfrac{3}{2}$

(B) -1

(C) $-\dfrac{1}{2}$

(D) $\dfrac{1}{2}$

Solution

- Rewrite the radical using the exponent rule for equare roots:

$$\frac{\sqrt{x^3}}{x^2} = x^n$$

$$\frac{\left(x^3\right)^{\frac{1}{2}}}{x^2} = x^n$$

$$\frac{x^{\frac{3}{2}}}{x^2} = x^n$$

- Use the quotient law of exponents:

$$x^{\frac{3}{2}-2} = x^n$$

- Simplify the exponent:

$$x^{\frac{3}{2}-\frac{4}{2}} = x^n$$

$$x^{-\frac{1}{2}} = x^n$$

$$n = -\frac{1}{2}$$

The correct choice is **(C)**.

RADICAL EQUATIONS

If an equation contains the square root of a variable term, isolate the radical in the usual way. Eliminate the radical by raising both sides of the equation to the second power. To solve $\sqrt{3x+1} - 2 = 3$ perform the following steps:

1. Isolate the radical:

$$\sqrt{3x+1} = 5$$

2. Raise both sides of the equation to the second power:

$$\left(\sqrt{3x+1}\right)^2 = (5)^2$$

3. Simplify:

$$3x + 1 = 25$$

$$\frac{3x}{3} = \frac{24}{3}$$

$$x = 8$$

> **MATH REFERENCE FACT**
>
> To solve an equation of the form $\sqrt[n]{x} = c$, raise both sides of the equation to the nth power:
>
> $$\left(\sqrt[n]{x}\right)^n = (c)^n \text{ so } x = c^n$$

EQUATIONS WITH FRACTIONAL EXPONENTS

Since the reciprocal of $\frac{3}{2}$ is $\frac{2}{3}$, raising both sides of the equation $x^{\frac{3}{2}} = 8$ to the $\frac{2}{3}$ power creates an equivalent equation in which the power of x is 1:

$$x^{\frac{3}{2} \cdot \frac{2}{3}} = (8)^{\frac{2}{3}}$$

$$x^1 = \left(\sqrt[3]{8}\right)^2$$

$$x = (2)^2 = 4$$

EXPONENTIAL EQUATIONS

If an equation contains a variable in an exponent, rewrite each side as a power of the same base. Then set the exponents equal to each other. For example, solve $8^{x-2} = 4^{x+1}$ as follows:

1. Rewrite each side as a power of 2: $(2^3)^{(x-2)} = (2^2)^{(x+1)}$
2. Simplify the exponents: $2^{3(x-2)} = 2^{2(x+1)}$
3. Set the exponents equal: $3(x-2) = 2(x+1)$

$$3x - 6 = 2x + 2$$

$$3x - 2x = 2 + 6$$

$$x = 8$$

Multiple-Choice

1. Which of the following is equal to $b^{-\frac{1}{2}}$ for all values of b for which the expression is defined?

   (A) $\dfrac{b}{b^2}$

   (B) $\dfrac{\sqrt{b}}{b}$

   (C) $\dfrac{1}{\sqrt{2b}}$

   (D) $\dfrac{1}{2}b$

2. Which expression is equivalent to $\left(9x^2y^6\right)^{-\frac{1}{2}}$?

   (A) $\dfrac{1}{3xy^3}$

   (B) $3xy^3$

   (C) $\dfrac{3}{xy^3}$

   (D) $\dfrac{xy^3}{3}$

3. If $4^y + 4^y + 4^y + 4^y = 16^x$, then $y =$

   (A) $2x - 1$
   (B) $2x + 1$
   (C) $x - 2$
   (D) $x + 2$

4. If $\sqrt{m} = 2p$, then $m^{\frac{3}{2}} =$

   (A) $\dfrac{p}{3}$

   (B) $2p^2$
   (C) $6p^3$
   (D) $8p^3$

5. If $3^x = 81$ and $2^{x+y} = 64$, then $\dfrac{x}{y} =$

   (A) 1

   (B) $\dfrac{3}{2}$

   (C) 2

   (D) $\dfrac{5}{2}$

6. Which of the following is equal to $y^{\frac{3}{2}}$ for all values of y for which the expression is defined?

   (A) $\sqrt[3]{y^2}$

   (B) $\sqrt{y^3}$

   (C) $\sqrt[3]{y^{\frac{1}{2}}}$

   (D) $3\sqrt{y}$

7. Which expression is equivalent to $\dfrac{(2xy)^{-2}}{4y^{-5}}$?

   (A) $-\dfrac{y^3}{x^2}$

   (B) $-\dfrac{y^3}{16x^2}$

   (C) $\dfrac{y^3}{x^2}$

   (D) $\dfrac{y^3}{16x^2}$

8. If $10^k = 64$, what is the value of $10^{\frac{k}{2}+1}$?

   (A) 18
   (B) 42
   (C) 80
   (D) 81

9. If x is a positive integer greater than 1, how much greater than x^2 is $x^{\frac{5}{2}}$?

(A) $x^2\left(1-x^{\frac{1}{2}}\right)$

(B) $x^{-\frac{1}{2}}$

(C) $x^2\left(x^{\frac{1}{2}}-1\right)$

(D) $x^{\frac{1}{2}}$

10. The expression $\dfrac{x^2}{\sqrt{x^3}}$ is equivalent to

(A) $\sqrt[3]{x}$

(B) $\dfrac{1}{\sqrt{x}}$

(C) $\sqrt{x}$

(D) $\dfrac{1}{\sqrt[3]{x^2}}$

11. If n and p are positive integers such that $8(2^p) = 4^n$, what is n in terms of p?

(A) $\dfrac{p+2}{3}$

(B) $\dfrac{2p}{3}$

(C) $\dfrac{p+3}{2}$

(D) $\dfrac{3p}{2}$

$$2\sqrt{x-k} = x - 6$$

12. If $k = 3$, what is the solution of the equation above?

(A) {4, 12}

(B) {3}

(C) {4}

(D) {12}

13. When $x^{-1} - 1$ is divided by $x - 1$, the quotient is

(A) -1

(B) $-\dfrac{1}{x}$

(C) $\dfrac{1}{x^2}$

(D) $\dfrac{1}{(x-1)^2}$

14. If n is a negative integer, which statement is *always* true?

(A) $6n^{-2} < 4n^{-1}$

(B) $\dfrac{n}{4} > -6n^{-1}$

(C) $6n^{-1} < 4n^{-1}$

(D) $4n^{-1} > (6n)^{-1}$

$$g(x) = a\sqrt{a(1-x)}$$

15. Function g is defined by the equation above. If $g(-8) = 375$, what is the value of a?

(A) 25

(B) 75

(C) 125

(D) 625

16. If $27^x = 9^{y-1}$, then

(A) $y = \dfrac{3}{2}x + 1$

(B) $y = \dfrac{3}{2}x + 2$

(C) $y = \dfrac{3}{2}x + \dfrac{1}{2}$

(D) $y = \dfrac{1}{2}x + \dfrac{2}{3}$

Grid-In

$$\sqrt{3p^2 - 11} - x = 0$$

1. If $p > 0$ and $x = 8$ in the equation above, what is the value of p?

2. If $x^{-\frac{1}{2}} = \frac{1}{8}$, what is the value of $x^{\frac{2}{3}}$?

3. If y is not equal to 0, what is the value of $\dfrac{6(2y)^{-2}}{(3y)^{-2}}$?

4. If $(2rs)^{-1} = 3s^{-2}$, what is the value of $\dfrac{r}{s}$?

5. If m and p are positive integers and $\left(2\sqrt{2}\right)^m = 32^p$, what is the value of $\dfrac{p}{m}$?

6. If a, b, and c are positive numbers such that $\sqrt{\dfrac{a}{b}} = 8c$ and $ac = b$, what is the value of c?

7. If $k = 8\sqrt{2}$ and $\dfrac{1}{2}k = \sqrt{3h}$, what is the value of h?

8. If $64^{2n+1} = 16^{4n-1}$, what is the value of n?

$$\frac{\sqrt[3]{a^8}}{\left(\sqrt{a}\right)^3} = a^x, \text{ where } a > 1$$

9. In the equation above, what is the value of x?

10. A meteorologist estimates how long a passing storm will last by using the function $t(d) = 0.08d^{\frac{3}{2}}$, where d is the diameter of the storm, in miles, and t is the time, in hours. If the storm lasts 16.2 minutes, find its diameter, in miles.

> ### OVERVIEW
> This section extends some previously covered algebraic skills related to factors, equations, and more complicated algebraic expressions.

FACTORING BY GROUPING PAIRS OF TERMS

Some polynomials with four terms can be factored by grouping appropriate pairs of terms together and then factoring out a binomial that is common to both pairs of terms.

- $ab - 4b + 3a - 12 = (ab - 4b) + (3a - 12)$
$$= b(a - 4) + 3(a - 4)$$
$$= (a - 4)(b + 3)$$

- $x^3 + 5x^2 - 9x - 45 = x^2(x + 5) - 9(x + 5)$
$$= (x + 5)(x^2 - 9)$$
$$= (x + 5)(x + 3)(x - 3)$$

ZEROS OF A FUNCTION

A **zero** of a function is any value of the variable that makes the function evaluate to zero. The zeros of a function correspond to the x-intercepts of the graph of the function.

> **MATH REFERENCE FACT**
>
> The following statements have the same meaning as "c is a zero of function f":
>
> - $x = c$ is a root or solution of the equation $f(x) = 0$.
> - $x - c$ is a factor of $f(x)$.
> - The graph of $y = f(x)$ intersects the x-axis at $(c, 0)$.

➡ Example

$$f(x) = 2x^3 - 5x^2 - 8x + 20$$

What are the zeros of function f defined by the above equation?

Solution

To find the zeros of function f, find the solutions of the equation $f(x) = 0$:

$$(2x^3 - 5x^2) + (-8x + 20) = 0$$
$$x^2(2x - 5) - 4(2x - 5) = 0$$
$$(2x - 5)(x^2 - 4) = 0$$
$$(2x - 5)(x - 2)(x + 2) = 0$$

Setting each factor equal to 0 yields $x = -2$, 2, and $\frac{5}{2}$.

The zeros of the function are **−2, 2, and $\frac{5}{2}$**. The zeros of function f correspond to the x-intercepts of its graph in the xy-plane.

➡ Example

$$f(x) = 3x^3 + kx^2 - 32x + 28$$

The function f is defined by the equation above where k is a nonzero constant. In the xy-plane, the graph of f intersects the x-axis at three points: $(-2, 0)$, $\left(\dfrac{3}{2}, 0\right)$, and $(c, 0)$. What is the value of k?

(A) -25
(B) -17
(C) 7
(D) 14

Solution

- Since the x-intercepts correspond to the factors of a polynomial function, $f(x) = (x + 2)(3x - 2)(x - c)$. If the factors of function f are multiplied out, the constant term would be equal to the product of the constant terms in the three factors:

$$(2)(-2)(-c) = 28$$
$$4c = 28$$
$$c = 7$$

- Multiply the factors of $f(x)$ together:

$$\begin{aligned}
f(x) &= (x + 2)(3x - 2)(x - 7) \\
&= (3x^2 + 4x - 4)(x - 7) \\
&= 3x^3 - 17x^2 - 32x + 28
\end{aligned}$$

- Since k is the coefficient of the x^2-term, $k = -17$.

The correct choice is **(B)**.

REMAINDERS AND FACTORS OF POLYNOMIALS

If $f(x) = x^2 - x - 6$, then $x - 3$ divides evenly into $f(x)$ since

$$\frac{x^2 - x - 6}{x - 3} = \frac{(x - 3)(x + 2)}{x - 3} = x + 2 \text{ with a remainder of } 0$$

Also, $f(3) = 3^2 - 3 - 6 = 0$. This illustrates that if $f(x)$ is divisible by $x - c$, then $f(c) = 0$.

REMAINDER AND FACTOR THEOREMS

If a polynomial $f(x)$ is divided by a binomial of the form $x - c$ where c is a constant, then the function value $f(c)$ is equal to the remainder.

- If $f(c) = 0$, then there is zero remainder so $f(x)$ is divisible by $x - c$ or, equivalently, $x - c$ is a factor of $f(x)$. This also means that the graph of $y = f(x)$ intersects the x-axis at $x = c$.
- If $f(c) \neq 0$, then $f(x)$ is *not* divisible by $x - c$ so it is *not* a factor of $f(x)$.

➡ Example

If $x + 3$ is a factor of $f(x) = px^2 + p^2x + 30$ and $p > 0$, what is the value of p?

Solution

Since $x + 3 = x - (-3)$ is a factor of the polynomial, $f(-3) = 0$:

$$f(-3) = p(-3)^2 + p^2(-3) + 30$$

$$0 = 9p - 3p^2 + 30$$

$$3p^2 - 9p - 30 = 0$$

$$p^2 - 3p - 10 = 0$$

$$(p - 5)(p + 2) = 0$$

$$p = 5 \quad or \quad p = -2 \leftarrow \text{Reject since } p > 0$$

The value of p is **5**.

SIMPLIFYING COMPLEX FRACTIONS

A **complex fraction** is a fraction in which its numerator, denominator, or both contain other fractions. To simplify a complex fraction, multiply its numerator and its denominator by the lowest common multiple (LCM) of all its denominators. To simplify

$$\frac{1 + \dfrac{2}{x}}{1 - \dfrac{4}{x^2}}$$

multiply the numerator and denominator by x^2:

$$\frac{1 + \dfrac{2}{x}}{1 - \dfrac{4}{x^2}} = \frac{x^2\left(1 + \dfrac{2}{x}\right)}{x^2\left(1 - \dfrac{4}{x^2}\right)}$$

$$= \frac{x^2 + 2x}{x^2 - 4}$$

$$= \frac{x\cancel{(x + 2)}}{\cancel{(x + 2)}(x - 2)}$$

$$= \frac{x}{x - 2}$$

➡ Example

When $x^{-1} - 1$ is divided by $x - 1$, the quotient is

(A) -1

(B) $-\dfrac{1}{x}$

(C) $\dfrac{1}{x^2}$

(D) $\dfrac{1}{(x-1)^2}$

Solution

$$\frac{x^{-1} - 1}{x - 1} = \frac{\dfrac{1}{x} - 1}{x - 1}$$

$$= \frac{x\left(\dfrac{1}{x} - 1\right)}{x(x - 1)}$$

$$= \frac{1 - x}{x(x - 1)}$$

$$= \frac{\overset{-1}{\cancel{1 - x}}}{x\cancel{(x - 1)}}$$

$$= -\frac{1}{x}$$

The correct choice is **(B)**.

SOLVING FRACTIONAL EQUATIONS

To solve an equation that contains algebraic fractions, clear the equation of its fractions by multiplying each term of the equation by the lowest common multiple of all of the denominators. To find the solution to the equation

$$\frac{4}{3} = \frac{-(3x + 13)}{3x} + \frac{5}{6x}$$

first remove the parentheses by taking the opposite of each term inside the parentheses:

$$\frac{4}{3} = \frac{-3x - 13}{3x} + \frac{5}{6x}$$

Eliminate the fractions by multiplying each term by $6x$, the lowest common multiple of all the denominators:

$$\overset{2x}{\cancel{6x}}\left(\frac{4}{\cancel{3}}\right)=\overset{2}{\cancel{6x}}\left(\frac{-3x-13}{\cancel{3x}}\right)+\overset{1}{\cancel{6x}}\left(\frac{5}{\cancel{6x}}\right)$$

$$2x(4)=2(-3x-13)+5$$

$$8x=-6x-26+5$$

$$8x+6x=-21$$

$$\frac{14x}{14}=\frac{-21}{14}$$

$$x=-\frac{3}{2}$$

➡ Example

$$\frac{2(n-1)}{3}-\frac{3(n+1)}{4}=\frac{n+3}{2}$$

In the equation above, what is the value of n^2?

Solution

Eliminate the fractional terms by multiplying each term of the equation by 12, the lowest common multiple of all the denominators:

$$\overset{4}{\cancel{12}}\left[\frac{2(n-1)}{\cancel{3}}\right]-\overset{3}{\cancel{12}}\left[\frac{3(n+1)}{\cancel{4}}\right]=\overset{6}{\cancel{12}}\left(\frac{n+3}{\cancel{2}}\right)$$

$$4\cdot 2(n-1)-3\cdot 3(n+1)=6(n+3)$$

$$8n-8-9n-9=6n+18$$

$$-n-17=6n+18$$

$$-n-6n=18+17$$

$$\frac{-7n}{-7}=\frac{35}{-7}$$

$$n=-5$$

Hence, $n^2=(-5)^2=\mathbf{25}$.

➡ Example

$$\frac{9}{y+1}+\frac{18}{y^2-1}=1$$

What is a possible solution to the equation above?

Solution

First factor the second denominator. Then eliminate the fractions by multiplying each term of the equation by the lowest common multiple of the denominators:

$$\frac{9}{y+1} + \frac{18}{(y+1)(y-1)} = 1$$

$$[(y+1)(y-1)]\frac{9}{y+1} + [(y+1)(y-1)]\frac{18}{(y+1)(y-1)} = [(y+1)(y-1)] \cdot 1$$

$$9(y-1)+18 = y^2 - 1$$

$$9y - 9 + 18 = y^2 - 1$$

$$y^2 - 9y - 10 = 0$$

$$(y-10)(y+1) = 0$$

$$y - 10 = 0 \quad or \quad y + 1 = 0$$

$$y = 10 \ or \qquad y = -1 \longleftarrow \text{reject!}$$

If $y = -1$, the denominator of the first fraction in the original equation evaluates to 0 so this solution is rejected. The only possible solution is $y = \mathbf{10}$.

Multiple-Choice

1. The polynomial $x^3 - 2x^2 - 9x + 18$ is equivalent to

   (A) $(x - 9)(x - 2)^2$
   (B) $(x - 2)(x - 3)(x + 3)$
   (C) $(x + 3)(x - 2)^2$
   (D) $(x - 2)(x + 2)(x - 3)$

2. When resistors R_1 and R_2 are connected in a parallel electric circuit, the total resistance is $\dfrac{1}{\dfrac{1}{R_1} + \dfrac{1}{R_2}}$. This fraction is equivalent to

   (A) $R_1 + R_2$
   (B) $\dfrac{R_1 + R_2}{R_1 R_2}$
   (C) $\dfrac{R_1}{R_2} + \dfrac{R_2}{R_1}$
   (D) $\dfrac{R_1 R_2}{R_1 + R_2}$

3. In how many different points does the graph of the function $f(x) = x^3 - 2x^2 + x - 2$ intersect the x-axis?

   (A) 0
   (B) 1
   (C) 2
   (D) 3

$$\frac{x^2 + 9x - 22}{x^2 - 121} \div (2 - x)$$

4. The expression above is equivalent to

   (A) $x - 11$
   (B) $\dfrac{1}{x - 11}$
   (C) $11 - x$
   (D) $\dfrac{1}{11 - x}$

5. If $p(x)$ is a polynomial function and $p(4) = 0$, then which statement is true?

   (A) $x + 4$ is a factor of $p(x)$.
   (B) $x - 4$ is a factor of $p(x)$.
   (C) The greatest power of x in $p(x)$ is 4.
   (D) $p(x)$ is divisible by 4.

$$\left(\frac{9}{4}x^2 - 1\right) - \left(\frac{3}{2}x - 1\right)^2$$

6. The expression above is equivalent to

   (A) $3x - 2$
   (B) $-3x$
   (C) $\dfrac{3}{4}x - 2$
   (D) 0

$$\frac{\dfrac{x - y}{y}}{y^{-1} - x^{-1}}$$

7. The expression above is equivalent to

   (A) x
   (B) y
   (C) $\dfrac{1}{y}$
   (D) $-\dfrac{x}{y}$

$$f(x) = 3x^3 - 5x^2 - 48x + 80$$

8. If the zeros of function f defined above are represented by r, s, and t, what is the value of the sum $r + s + t$?

(A) $\dfrac{3}{5}$

(B) $\dfrac{5}{3}$

(C) $\dfrac{17}{3}$

(D) 8

$$\frac{y^3 + 3y^2 - y - 3}{y^2 + 4y + 3}$$

9. The expression above is equivalent to

(A) $y - 1$

(B) $y + 1$

(C) $\dfrac{y-1}{y+3}$

(D) $y^2 - 1$

x	f(x)	g(x)
–3	3	0
–1	0	3
0	–4	4
2	0	–2

10. Several values of x, and the corresponding values for polynomial functions f and g are shown in the table above. Which of the following statements is true?

  I. $f(0) + g(0) = 0$
 II. $f(x)$ is divisible by $x + 2$
III. $g(x)$ is divisible by $x + 3$

(A) I, II, and III

(B) I and II, only

(C) II and III, only

(D) I and III, only

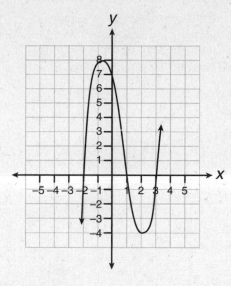

11. Which equation(s) represent(s) the graph above?

  I. $y = (x + 2)(x^2 - 4x - 12)$
 II. $y = (x - 3)(x^2 + x - 2)$
III. $y = (x - 1)(x^2 - 5x - 6)$

(A) I only

(B) II only

(C) I and II

(D) II and III

12. Which of the following functions have zeros –1, 1, and 4?

(A) $f(x) = (x - 4)(1 + x^2)$

(B) $f(x) = (x + 4)(1 - x^2)$

(C) $f(x) = (x - 1)(x^2 - 3x - 4)$

(D) $f(x) = (x - 1)(x^2 + 3x - 4)$

$$\left(\frac{10x^2y}{x^2 + xy}\right) \times \left(\frac{(x+y)^2}{2xy}\right) \div \left(\frac{x^2 - y^2}{y^2}\right)$$

13. Which of the following is equivalent to the expression above?

(A) $\dfrac{5y^2}{x - y}$

(B) $\dfrac{y^2}{x - y}$

(C) $\dfrac{xy}{x - y}$

(D) $\dfrac{x + y}{xy}$

$$f(x) = (2 - 3x)(x + 3) + 4(x^2 - 6)$$

14. What is the sum of the zeros of function f defined by the equation above?

    (A) 3
    (B) 6
    (C) 7
    (D) 11

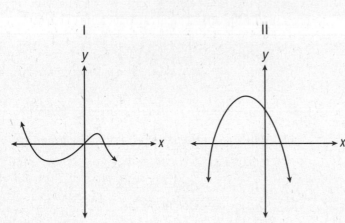

I   II

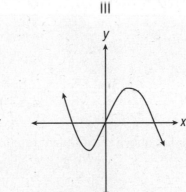

III

15. A polynomial function contains the factors x, $x - 2$, and $x + 5$. Which of the graph(s) above could represent the graph of this function?

    (A) I only
    (B) II only
    (C) III only
    (D) I and III

16. What is the greatest integer value of k for which $f(x) = k$ has exactly 3 real solutions?

    (A) −5
    (B) 0
    (C) 6
    (D) 7

17. What is the best estimate of the remainder when $f(x)$ is divided by $x + 3$?

    (A) −6.0
    (B) 0
    (C) 6.5
    (D) It cannot be determined.

18. What is the maximum number of points a circle whose center is at the origin can intersect the graph of $y = f(x)$?

    (A) 2
    (B) 3
    (C) 4
    (D) 6

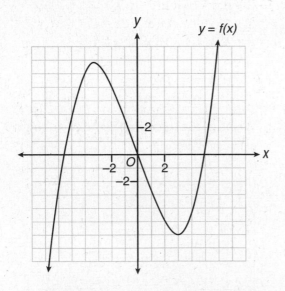

$y = f(x)$

16–18 The graph of polynomial function f is shown above.

$$(y^2 + ky - 3)(y - 4) = y^3 + by^2 + 5y + 12$$

19. In the equation above, k is a nonzero constant. If the equation is true for all values of y, what is the value of k?

    (A) $-\dfrac{1}{2}$

    (B) -2

    (C) 4

    (D) 6

$$\dfrac{16a^4 - 81b^4}{8a^3 + 12a^2b + 18ab^2 + 27b^3}$$

20. Which of the following expressions is equivalent to the expression above?

    (A) $4a^2b + 9ab^2 - a^2b^2$

    (B) $4a^2b - 9ab^2$

    (C) $2a + 3b$

    (D) $2a - 3b$

Grid-In

$$\dfrac{k}{6} + \dfrac{3(1 - k)}{4} = \dfrac{k - 5}{2}$$

1. What is the solution for k in the equation above?

$$\dfrac{3}{2} = \dfrac{-(5m - 3)}{3m} + \dfrac{7}{12m}$$

2. What is the solution for m in the equation above?

$$f(x) = x^3 + 5x^2 - 4x - 20$$

3. How many of the zeros of function f defined by the equation above are located in the interval $-4 \le x \le 4$?

$$\dfrac{t}{t - 3} - \dfrac{t - 2}{2} = \dfrac{5t - 3}{4t - 12}$$

4. If x and y are solutions of the equation above and $y > x$, what is the value of $y - x$?

$$x^3 + 150 = 6x^2 + 25x$$

5. What is the sum of all values of x that satisfy the equation above?

$$p(t) = t^5 - 3t^4 - kt + 7k^2$$

6. In the polynomial function above, k is a nonzero constant. If $p(t)$ is divisible by $t - 3$, what is the value of k?

> ### OVERVIEW
>
> The **imaginary unit** i is defined such that $i^2 = -1$ so that $i = \sqrt{-1}$. An **imaginary number** is the product of a nonzero real number and the imaginary unit i, as in $2i$.

COMPLEX NUMBERS

When an imaginary number is combined with a real number, as in $3 + 2i$, the result is called a *complex number*. A **complex number** is a number that has the form $a + bi$ where a and b are real numbers and $i = \sqrt{-1}$. Since any real number a can be written in the form $a + 0 \cdot i$, the set of complex numbers includes the set of real numbers. For example, 5 is a complex number since $5 = 5 + 0 \cdot i$.

Properties of Complex Numbers

All of the properties of arithmetic that work for real numbers also hold for complex numbers.

- When performing arithmetic operations with complex numbers, terms involving i are treated as if they are monomials:

  1. $3i + 5i = 8i$

  2. $6i - i = 5i$

  3. $i \cdot i^2 = i^{1+2} = i^3$

  4. $\dfrac{i^{13}}{i^4} = i^{13-4} = i^9$

- To add or subtract complex numbers of the form $a + bi$, combine the real parts and then combine the imaginary parts:

  1. $(2 + 3i) + (4 - 5i) = (2 + 4) + (3i - 5i) = 6 - 2i$

  2. $(1 - 2i) - (-4 + 6i) = (1 - 2i) + (4 - 6i)$
  $$= (1 + 4) + (-2i - 6i)$$
  $$= 5 - 8i$$

  3. $\sqrt{-64} + 2\sqrt{-9} = \sqrt{64} \cdot \sqrt{-1} + 2\sqrt{9} \cdot \sqrt{-1}$
  $$= 8i + 6i$$
  $$= 14i$$

  4. $\sqrt{-50} + 4\sqrt{-18} = \left(\sqrt{25} \cdot \sqrt{2} \cdot \sqrt{-1}\right) + \left(4\sqrt{9} \cdot \sqrt{2} \cdot \sqrt{-1}\right)$
  $$= 5\sqrt{2}i \qquad\quad + 12\sqrt{2}i$$
  $$= 17\sqrt{2}\,i$$

- To multiply complex numbers of the form $a + bi$, treat the complex numbers as binomials and use the FOIL method.

$$\overset{\textbf{F}\quad\textbf{O}\quad\textbf{I}\quad\textbf{L}}{(3 + 2i)(5 - 4i) = (5 \cdot 3) + (3)(-4i) + (2i)(5) + (2i)(-4i)}$$
$$= 15 - 12i + 10i - 8i^2$$
$$= 15 - 2i - 8(-1)$$
$$= 23 - 2i$$

Simplifying Higher Powers of i

The first few powers of i are worth remembering:

$$i^0 = 1$$
$$i^1 = i$$
$$i^2 = -1$$
$$i^3 = -i$$

You can simplify a higher power of i by rewriting it so that it is expressed in terms of a power of i^2:

$$i^{35} = i^{34} \cdot i \quad \leftarrow \text{Factor } i^{35} \text{ so that one of its factors has the greatest even exponent}$$
$$= (i^2)^{17} \cdot i \quad \leftarrow \text{Rewrite } i^{34} \text{ as a power of } i^2$$
$$= (-1)^{17} \cdot i \quad \leftarrow \text{Let } i^2 = -1$$
$$= -i \quad \leftarrow \text{Simplify}$$

TIP

In simplifying higher powers of i, make use of these facts: $i^2 = -1$, -1 raised to an even power is 1, and -1 raised to an odd power is -1.

To illustrate further,

$$i^{82} = (i^2)^{41}$$
$$= (-1)^{41}$$
$$= -1$$

and

$$i^{13} = i^{12} \cdot i^1$$
$$= (i^2)^6 \cdot i$$
$$= (-1)^6 \cdot i$$
$$= i$$

COMPLEX CONJUGATES: $a \pm bi$

The numbers $3 + 2i$ and $3 - 2i$ are *complex conjugates*. A pair of complex numbers having the form $a + bi$ and $a - bi$ are **complex conjugates**. The sum and product of a pair of complex conjugates are always positive real numbers. For example,

$$(3 + 2i)(3 - 2i) = 9 - 6i + 6i - 4i^2$$
$$= 9 - 4(-1)$$
$$= 9 + 4$$
$$= 13$$

MATH REFERENCE FACT

The product of two complex conjugates is always a positive real number:

$$(a + bi)(a - bi) = a^2 + b^2$$

➡ Example

Which of the following complex numbers is equivalent to $\dfrac{8-4i}{5+3i}$? (Note: $i = \sqrt{-1}$)

(A) $\dfrac{14}{17} + \dfrac{22}{17}i$

(B) $\dfrac{14}{17} - \dfrac{22}{17}i$

(C) $\dfrac{13}{8} - \dfrac{24}{17}i$

(D) $\dfrac{13}{8} + \dfrac{24}{17}i$

Solution

In order to produce an equivalent fraction with a real number in the denominator, multiply the numerator and the denominator by $5 - 3i$, the complex conjugate of $5 + 3i$:

$$
\begin{aligned}
\frac{8-4i}{5+3i} \cdot \frac{5-3i}{5-3i} &= \frac{(8-4i)(5-3i)}{25-9i^2} \\
&= \frac{40-24i-20i+12i^2}{25+9} \\
&= \frac{40-44i-12}{34} \\
&= \frac{28-44i}{34} \\
&= \frac{2(14-22i)}{34} \\
&= \frac{14-22i}{17} \\
&= \frac{14}{17} - \frac{22}{17}i
\end{aligned}
$$

The correct choice is (**B**).

Multiple-Choice

NOTE: *Unless indicated otherwise, $i = \sqrt{-1}$ for each problem.*

1. Which of the following is equal to $i^{50} + i^{0}$?

   (A) 1
   (B) 2
   (C) −1
   (D) 0

2. Which of the following is equivalent to $2i^2 + 3i^3$?

   (A) −2 − 3i
   (B) 2 − 3i
   (C) −2 + 3i
   (D) 2 + 3i

3. Expressed in simplest form, $2\sqrt{-50} - 3\sqrt{-8}$ is equivalent to

   (A) $16i\sqrt{2}$
   (B) $3i\sqrt{2}$
   (C) $4i\sqrt{2}$
   (D) $-i\sqrt{2}$

4. If $x = 3i$, $y = 2i$, $z = m + i$, and $i = \sqrt{-1}$, then the expression $xy^2z =$

   (A) −12 − 12mi
   (B) −6 − 6mi
   (C) 12 − 12mi
   (D) 6 − 6mi

5. If $g(x) = \left(x\sqrt{1-x}\right)^2$, what is $g(10)$?

   (A) −30
   (B) −900
   (C) 30i
   (D) 900i

6. Which of the following is equal to $(x + i)^2 - (x - i)^2$?

   (A) 0
   (B) −2
   (C) −2 + 4xi
   (D) 4xi

$$i^{13} + i^{18} + i^{31} + n = 0$$

7. In the equation above, what is the value of n in simplest form?

   (A) −i
   (B) −1
   (C) 1
   (D) i

8. Which of the following is equivalent to $2i(xi - 4i^2)$?

   (A) 2x − 8i
   (B) −2x + 8i
   (C) −6xi
   (D) −8xi

9. If $x = 2i$, $y = -4$, $z = 3i$, and $i = \sqrt{-1}$, then $\sqrt{x^3yz} =$

   (A) $4\sqrt{6}i$
   (B) 24i
   (C) $-4\sqrt{6}$
   (D) −24

10. Which of the following is equal to $(13 + 17i)(4 - 9i)$?

   (A) −12
   (B) 116
   (C) 115 − 89i
   (D) 52 − 126i

11. If $(x - yi) + (a + bi) = 2x$ and $i = \sqrt{-1}$, then $(x + yi)(a + bi) =$

(A) $x^2 + y^2$
(B) $x^2 - y^2$
(C) $4x^2 + y^2$
(D) $5x^2$

12. Which of the following complex numbers is equivalent to $\dfrac{3+i}{4-7i}$?

(A) $\dfrac{17}{28}$

(B) $-\dfrac{19}{33} - \dfrac{25}{33}i$

(C) $\dfrac{1}{13} - \dfrac{5}{13}i$

(D) $\dfrac{1}{13} + \dfrac{5}{13}i$

13. In an electrical circuit, the voltage, E, in volts, the current, I, in amps, and the opposition to the flow of current, called impedance, Z, in ohms, are related by the equation $E = IZ$. What is the impedance, in ohms, of an electrical circuit that has a current of $(3 + i)$ amps and a voltage of $(-7 + i)$ volts?

(A) $-2 + i$
(B) $1 - 2i$
(C) $\dfrac{-11}{25} - \dfrac{1}{5}i$
(D) $-\dfrac{16}{25}i$

$$(9 + 2i)(4 - 3i) - (5 - i)(4 - 3i)$$

14. The expression above is equivalent to which of the following?

(A) 7
(B) $14 - 18i$
(C) 25
(D) $16 + 18i$

Grid-In

NOTE: *Unless indicated otherwise, $i = \sqrt{-1}$ for each problem.*

1. What is the value of $\left(\dfrac{1}{2} + i\sqrt{5}\right)\left(\dfrac{1}{2} - i\sqrt{5}\right)$?

$$\left(2 - \sqrt{-25}\right)\left(-7 + \sqrt{-4}\right) = x + yi$$

2. In the equation above, what is the value of y?

3. If $(1 - 3i)(7 + 5i + i^2) = a + bi$, what is the value of $a + b$?

4. If $\dfrac{6 + 4i}{1 - 3i} = a + bi$, what is the value of $a + b$?

$$g(x) = a\sqrt{41 - x^2}$$

5. Function g is defined by the equation above where a is a nonzero real constant. If $g(2i) = \sqrt{5}$, where $i = \sqrt{-1}$, what is the value of a?

OVERVIEW

This section explains how to complete the square and shows how it can be used to solve quadratic equations including those that cannot be solved by factoring.

HOW TO COMPLETE THE SQUARE

Sometimes it is useful to complete the square within a quadratic expression so that it includes the square of a binomial.

- To complete the square of a quadratic trinomial of the form $x^2 + bx + c$, add and then subtract $\left(\dfrac{b}{2}\right)^2$. To complete the square within $x^2 + 10x + 7$, add and then subtract $\left(\dfrac{10}{2}\right)^2 = 25$:

$$x^2 + 10x + 7 = \underbrace{\left(x^2 + 10x + \boxed{25}\right)}_{\text{Perfect square}} + 7 - 25$$

$$= (x + 5)^2 - 18$$

- To complete the square within a quadratic expression having the form $ax^2 + bx + c$, first factor out a from the variable terms:

$$2x^2 - 12x + 5 = 2\left(x^2 - 6x + \boxed{?}\right) + 5$$

$$= 2\left(x^2 - 6x + \boxed{9}\right) + 5 - 18$$

$$= 2(x - 3)^2 - 13$$

Since 2×9 is being added to complete the square, 18 also needs to be subtracted in order to produce an equivalent expression.

SOLVING NONFACTORABLE QUADRATIC EQUATIONS

Any quadratic equation, including those that are not factorable, can be solved by putting them in the form $(x + p)^2 = k$ and then taking the square root of both sides of the equation. To solve $x^2 - 8x + 9 = 0$ by completing the square,

- Rewrite the equation as $x^2 - 8x = -9$ so that only terms involving x are on one side of the equation.

- Complete the square by adding $\left(-\dfrac{8}{2}\right)^2 = 16$ to both sides of the equation:

$$x^2 - 8x + 16 = -9 + 16$$

$$x^2 - 8x + 16 = 7$$

$$(x - 4)^2 = 7$$

- Take the square root of both sides of the equation:

$$\sqrt{(x-4)^2} = \pm\sqrt{7}$$

$$x - 4 = +\sqrt{7} \quad or \quad x - 4 = -\sqrt{7}$$

$$x = 4 + \sqrt{7} \qquad\qquad x = 4 - \sqrt{7}$$

SOLVING A QUADRATIC EQUATION BY COMPLETING THE SQUARE

To solve a quadratic equation by completing the square,

- Rewrite the quadratic equation in the form $x^2 + bx = k$.
- Add the number that completes the square to both sides of the equation.
- Take the square root of both sides of the equation and solve for x.

The Quadratic Formula

If you need to find the solutions to a quadratic equation, you may find it easier to use the quadratic formula.

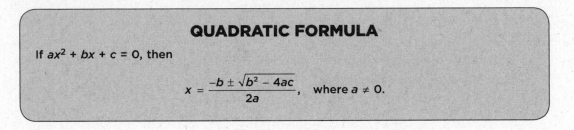

QUADRATIC FORMULA

If $ax^2 + bx + c = 0$, then

$$x = \frac{-b \pm \sqrt{b^2 - 4ac}}{2a}, \quad \text{where } a \neq 0.$$

Since the quadratic formula is *not* included on the SAT formula reference sheet, you should memorize it. The quantity underneath the radical sign, $b^2 - 4ac$, is called the **discriminant**. The discriminant is the part of the quadratic formula that determines whether the solutions are real or not real, equal or unequal, rational or irrational.

- If $b^2 - 4ac < 0$, the solutions are imaginary (not real).
- If $b^2 - 4ac = 0$, the solutions are real, equal, and rational.
- If $b^2 - 4ac > 0$, the solutions are real and unequal. If $b^2 - 4ac$ is a perfect square, then the solutions are rational; otherwise, the solutions are irrational.

➡ Example

What are the solutions to $y^2 + 6y + 7 = 0$?

Solution

Since the quadratic trinomial $y^2 + 6y + 7$ is not factorable, solve by either completing the square or by using the quadratic formula:

METHOD 1: Rewrite the equation as $y^2 + 6y = -7$ and solve it by completing the square.

- Add $\left(\dfrac{6}{2}\right)^2 = 9$ to both sides of the equation:

$$y^2 + 6y + 9 = -7 + 9 \text{ so } (y+3)^2 = 2$$

- Take the square root of both sides of the equation:

$$y + 3 = \pm\sqrt{2}$$

- Solve for y:

$$y = -3 + \sqrt{2} \quad \text{or} \quad y = -3 - \sqrt{2}$$

METHOD 2: Solve using the quadratic formula where $a = 1$, $b = 6$ and $c = 7$:

$$\begin{aligned}
x &= \frac{-b \pm \sqrt{b^2 - 4ac}}{2a} \\
&= \frac{-6 \pm \sqrt{6^2 - 4(1)(7)}}{2(1)} \\
&= \frac{-6 \pm \sqrt{36 - 28}}{2} \\
&= \frac{-6 \pm \sqrt{8}}{2} \\
&= \frac{-6 \pm 2\sqrt{2}}{2} \\
&= \frac{2\left(-3 \pm \sqrt{2}\right)}{2} \\
&= \mathbf{-3 \pm \sqrt{2}}
\end{aligned}$$

➥ Example

Solve $4x^2 - 8x - 3 = 0$ by completing the square.

Solution

- Transpose the constant term, -3, to the right side of the equation and factor out 4 from the left side:

$$4(x^2 - 2x) = 3$$

- Complete the square by adding $\left(\dfrac{-2}{2}\right)^2 = 1$ inside the parentheses. Since this is being multiplied by 4, you must add 4×1 to the right side of the equation:

$$4\left(x^2 - 2x + 1\right) = 3 + 4$$
$$(x - 1)^2 = \frac{7}{4}$$

- Solve for x by taking the square root of both sides of the equation:

$$\sqrt{(x-1)^2} = \pm\sqrt{\frac{7}{4}}$$

$$x - 1 = \frac{\sqrt{7}}{2} \qquad \text{or} \quad x - 1 = -\frac{\sqrt{7}}{2}$$

$$x = 1 + \frac{\sqrt{7}}{2} \qquad\qquad x = 1 - \frac{\sqrt{7}}{2}$$

➥ Example

What is the smallest integral value of k for which the roots of $3x^2 + 8x - k = 0$ are real?

(A) −6
(B) −5
(C) 0
(D) 6

Solution

The roots of a quadratic equation are real when the discriminant is greater than or equal to 0. If $3x^2 + 8x - k = 0$, then $a = 3$, $b = 8$, and $c = -k$:

$$b^2 - 4ac \geq 0$$
$$8^2 - 4(3)(-k) \geq 0$$
$$64 + 12k \geq 0$$
$$12k \geq -64$$
$$k \geq -\frac{64}{12}$$
$$k \geq -5\frac{1}{3}$$

The smallest integral value of k that satisfies the inequality is −5.

The correct choice is **(B)**.

SUM AND PRODUCT OF ROOTS

A quick way of finding the sum or product of the solutions to a quadratic equation without actually solving the equation is to use these formulas:

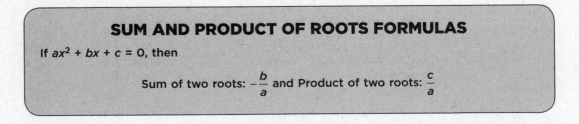

SUM AND PRODUCT OF ROOTS FORMULAS

If $ax^2 + bx + c = 0$, then

Sum of two roots: $-\dfrac{b}{a}$ and Product of two roots: $\dfrac{c}{a}$

➡ Example

By what amount does the product of the solutions of $3x^2 - 10x + 13 = 0$ exceed the sum of its solutions?

Solution

- Sum of the solutions:

$$-\frac{b}{a} = -\left(\frac{-10}{3}\right) = \frac{10}{3}$$

- Product of the solutions:

$$\frac{c}{a} = \frac{13}{3}$$

- Product of the solutions exceeds their sum by $\dfrac{13}{3} - \dfrac{10}{3} = \dfrac{3}{3} = \mathbf{1}$.

Multiple-Choice

1. What are the solutions to $3x^2 - 33 = 18x$?

   (A) $x = 3 \pm 2\sqrt{5}$

   (B) $x = \dfrac{3 \pm \sqrt{5}}{2}$

   (C) $x = 3 \pm 4\sqrt{5}$

   (D) $x = 3 \pm \dfrac{\sqrt{5}}{2}$

2. If the solutions to $2x^2 - 8x - 5 = 0$ are p and q with $p > q$, what is the value of $p - q$?

   (A) $\sqrt{26}$

   (B) $\dfrac{7}{2}$

   (C) $2\sqrt{13}$

   (D) $\dfrac{11}{2}$

$$\frac{x+5}{4} = \frac{1-x}{3x-4}$$

3. If the solutions to the equation above are r and s with $r > s$, what is the value of $r - s$?

   (A) $\sqrt{7}$

   (B) $\dfrac{5}{2}$

   (C) $\sqrt{57}$

   (D) 5

4. If the equation $y = 3x^2 + 18x - 13$ is written in the form $y = a(x - h)^2 + k$, what are the values of h and k?

   (A) $h = 3, k = 14$
   (B) $h = -3, k = -40$
   (C) $h = 3, k = -13$
   (D) $h = -3, k = -22$

$$x^2 + 6x + y^2 - 8y = 56$$

5. If the above equation is written in the form $(x - h)^2 + (y - k)^2 = r^2$, what is the value of r?

   (A) 6
   (B) 8
   (C) 9
   (D) $\sqrt{31}$

$$\frac{4}{x-3} + \frac{2}{x-2} = 2$$

6. If the solutions of the equation above in simplest radical form are $x = a \pm \sqrt{b}$, what are the values of a and b?

   (A) $a = 4, b = 3$
   (B) $a = -4, b = 5$
   (C) $a = 3, b = 5$
   (D) $a = -3, b = 5$

7. Which quadratic equation has $2 + 3i$ and $2 - 3i$ as its solutions?

   (A) $x^2 + 4x - 13 = 0$
   (B) $x^2 - 4x + 13 = 0$
   (C) $x^2 + 13x - 4 = 0$
   (D) $x^2 - 13x - 4 = 0$

8. The equation $ax^2 + 6x - 9 = 0$ will have imaginary roots if

   (A) $a < -1$
   (B) $a \geq -1$
   (C) $a \leq 1$
   (D) $-1 < a < 1$

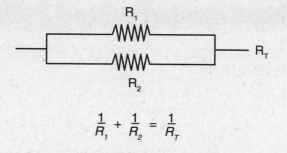

$$\frac{1}{R_1} + \frac{1}{R_2} = \frac{1}{R_T}$$

9. If electrical circuits are hooked up in parallel, the reciprocal of the total resistance in the series is found by adding the reciprocals of each resistance as shown in the diagram above. In a certain circuit, R_2 exceeds the resistance of R_1 by 2 ohms, and the total resistance, R_T, is 1.5 ohms. Which expression represents the number of ohms in R_1?

(A) $\sqrt{13} - 1$

(B) $\sqrt{11} - 1$

(C) $\dfrac{1 + \sqrt{11}}{2}$

(D) $\dfrac{1 + \sqrt{13}}{2}$

10. The amount of water remaining in a certain bathtub as it drains when the plug is pulled is represented by the equation $L = -4t^2 - 8t + 128$, where L represents the number of liters of water in the bathtub and t represents the amount of time, in minutes, since the plug was pulled. Which expression represents the number of minutes it takes for half of the water that was in the bathtub before the plug was pulled to drain?

(A) $-1 + \sqrt{33}$

(B) $-1 + \sqrt{17}$

(C) $\dfrac{-1 + \sqrt{33}}{2}$

(D) $\dfrac{-1 + 2\sqrt{17}}{2}$

OVERVIEW

The graph of the quadratic function $y = ax^2 + bx + c$ is a **U**-shaped curve called a **parabola** with c as its y-intercept.

- If the constant $a > 0$, the parabola opens up and the turning point or vertex of the parabola is the lowest point on the curve.

- If the constant $a < 0$, the parabola opens down and the vertex of the parabola is the highest point on the curve.

VERTEX OF A PARABOLA

A line can be drawn through a parabola that divides it into two mirror image parts. The line of symmetry intersects the parabola at its **vertex**. The vertex is either the highest or the lowest point on the parabola. For the parabola $y = ax^2 + bx + c$,

- If $a > 0$, the parabola opens up and the vertex is the lowest (minimum) point on the curve. See Figure 5.1. The graph of $y = x^2 - x - 6$ is a parabola that opens up since the coefficient of the x^2-term is positive.

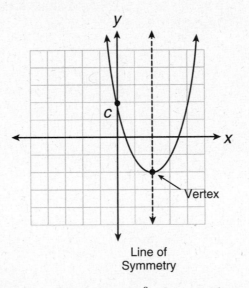

Figure 5.1 Graph of $y = ax^2 + bx + c$ with $a > 0$

- If $a < 0$, the parabola opens down and the vertex is the highest (maximum) point on the curve. See Figure 5.2. The graph of $y = -2x^2 - x + 8$ is a parabola that opens down since the coefficient of the x^2-term is negative.

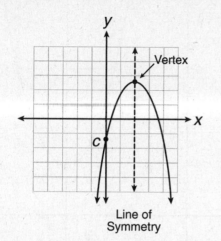

Figure 5.2 Graph of $y = ax^2 + bx + c$ with $a < 0$

EQUATION OF LINE OF SYMMETRY

If a parabola equation is given in the standard $y = ax^2 + bx + c$ form, you can use the formula $x = -\dfrac{b}{2a}$ to find an equation of the line of symmetry, which also gives the x-coordinate of the vertex. If $y = 2x^2 - 12x + 3$, then an equation of the line of symmetry is

$$x = -\frac{b}{2a}$$
$$= -\frac{-12}{2(2)}$$
$$= 3$$

Since the x-coordinate of the vertex is 3, you can find the y-coordinate of the vertex by substituting 3 for x in the parabola equation:

$$y = 2x^2 - 12x + 3$$
$$= 2(3)^2 - 12(3) + 3$$
$$= 18 - 36 + 3$$
$$= -15$$

The vertex is at $(3, -15)$. Since the vertex is a minimum point on the parabola, -15 represents the minimum value of the function $f(x) = 2x^2 - 12x + 3$.

MATCHING UP PAIRS OF PARABOLA POINTS

Each point on the parabola, other than the vertex, has a matching point on the other side of the line of symmetry that is the same distance from it. Figure 5.3 shows the graph of $f(x) = x^2 - 10x + 27$ with $x = 5$ as its line of symmetry. The point $(3, 6)$ is on the parabola and 2 units from the line of symmetry. The matching point on the parabola is $(7, 6)$, which lies on the opposite side of the line of symmetry and is also 2 units from it. Using function notation, $f(3) = f(7)$. Similarly, $f(4) = f(6)$.

 TIP

A parabola's line of symmetry is the perpendicular bisector of every horizontal segment that connects two points on the parabola.

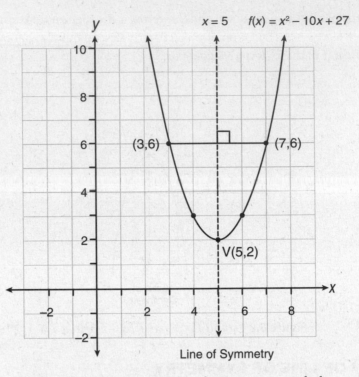

Figure 5.3 Matching up pairs of points on a parabola

VERTEX FORM OF PARABOLA EQUATION: $y = a(x - h)^2 + k$

The parabola equation $y = ax^2 + bx + c$ can also be written in the **vertex form** $y = a(x - h)^2 + k$ where (h, k) are the coordinates of the vertex. The constant a has the same meaning in both forms of the parabola equation. If $y = 3(x - 4)^2 + 7$, then you know the parabola opens up since $3 > 0$; the vertex is at $(4, 7)$ and is a minimum point on the graph. An equation of the line of symmetry is $x = 4$.

REWRITING PARABOLA EQUATIONS

The SAT may ask you to rewrite the equation of a parabola in order to reveal certain information about it.

- To write an equation of a parabola so that its x-intercepts appear as constants in the equation, write the parabola equation in factored form. By rewriting the parabola equation $y = x^2 - 2x + 8$ as $y = (x - 2)(x - 4)$, its x-intercepts appear as constants.
- To write an equation of a parabola so that its minimum or maximum value appears as a constant, write the parabola equation in vertex form. When the parabola equation $y = x^2 - 2x + 8$ is written in the equivalent form $y = (x - 1)^2 + 7$, the minimum value of the function, 7, appears as a constant.

➡ Example

The graph of the function $f(x) = -\frac{1}{2}(x+4)(x+8)$ in the xy-plane is a parabola. Which of the following is an equivalent form of function f in which the maximum value of the function appears as a constant?

(A) $f(x) = (x+4)\left(4 - \frac{1}{2}x\right)$

(B) $f(x) = -\frac{1}{2}(x+6)^2 + 2$

(C) $f(x) = -\frac{1}{2}x(x+12) + 16$

(D) $f(x) = \frac{1}{2}(x+6)^2 + 20$

Solution

The maximum (or minimum) value of a quadratic function is the y-coordinate of the vertex of its graph which appears as the constant k in the vertex form of the equation $f(x) = a(x-h)^2 + k$.

- Using FOIL, multiply the binomial factors of $f(x)$, which gives

$$f(x) = -\frac{1}{2}(x^2 + 12x + 32)$$

$$= -\frac{1}{2}(x^2 + 12x) - \frac{1}{2}(32)$$

$$= -\frac{1}{2}(x^2 + 12x) - 16$$

- Change to the vertex form of the parabola equation by completing the square:

$$f(x) = -\frac{1}{2}(x^2 + 12x + 36) - 16 + \boxed{18} \leftarrow \text{Add 18 to balance}$$

$$\text{adding}\left(-\frac{1}{2} \times 36\right)$$

$$= -\frac{1}{2}(x+6)^2 + 2$$

- By comparing $f(x) = -\frac{1}{2}(x+6)^2 + 2$ to $f(x) = a(x-h)^2 + k$, you know that $a = -\frac{1}{2}$, $h = -6$, and $k = 2$. Since $a < 0$, the vertex of the parabola is a maximum point so that the maximum value of f is 2.

The correct choice is **(B)**.

➡ **Example**

$$y = -x^2 + 120x - 2,000$$

The equation above gives the profit in dollars, y, a coat manufacturer earns each day where x is the number of coats sold. What is the maximum profit he earns in dollars?

Solution

- Because the leading coefficient of the parabola equation is negative, the vertex of the parabola is the highest point on the graph.
- First, determine the value of x that maximizes the profit by determining the coordinate of the vertex:

$$x = -\frac{b}{2a}$$
$$= -\frac{120}{2(-1)}$$
$$= 60$$

- The coat manufacturer must sell 60 coats each day to earn a maximum profit for that day. Because the maximum profit corresponds to the y-coordinate of the vertex, substitute 60 for x in the parabola equation to find the corresponding value of y:

$$y = -60^2 + 120(60) - 2,000$$
$$= -3,600 + 7,200 - 2,000$$
$$= 1,600$$

The maximum profit is **$1,600**. You could also have arrived at the correct answer by changing the parabola equation to vertex form.

➡ **Example**

Stacy has 30 meters of fencing that she wishes to use to enclose a rectangular garden. If all of the fencing is used, what is the maximum area of the garden, in square meters, that can be enclosed?

(A) 48.75
(B) 56.25
(C) 60.50
(D) 168.75

Solution

- If x represents the length of the enclosed rectangular garden and w represents its width, then all 30 meters of fencing are used when $2x + 2w = 30$. Simplifying makes $x + w = 15$ so $w = 15 - x$. If $A(x)$ represents the area of the enclosed rectangle as a function of x, then

$$A(x) = xw = x(15 - x) = -x^2 + 15x$$

- Find the x-coordinate of the vertex:

$$x = -\frac{b}{2a} = -\frac{15}{2(-1)} = 7.5$$

- The maximum value of $A(x)$ occurs at $x = 7.5$:

$$
\begin{aligned}
A(7.5) &= -(7.5)^2 + (15)(7.5) \\
&= -56.25 + 112.5 \\
&= 56.25 \text{ m}^2
\end{aligned}
$$

The correct choice is **(B)**. You could also have arrived at the correct answer by changing the parabola equation to vertex form.

Table 5.2 summarizes some parabola facts you need to remember. In each parabola equation, the sign of the x^2-term determines whether the vertex at (h, k) is a maximum point (if $a < 0$) or a minimum point (if $a > 0$).

Table 5.2 Some Parabola Facts

Parabola Facts	Parabola ($a > 0$) $y = ax^2 + bx + c$ or $y = a(x - h)^2 + k$
• If $y = ax^2 + bx + c$, an equation of the line of symmetry is $$x = -\frac{b}{2a}$$ • If $f(x) = a(x - h)^2 + k$, the vertex is (h, k), and an equation of the line of symmetry is $x = h$. If $a > 0$, k is the minimum value of the function; if $a < 0$, k is the maximum value of the function. • The line of symmetry is the perpendicular bisector of any horizontal segment whose endpoints are on the parabola: $$h = \frac{p+q}{2} \quad \text{and} \quad h = \frac{r+t}{2}$$	

MODELING PROJECTILE MOTION

Projectile motion that is under the influence of gravity, as when a ball is tossed in the air, has a parabola-shaped flight path in which the x-coordinate of each point on the curve represents how much time has elapsed after the object was launched and the corresponding y-coordinate gives the height of the object at that instant of time. The vertex of the parabola corresponds to the point at which the object reaches its maximum height.

Example

$$h(t) = 144t - 16t^2$$

The function above represents the height, in feet, a ball reaches t seconds after it is tossed in the air from ground level.

a. What is the maximum height of the ball?
b. After how many seconds will the ball hit the ground before rebounding?

Solution

a. The maximum height of the ball corresponds to the y-coordinate of the vertex.

- The x-coordinate of the vertex is

$$-\frac{b}{2a} = \frac{-144}{2(-16)} = 4.5$$

- The y-coordinate of the vertex is

$$h(4.5) = 144(4.5) - 16(4.5)^2$$
$$= 324$$

The maximum height is **324** feet.

b. When the ball hits the ground, $h(t) = 0$:

$$144t - 16t^2 = 0$$
$$4t(36 - 4t) = 0$$
$$4t = 0 \quad or \quad 36 - 4t = 0$$
$$t = \frac{0}{4} \qquad\qquad -4t = -36$$
$$t = 0 \qquad\qquad t = \frac{-36}{-4} = 9 \text{ seconds}$$

The solution $t = 0$ seconds represents the instant of time at which the ball is tossed in the air. The solution $t = 9$ seconds is the number of seconds it takes for the ball to reach the ground after it is launched.

SOLVING NONLINEAR SYSTEMS OF EQUATIONS

A linear quadratic system can be solved algebraically by solving the first degree equation for either variable and then substituting for that variable in the second degree equation. To solve the system $y = -x^2 + 4x - 3$ and $x + y = 1$ algebraically,

- Solve the linear equation for y that gives $y = 1 - x$. Eliminate y in the quadratic equation by replacing it with $1 - x$. The result is $1 - x = -x^2 + 4x - 3$, which simplifies to $x^2 - 5x + 4 = 0$.
- Solve $x^2 - 5x + 4 = 0$ by factoring:

$$(x - 1)(x - 4) = 0$$
$$x - 1 = 0 \quad or \quad x - 4 = 0$$
$$x = 1 \qquad\qquad x = 4$$

Find the corresponding values of y by substituting each solution for x in the linear equation:

- If $x = 1$, then $1 + y = 1$ and $y = 0$ so $(1, 0)$ is a solution.
- If $x = 4$, then $4 + y = 1$ and $y = -3$ so $(4, -3)$ is a solution.

When a nonlinear system of equations is graphed in the xy-plane, the points of intersection common to *all* of the equations, if any, represent the solutions to the system of equations.

➡ Example

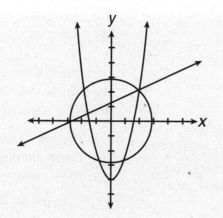

A system of three equations whose graphs in the xy-plane are a line, a circle, and a parabola are shown above. How many solutions does the system have?

(A) 0
(B) 1
(C) 2
(D) 4

Solution

The solution to the system of equations is the common point or points at which the line, circle, and parabola all intersect. There is exactly one such point which is in the first quadrant.

The correct choice is **(B)**.

Multiple-Choice

1. An archer shoots an arrow into the air such that its height at any time, t, is given by the function $h(t) = -16t^2 + kt + 3$. If the maximum height of the arrow occurs at 4 seconds after it is launched, what is the value of k?

   (A) 128
   (B) 64
   (C) 8
   (D) 4

2. A model rocket is launched vertically into the air such that its height at any time, t, is given by the function $h(t) = -16t^2 + 80t + 10$. What is the maximum height attained by the model rocket?

   (A) 140
   (B) 110
   (C) 85
   (D) 10

3. When a ball is thrown straight up at an initial velocity of 54 feet per second. The height of the ball t seconds after it is thrown is given by the function $h(t) = 54t - 12t^2$. How many seconds after the ball is thrown will it return to the ground?

   (A) 9.2
   (B) 6
   (C) 4.5
   (D) 4

4. The graph of $y + 3 = (x - 4)^2 - 6$ is a parabola in the xy-plane. What are the x-intercepts of the parabola?

   (A) 1 and 7
   (B) −1 and −7
   (C) 4 and −6
   (D) 4 and −9

5. The graph of $y = (2x - 4)(x - 8)$ in the xy-plane is a parabola. Which of the following are true?

   I. The graph's line of symmetry is $x = 5$
   II. The minimum value of y is −7
   III. The y-intercept of the graph is 32

   (A) I and II only
   (B) I and III only
   (C) II and III only
   (D) I, II, and III

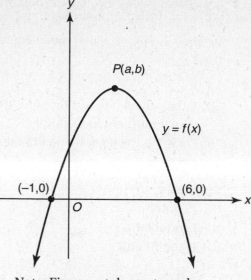

Note: Figure not drawn to scale

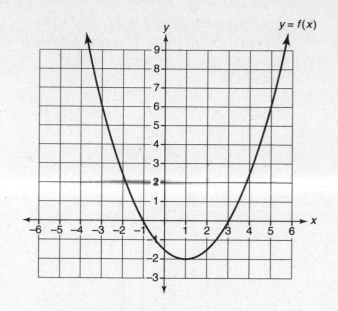

6. The graph of a quadratic function f is shown in the above figure. If $f(x) \leq b$ for all values of x, which of the following could be the coordinates of point P?

(A) $(1.5, 2)$
(B) $(2.25, 3.5)$
(C) $(2.5, 4)$
(D) $(2.75, 5)$

7. The figure above shows the graph of a quadratic function f with a minimum point at $(1, -2)$. If $f(5) = f(c)$, then which of the following could be the value of c?

(A) -5
(B) -3
(C) 0
(D) 6

8. The graph of a quadratic function f intersects the x-axis at $x = -2$ and $x = 6$. If $f(8) = f(p)$, which could be the value of p?

(A) -6
(B) -4
(C) -2
(D) 0

9. If in the quadratic function $f(x) = ax^2 + bx + c$, a and c are both negative constants, which of the following could be the graph of function f?

(A)

(B)

(C)

(D)

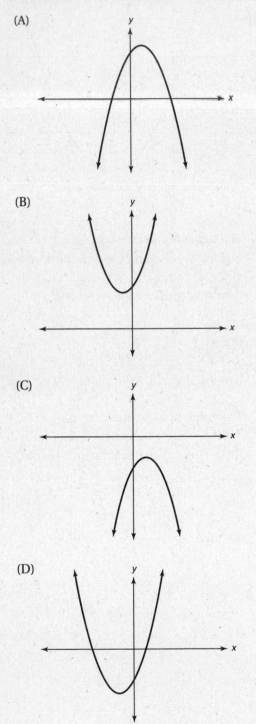

10. A parabola passes through the points $(0, 0)$ and $(6, 0)$. If the turning point of the parabola is $T(h, 4)$, which statement must be true?

   I. $h = 2$

   II. If the parabola passes through $(1, 2)$, then it must also pass through $(5, 2)$

   III. Point T is the highest point of the parabola

(A) II only
(B) III only
(C) I and II only
(D) II and III only

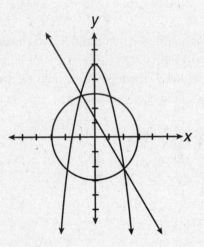

11. A system of three equations whose graphs in the xy-plane are a line, a circle, and a parabola is shown above. How many solutions does the system have?

(A) 1
(B) 2
(C) 3
(D) 4

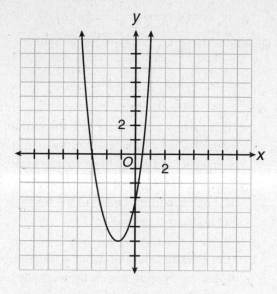

12. Which of the following could be the equation of the graph above?

(A) $y = (x - 3)(2x + 1)$

(B) $y = (x + 3)(2x - 1)$

(C) $y = -(x - 3)(1 + 2x)$

(D) $y = \frac{1}{2}(x + 3)(x - 1)$

$$y = 2x^2 - 12x + 11$$

13. The graph of the equation above is a parabola in the xy-plane. What is the distance between the vertex of the parabola and the point (3, 1)?

(A) 1

(B) 8

(C) 10

(D) 12

$$f(x) = ax^2 + bx + c, \ a > 0$$

14. The coordinates of the lowest point on the graph of the function defined by the equation above is (3, 2). If $f(-1) = p$, then which of the following represents the value of p?

(A) $f(-5)$

(B) $f(-4)$

(C) $f(6)$

(D) $f(7)$

15. The parabola whose equation is $y = ax^2 + bx + c$ passes through the points (−3, −40), (0, 29), and (−1, 10). What is an equation of the line of symmetry?

(A) $x = \dfrac{17}{4}$

(B) $x = \dfrac{9}{2}$

(C) $x = 5$

(D) $x = 6$

$$x^2 + y^2 = 416$$
$$y + 5x = 0$$

16. If (x, y) is a solution to the system of equations above and $x > 0$, what is the value of the difference $x - y$?

(A) 4

(B) 16

(C) 20

(D) 24

$$h(t) = -4.9t^2 + 68.6t$$

17. The function above gives the height of a model rocket, in meters, t seconds after it is launched from ground level. What is the maximum height, to the *nearest meter*, attained by the model rocket?

(A) 90

(B) 120

(C) 180

(D) 240

$$y = k(x - 1)(x + 9)$$

18. The graph of the equation above is a parabola in the xy-plane. If $k > 0$, what is the minimum value of y expressed in terms of k?

(A) $-7k$

(B) $-16k$

(C) $-25k$

(D) $-73k$

Grid-In

$$x^2 - y^2 = 18$$

$$y = x - 4$$

1. In the above system of equations, what is the value of $x + y$?

$$d(t) = -16t^2 + 40t + 24$$

2. A swimmer dives from a diving board that is 24 feet above the water. The distance, in feet, that the diver travels after t seconds have elapsed is given by the function above. What is the maximum height above the water, in feet, the swimmer reaches during the dive?

3. The marketing department at Sports Stuff found that approximately 600 pairs of running shoes will be sold monthly when the average price of each pair of running shoes is $90. It was observed that for each $5 reduction in price, an additional 50 pairs of running shoes will be sold monthly. What price per pair of running shoes will maximize the store's monthly revenue from the sale of running shoes?

Questions 4–6 refer to the equation below.

$$h(x) = -\frac{1}{225}x^2 + \frac{2}{3}x$$

The function h above models the path of a football when it is kicked during an attempt to make a field goal where x is the horizontal distance, in feet, from the kick, and $h(x)$ is the corresponding height of the football, in feet above the ground.

4. After the ball is kicked, what is the number of feet the football travels horizontally before it hits the ground?

5. What is the number of feet in the maximum height of the football?

6. The goal post is 10 feet high and a horizontal distance of 45 yards from the point at which the ball is kicked. By how many feet will the football fail to pass over the goal post?

OVERVIEW

Changing the equation of a function in a certain way can have predictable effects on moving its graph in the *xy*-plane without changing its size or shape. A **reflection** flips a graph over a line so that the original and reflected graphs are "mirror images." A **translation** moves a graph sideways or up/down, or both sideways and up/down.

REFLECTING GRAPHS OF FUNCTIONS

Inserting a negative sign within the parentheses of $y = f(x)$ flips its graph over the *y*-axis, whereas writing a negative sign outside the parentheses in front of the function flips its graph upside down over the *x*-axis. Reflecting a graph does not change its size or shape. See Figure 5.4.

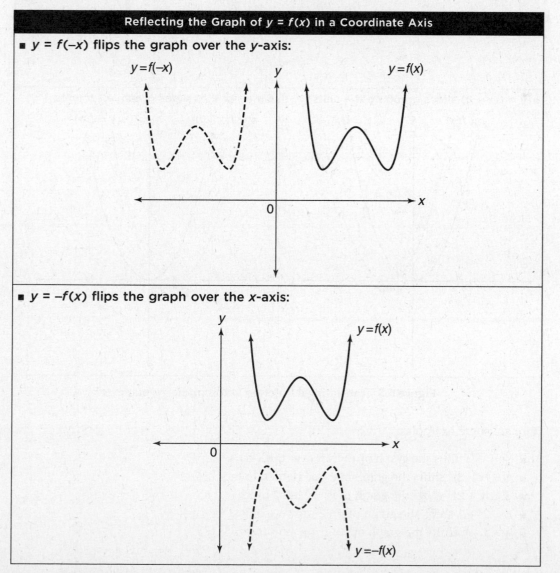

Reflecting the Graph of $y = f(x)$ in a Coordinate Axis

- $y = f(-x)$ flips the graph over the *y*-axis:

- $y = -f(x)$ flips the graph over the *x*-axis:

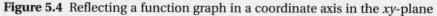

Figure 5.4 Reflecting a function graph in a coordinate axis in the *xy*-plane

TRANSLATING GRAPHS OF FUNCTIONS

Translating a graph in the *xy*-plane moves the graph sideways or up/down without rotating it, or changing its size or shape. See Figure 5.5.

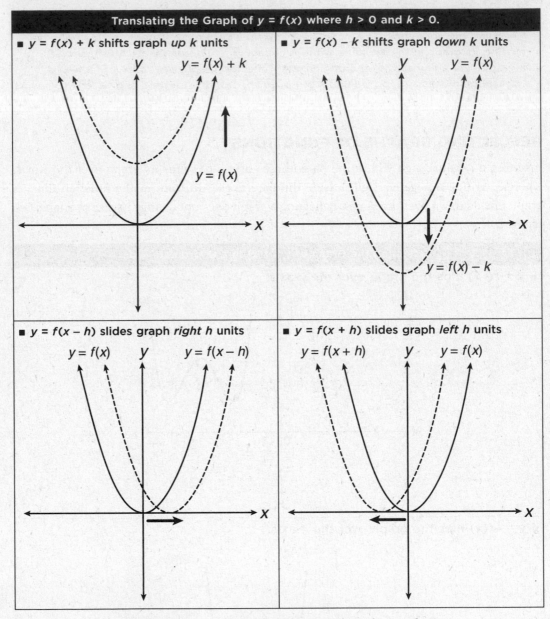

Figure 5.5 Translating a function graph in the *xy*-plane

Here are some examples:

- $y = -\sqrt{x}$ flips the graph of $y = \sqrt{x}$ over the *x*-axis.
- $y = (x - 3)^2$ shifts the graph of $y = x^2$ right 3 units.
- $y = (x + 2)^3$ shifts the graph of $y = x^3$ left 2 units.
- $y = 2^x - 1$ shifts the graph of $y = 2^x$ down 1 unit.
- $y = x + 4$ shifts the graph of $y = x^2$ up 4 units.

TIP

- $y = f(x) + k$:
  Adding *k* outside the parentheses shifts the graph of function *f* vertically *k* units.
- $y = f(x + h)$:
  Adding *h* *inside* the parentheses shifts the graph of function *f* sideways *h* units.

Graph of $y = |x|$

Because of the definition of absolute value of x, the graph of $y = |x|$ is comprised of the portion of the graphs of $y = x$ $(x \geq 0)$ and $y = -x$ $(x < 0)$ for which y is nonnegative as shown in Figure 5.6. Each ray of the graph is a reflection of the other in the y-axis.

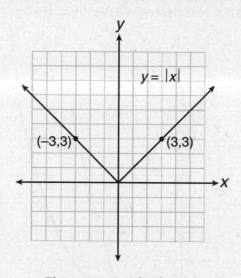

Figure 5.6 Graph of $y = |x|$.

As illustrated in Figure 5.7,

- The graph of $y = |x| + 2$ shifts the graph of $y = |x|$ up 2 units.
- The graph of $y = |x + 2|$ shifts the graph of $y = |x|$ left 2 units.

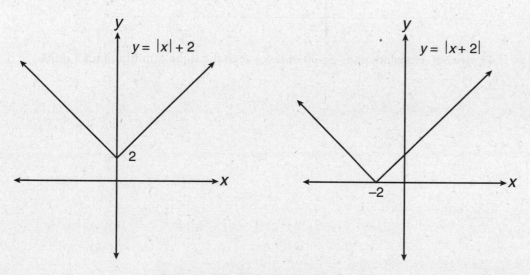

Figure 5.7 Shifting the graph of $y = |x|$

COMBINING HORIZONTAL AND VERTICAL SHIFTS

A graph may be shifted both vertically and horizontally as in Figure 5.8. The graph of $y = f(x - 2) + 3$ is the graph of $y = f(x)$ shifted sideways to the right 2 units and straight up 3 units. Each point (x, y) of the original graph corresponds to $(x + 2, y + 3)$ of the new graph. Since $(0, 0)$ is the vertex of $f(x) = x^2$, the vertex of the translated graph is $(0 + 2, 0 + 3) = (2, 3)$.

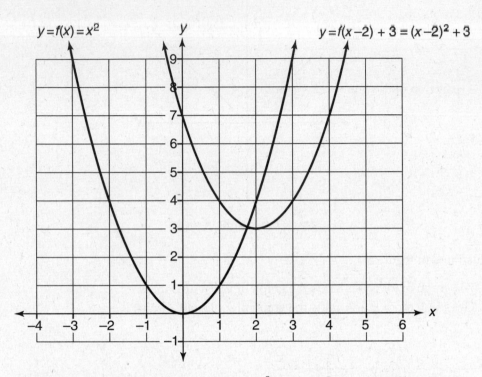

Figure 5.8 Translating the graph of $f(x) = x^2$ to the right 2 units and up 3 units

Multiple-Choice

1. The graph of $y = 2^{x-3}$ can be obtained by shifting the graph of $y = 2^x$

   (A) 3 units to the right
   (B) 3 units to the left
   (C) 3 units up
   (D) 3 units down

2. Which equation represents the line that is the reflection of the line $y = 2x - 3$ in the y-axis?

   (A) $y = -2x - 3$
   (B) $y = -2x + 3$
   (C) $y = 2x + 3$
   (D) $y = 3x - 2$

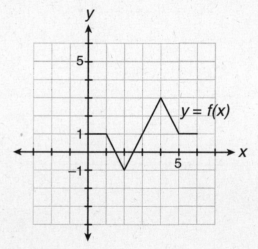

3. The figure above shows part of the graph of function f. If $f(x - 5) = f(x)$ for all values of x, what is the value of $f(19)$?

   (A) −1
   (B) 0
   (C) 1
   (D) 3

4. The endpoints of $\overline{AB}$ are $A(0, 0)$ and $B(9, -6)$. What is an equation of the line that contains the reflection of $\overline{AB}$ in the y-axis?

   (A) $y = -\dfrac{3}{2}x$

   (B) $y = -\dfrac{2}{3}x$

   (C) $y = -x + 3$

   (D) $y = \dfrac{2}{3}x$

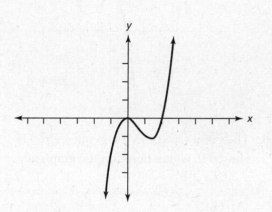

5. The figure above shows the graph of function f. If $g(x) = -f(x)$, which graph represents function g?

(A)

(B)

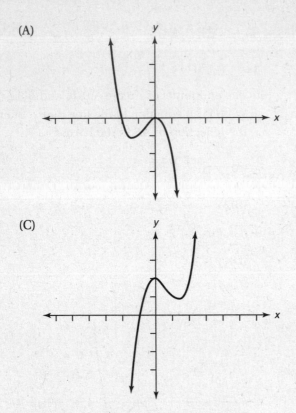

(C)

(D)

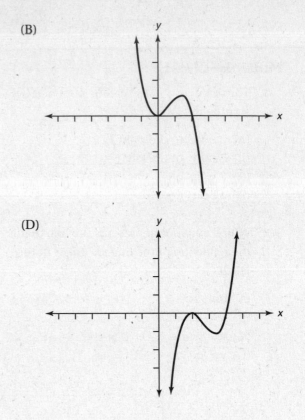

6. The point $(2, -1)$ on the graph $y = f(x)$ is shifted to which point on the graph of $y = f(x + 2)$?

(A) $(4, 1)$
(B) $(4, -1)$
(C) $(0, -1)$
(D) $(0, -3)$

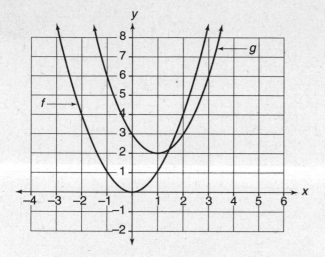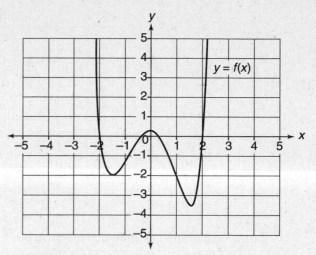

7. The accompanying figure shows the graphs of functions f and g. If f is defined by $f(x) = x^2$ and g is defined by $g(x) = f(x + h) + k$, where h and k are constants, what is the value of $h + k$?

(A) -3

(B) -2

(C) -1

(D) 1

8. If $g(x) = -2$ intersects the graph of $y = f(x) + k$ at one point, which of these choices could be the value of k?

(A) -1.5

(B) -0.5

(C) 0

(D) 1.5

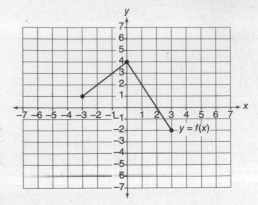

9. If the accompanying figure above shows the graph of function f, which of the following could represent the graph of $y = f(x + 1)$?

(A)

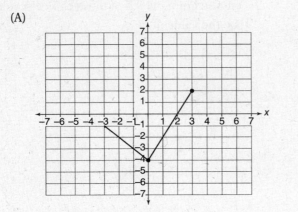

(B)

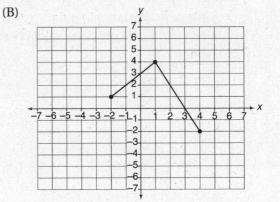

(C)

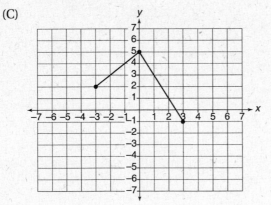

(D)

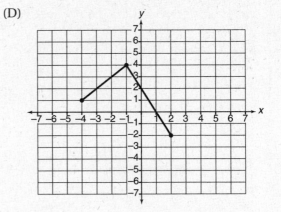

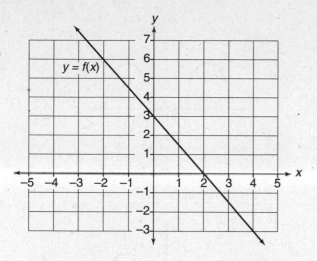

y = f(x)

10. A linear function *f* is shown in the accompanying figure. If function *g* is the reflection of function *f* in the *x*-axis (not shown), what is the slope of the graph of function *g*?

(A) $-\dfrac{3}{2}$

(B) $-\dfrac{2}{3}$

(C) $\dfrac{2}{3}$

(D) $\dfrac{3}{2}$

11. The graph of $y = f(x)$ is shown below.

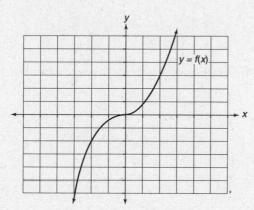

y = f(x)

Which of the following could represent the graph of $y = f(x - 2) + 1$?

(A)

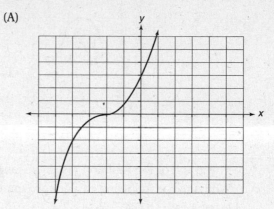

(B)

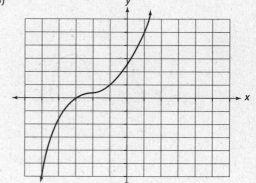

(C)

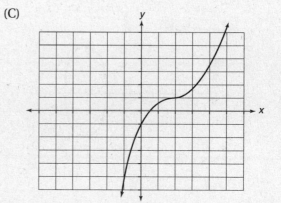

(D)

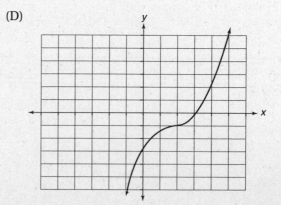

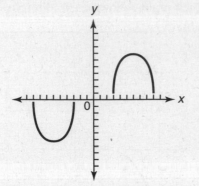

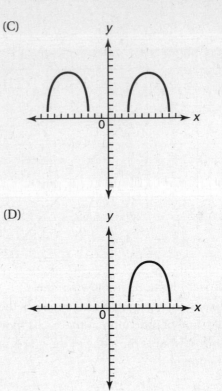

(C)

(D)

12. The graph of the function f is shown above. Which of the following could represent the graph of $y = |f(x)|$?

(A)

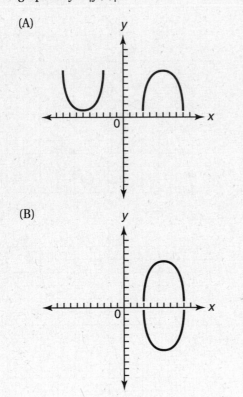

(B)

Grid-In

Questions 1 and 2 refer to the information and graph below.

Let function f be defined by the graph in the accompanying figure.

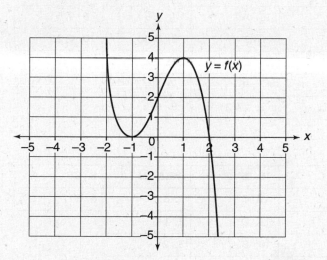

1. For what positive integer k is $(1, 0)$ an x-intercept of the graph of $y = f(x - k)$?

2. Let m represent the number of points at which the graphs of $y = f(x)$ and $g(x) = 3$ intersect. Let n represent the number of points at which the graphs of $y = f(x) - 1$ and $g(x) = 3$ intersect. What is the value of $m + n$?

NOTE: See pages 442–455 for worked out solutions.

Lesson 5–1

1. B	6. B	11. C	16. A	4. 1/6	9. 7/6
2. A	7. D	12. D	GRID-IN	5. 3/10	10. 2.25
3. A	8. C	13. B	1. 5	6. 1/4	
4. D	9. C	14. C	2. 16	7. 32/3	
5. C	10. C	15. A	3. 27/2	8. 5/2	

Lesson 5–2

1. B	6. A	11. B	16. C	GRID-IN	5. 6
2. D	7. A	12. C	17. C	1. 3	6. 3/7
3. B	8. B	13. A	18. D	2. 1/2	
4. D	9. A	14. C	19. B	3. 2	
5. B	10. D	15. A	20. D	4. 3/2	

Lesson 5–3

1. D	5. B	9. A	13. A	2. 39
2. A	6. D	10. C	14. C	3. 8
3. C	7. C	11. A	GRID-IN	4. 8/5
4. C	8. B	12. D	1. 21/4	5. 1/3

Lesson 5–4

1. A	3. C	5. C	7. B	9. D
2. A	4. B	6. A	8. A	10. B

Lesson 5–5

1. A	6. C	11. B	16. D	2. 49
2. B	7. B	12. B	17. D	3. 75
3. C	8. B	13. B	18. C	4. 150
4. A	9. C	14. D	GRID-IN	5. 25
5. B	10. D	15. A	1. 9/2	6. 1

Lesson 5–6

1. A	4. D	7. D	10. D	GRID-IN
2. A	5. B	8. D	11. C	1. 2
3. D	6. C	9. D	12. C	2. 5

Additional Topics in Math

6

In addition to algebra, you are also expected to know and to be able to apply key facts and relationships from geometry and trigonometry.

You not only need to know right triangle trigonometry but also need to be familiar with the general angle, angles measured in radians as well as degrees, the unit circle, and the coordinate definitions of the three basic trigonometric functions.

"Additional Topics in Math" represents the last of the four major mathematics content groups tested by the redesigned SAT.

LESSONS IN THIS CHAPTER

OVERVIEW

This lesson summarizes some important relationships from your study of Geometry that you should already know.

ANGLES AND LINES

When two lines intersect, vertical angles (opposite angles) have the same measure. See Figure 6.1.

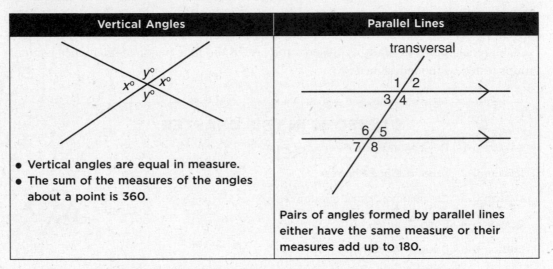

Vertical Angles	Parallel Lines
• Vertical angles are equal in measure. • The sum of the measures of the angles about a point is 360.	Pairs of angles formed by parallel lines either have the same measure or their measures add up to 180.

Figure 6.1 Angle relationships for intersecting and parallel lines

When two parallel lines are cut by another line, called a *transversal*, every pair of angles formed are either congruent (have the same measure) or are supplementary (have measures that add up to 180). In Figure 6.1, since the lines are parallel,

- Alternate interior angles 3 and 5 are equal in measure as are alternate interior angles 4 and 6.
- Corresponding pairs of angles 1 and 2, 3 and 7, 2 and 5, as well as 4 and 8 have equal measures.

Angles 2 and 8 look like they have different measures (one acute angle and one obtuse angle) so you can assume because the lines are parallel that their measures add up to 180.

TRIANGLES AND POLYGONS

For triangles (see Figure 6.2),

- The sum of the measures of the three angles is 180.
- The measure of an exterior angle is equal to the sum of the two nonadjacent interior angles of the triangle.

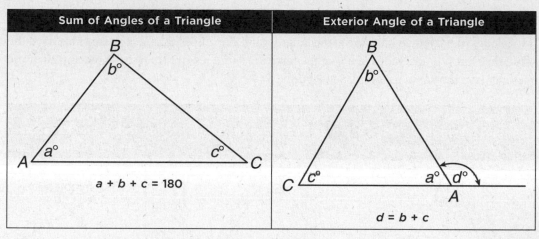

Figure 6.2 Angle relationships in a triangle

For any polygon with n sides (see Figure 6.3),

- The sum of the measures of the interior angles is $(n - 2) \cdot 180$. The sum of the measures of the four angles of a quadrilateral is $(4 - 2) \cdot 180 = 2 \cdot 180 = 360$.
- The sum of the exterior angles, one angle at each vertex, is 360.

> **MATH REFERENCE FACT**
>
> In a **regular polygon**, all of the sides have the same length and all of the angles have the same measure. For a *regular* polygon with n sides,
>
> - The measure of each exterior angle is $\dfrac{360}{n}$.
> - The measure of each interlor angle is $180 - \dfrac{360}{n}$.

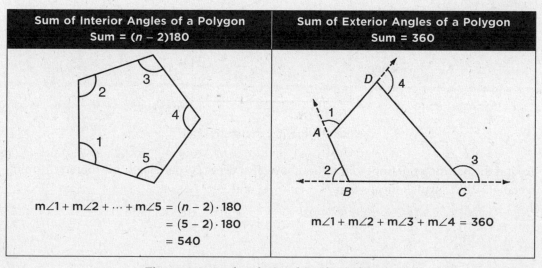

Figure 6.3 Angle relationships for polygons

ISOSCELES, EQUILATERAL, AND RIGHT TRIANGLES

If two sides of a triangle have the same length, then the angles that face these sides have the same measures. If all 3 sides of a triangle have the same length, then the three angles of the triangle have the same measure. See Figure 6.4.

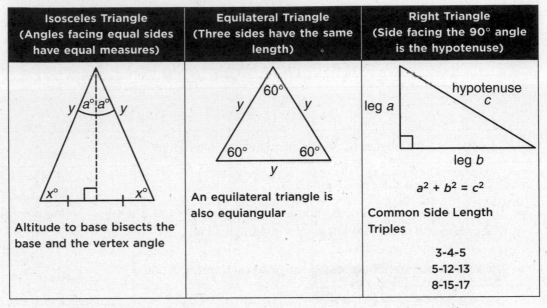

Isosceles Triangle (Angles facing equal sides have equal measures)	Equilateral Triangle (Three sides have the same length)	Right Triangle (Side facing the 90° angle is the hypotenuse)

Figure 6.4 Relationships in isosceles, equilateral, and right triangles

➡ **Example**

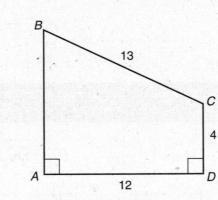

Note: Figure not drawn to scale

To find the perimeter of trapezoid *ABCD* above, first draw $\overline{EC}$ parallel to $\overline{AD}$ thereby forming rectangle *AECD*, which means that $EC = AD = 12$ and $EA = CD = 4$.

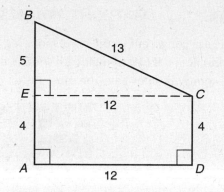

The lengths of the sides of right triangle BEC form a 5-12-13 Pythagorean triple with $BEC = 5$. Hence, the perimeter of trapezoid $ABCD$ is $12 + 9 + 13 + 4 = 38$.

SPECIAL RIGHT TRIANGLE RELATIONSHIPS

The lengths of the sides of a 45-45 right triangle and of a 30-60 right triangle are in fixed ratios as illustrated in Figure 6.5. These relationships will be provided in the reference page in your SAT test booklet.

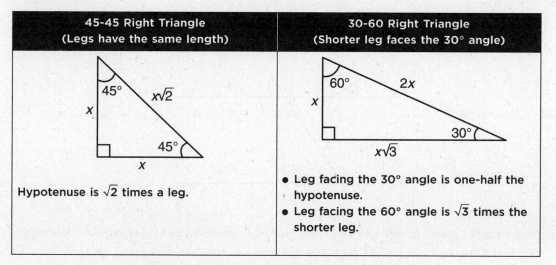

Figure 6.5 Special right triangle relationships

Here are some examples:

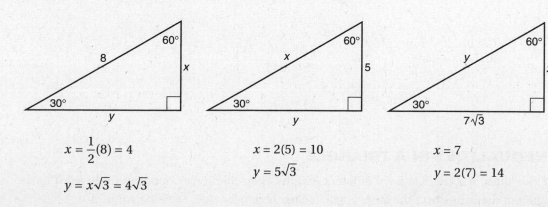

$x = \dfrac{1}{2}(8) = 4$

$y = x\sqrt{3} = 4\sqrt{3}$

$x = 2(5) = 10$

$y = 5\sqrt{3}$

$x = 7$

$y = 2(7) = 14$

SIMILAR TRIANGLES

If two angles of one triangle are congruent to the corresponding angles of another triangle, then the two triangles are similar (:). If two triangles are similar, as shown in Figure 6.6, then the lengths of pairs of corresponding sides have the same ratio:

$$\frac{a}{x} = \frac{b}{y} = \frac{c}{z}$$

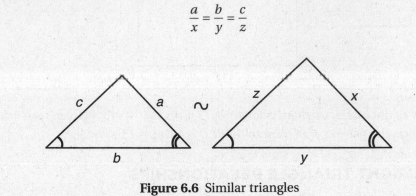

Figure 6.6 Similar triangles

➥ Example

In the accompanying figure, $\overline{AB} \| \overline{CD}$. If $AB = 40$, $CD = 16$, and $BC = 49$, what is the length of $\overline{BE}$?

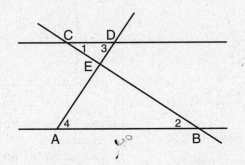

Angles pairs 1 and 2, as well as 3 and 4, are formed by parallel lines (alternate interior angles) and look congruent, so for SAT purposes, you can assume that they are congruent: $\angle 1 \cong \angle 2$ and $\angle 3 \cong \angle 4$. Hence, $\triangle ABE \sim \triangle DCE$. If x represents the length of $\overline{BE}$, then:

$$\frac{x}{49 - x} = \frac{40}{16}$$
$$\frac{x}{49 - x} = \frac{5}{2}$$
$$2x = 245 - 5x$$
$$\frac{7x}{7} = \frac{245}{7}$$
$$x = 35$$

INEQUALITIES IN A TRIANGLE

If two sides of a triangle have different lengths, then the angles opposite these sides have different measures with the larger angle facing the longer side. See Figure 6.7.

Unequal Sides Imply Unequal Angles	Side Length Restrictions in a Triangle
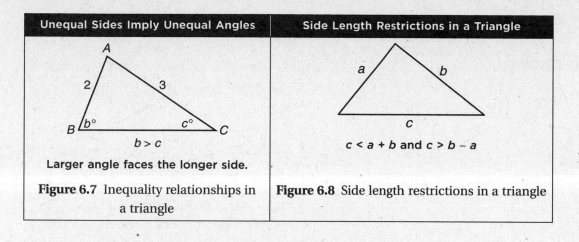	

b > c

Larger angle faces the longer side.

Figure 6.7 Inequality relationships in a triangle

c < a + b and c > b − a

Figure 6.8 Side length restrictions in a triangle

SIDE LENGTH RESTRICTIONS IN A TRIANGLE

In a triangle, the length of each side must be less than the sum of the lengths of the other two sides and greater than their difference. See Figure 6.8.

➡ Example

If 3, 8, and x represent the lengths of the sides of a triangle, how many integer values for x are possible?

Solution

- The value of x must be greater than $8 - 3 = 5$ *and* less than $8 + 3 = 11$.
- Since $x > 5$ and $x < 11$, x must be between 5 and 11.
- Because it is given that x is an integer, x can be equal to 6, 7, 8, 9, or 10.
- Hence, there are five possible integer values for x.

➡ Example

If the lengths of two sides of an isosceles triangle are 3 and 7, what is the perimeter of the triangle?

Solution

The length of the third side of this isosceles triangle could be 3 or 7. If it were 3, then the third side, 7, would be greater than the sum of the lengths of the other two sides, which is not possible. Hence, the length of the third side is 7 so the perimeter of the triangle is $7 + 7 + 3 = \textbf{17}$.

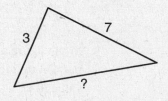

Multiple-Choice

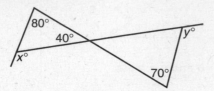

1. In the figure above, $x + y =$

   (A) 270
   (B) 230
   (C) 210
   (D) 190

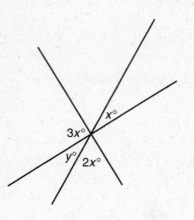

2. In the figure above, what is the value of y?

   (A) 20
   (B) 30
   (C) 45
   (D) 60

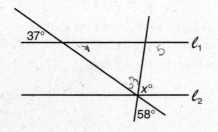

3. In the figure above, if $\ell_1 \parallel \ell_2$, what is the value of x?

   (A) 90
   (B) 85
   (C) 75
   (D) 70

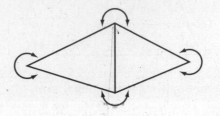

4. In the figure above, what is the sum of the degree measures of all of the angles marked?

   (A) 540
   (B) 720
   (C) 900
   (D) 1080

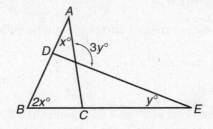

5. In the figure above, what is y in terms of x?

   (A) $\dfrac{3}{2}x$

   (B) $\dfrac{4}{3}x$

   (C) x

   (D) $\dfrac{3}{4}x$

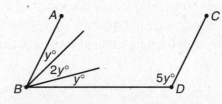

6. In the figure above, if line segment AB is parallel to line segment CD, what is the value of y?

   (A) 12
   (B) 15
   (C) 18
   (D) 20

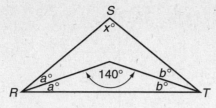

Note: Figure not drawn to scale.

7. In △RST above, what is the value of x?

(A) 80
(B) 90
(C) 100
(D) 110

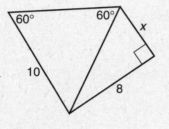

8. In the figure above, x =

(A) 4
(B) 6
(C) $4\sqrt{2}$
(D) $4\sqrt{3}$

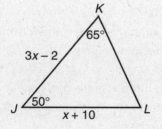

9. In △JKL above, what is the value of x?

(A) 2
(B) 3
(C) 4
(D) 6

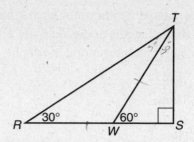

Note: Figure not drawn to scale.

10. In the figure above, what is the ratio of RW to WS?

(A) $\sqrt{2}$ to 1
(B) $\sqrt{3}$ to 1
(C) 2 to 1
(D) 3 to 1

11. Katie hikes 5 miles north, 7 miles east, and then 3 miles north again. What number of miles, measured in a straight line, is Katie from her starting point?

(A) $\sqrt{83}$
(B) 10
(C) $\sqrt{113}$
(D) 13

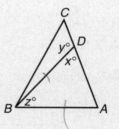

Note: Figure not drawn to scale.

12. In △ABC, if AB = BD, which of the following statements must be true?

 I. x > z
 II. y > x
 III. AB > BC

(A) I only
(B) II only
(C) I and II only
(D) II and III only

13. How many different triangles are there for which the lengths of the sides are 3, 8, and n, where n is an integer and $3 < n < 8$?

   (A) Two
   (B) Three
   (C) Four
   (D) Five

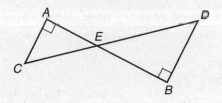

14. If, in the figure above, $AC = 3$, $DB = 4$, and $AB = 14$, then $AE =$

   (A) 4.5
   (B) 6
   (C) 8
   (D) 10.5

15. What is the number of sides of a polygon in which the sum of the degree measures of the interior angles is 4 times the sum of the degree measures of the exterior angles?

   (A) 10
   (B) 12
   (C) 14
   (D) No such polygon exists.

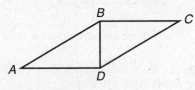

16. For parallelogram $ABCD$ above, if $AB > BD$, which of the following statements must be true?

   I. $CD < BD$
   II. $\angle ADB > \angle C$
   III. $\angle CBD > \angle A$

   (A) None
   (B) I only
   (C) II and III only
   (D) I and III only

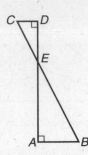

17. If, in the figure above, $CD = 1$, $AB = 2$, and $AD = 6$, then $BC =$

   (A) 5
   (B) 9
   (C) $2 + \sqrt{5}$
   (D) $3\sqrt{5}$

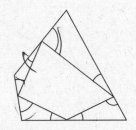

18. In the figure above, what is the sum of the degree measures of the marked angles?

   (A) 120
   (B) 180
   (C) 360
   (D) It cannot be determined from the information given.

19. If each interior angle of a regular polygon measures 140°, how many side does the polygon have?

   (A) 5 sides
   (B) 6 sides
   (C) 9 sides
   (D) 10 sides

Grid-In

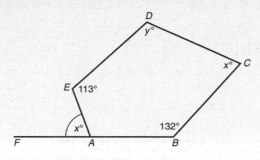

1. In the accompanying figure of pentagon *ABCDE*, points *F*, *A*, and *B* lie on the same line. What is the value of *y*?

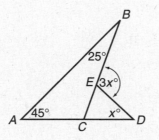

2. In the figure above, what is the value of *x*?

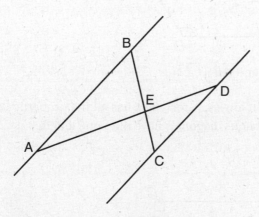

3. In the figure above, $\overline{AB} \parallel \overline{CD}$, *AD* = 30, *AB* = 21, and *CD* = 15. What is the length of $\overline{DE}$?

4. In the accompanying diagram of triangle *ABC*, *AC* = *BC*, *D* is a point on $\overline{AC}$, $\overline{AB}$ is extended to *E*, and $\overline{DFE}$ is drawn so that △*ADE* ~ △*ABC*. If m∠*C* = 30, what is the value of *x*?

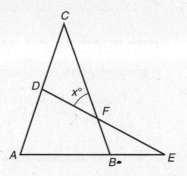

5. Two hikers started at the same location. One traveled 2 miles east and then 1 mile north. The other traveled 1 mile west and then 3 miles south. At the end of their hikes, how many miles apart were the two hikers?

> ### OVERVIEW
>
> This lesson reviews some key properties of parallelograms as well as the formulas to find the areas of parallelograms, triangles, and trapezoids.

PROPERTIES OF PARALLELOGRAMS AND RECTANGLES

A **parallelogram** is a quadrilateral whose opposite sides are parallel. In addition,

- Opposite sides have the same length:

$$AB = CD \text{ and } AD = BC$$

- Opposite angles have the same measure:

$$\angle A = \angle C \text{ and } \angle B = \angle D$$

- Diagonals bisect each other:

$$AE = EC \text{ and } BE = ED$$

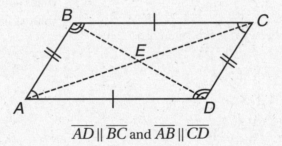

$$\overline{AD} \parallel \overline{BC} \text{ and } \overline{AB} \parallel \overline{CD}$$

A **rectangle** is a parallelogram with four right angles. A rectangle has all the properties of a parallelogram and the additional property that its diagonals have the same length:

$$AC = BD$$

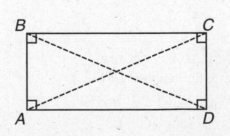

AREA OF A PARALLELOGRAM AND RECTANGLE

The area of a parallelogram and the area of a rectangle are each obtained by multiplying the base and altitude together as shown in Figure 6.9.

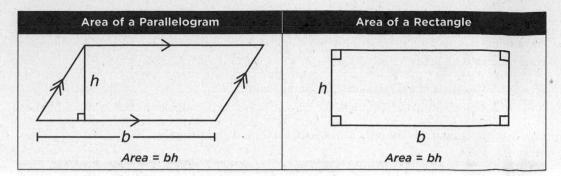

Area of a Parallelogram	Area of a Rectangle
Area = bh	Area = bh

Figure 6.9 Area formulas for a parallelogram and rectangle

➡ Example

To find the area of parallelogram *ABCD*, draw perpendicular segment *BH*, as shown. Since *BH* is the side opposite a 45° angle in a right triangle:

$$h = 6 \times \frac{\sqrt{2}}{2} = 3\sqrt{2}$$

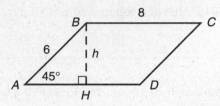

Opposite sides of a parallelogram are equal, so *AD* = 8. Hence,

$$
\begin{aligned}
\text{Area of parallelogram } ABCD &= bh \\
&= AD \times h \\
&= 8 \times 3\sqrt{2} \\
&= 24\sqrt{2}
\end{aligned}
$$

➡ Example

To find the area of rectangle *ABCD*, note that the diagonal forms a (5-12-13) right triangle **so** the width of the rectangle is 5, Hence:

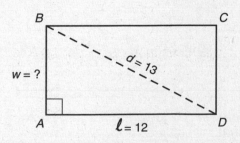

Area of rectangle = $\ell w = 12 \times 5 = 60$

AREA OF A RHOMBUS AND A SQUARE

A **rhombus** is a parallelogram in which all sides have the same length. A **square** is a rhombus with four right angles.

- The diagonals of a square have the same length.
- In both a rhombus and square, the diagonals are perpendicular bisectors of each other.

Figure 6.10 summarizes the area formulas for a square and rhombus.

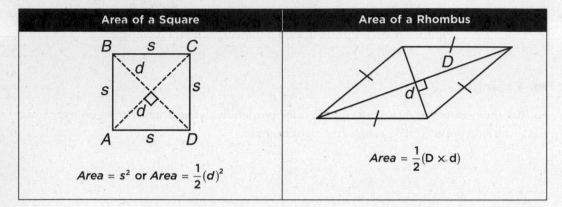

Figure 6.10 Area of a square and area of a rhombus

➡ Example

If rhombus *ABCD* has a side length of 10 and the longer diagonal measures 16, then the shorter diagonal measures 12 since

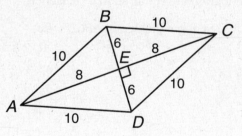

- diagonals $\overline{AC}$ and $\overline{BD}$ are perpendicular to each other and bisect each other, which makes $\triangle AED$ a 6-8-10 right triangle.
- $BD = 6 + 6 = 12$

Hence,

$$\text{Area rhombus } ABCD = \frac{1}{2}(16)(12) = 96$$

AREA OF A TRIANGLE

Figure 6.11 shows the area formulas for different type of triangles.

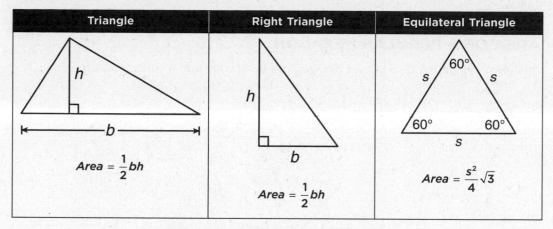

Triangle	Right Triangle	Equilateral Triangle

Figure 6.11 Area formulas for triangles

➼ Example

To find the area of $\triangle ABC$, note that the lengths of the sides of $\triangle ABC$ form a (3-4-5) Pythagorean triple, where $AC = 4$. Hence,

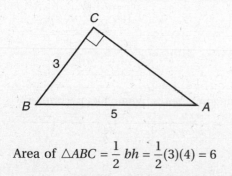

$$\text{Area of } \triangle ABC = \frac{1}{2}\,bh = \frac{1}{2}(3)(4) = 6$$

Note that in a right triangle either leg is the base and the other leg is the height.

➼ Example

To find the area of $\triangle JKL$, drop a perpendicular segment from vertex J to side KL, extending it as necessary. Since $\angle JKH$ measures 30°:

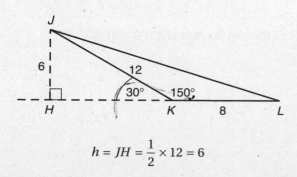

$$h = JH = \frac{1}{2} \times 12 = 6$$

and

$$\text{Area of } \triangle JKL = \frac{1}{2}bh = \frac{1}{2}(8)(6) = 24$$

AREA OF A REGULAR HEXAGON

As shown in Figure 6.12, a regular hexagon can be divided into 6 equilateral triangles:

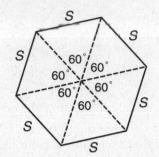

Figure 6.12 Dividing a regular hexagon into 6 equilateral triangles

Since the area of each equilateral triangle is $\frac{s^2}{4}\sqrt{3}$, the area of a regular hexagon is

$$6\left(\frac{s^2}{4}\sqrt{3}\right)$$

➥ Example

If the area of a regular hexagon is $96\sqrt{3}$, what is its perimeter?

Solution

Find the length, s, of each side of the regular hexagon:

$$6\left(\frac{s^2}{4}\sqrt{3}\right) = 96\sqrt{3}$$

$$s^2\sqrt{3} = \frac{2}{3}\left(96\sqrt{3}\right)$$

$$s^2 = 64$$

$$s = \sqrt{64} = 8$$

Hence, the perimeter of the regular hexagon is $6 \times 8 = 48$.

AREA OF A TRAPEZOID

A **trapezoid** is a quadrilateral with one pair of parallel sides, as shown in Figure 6.13.

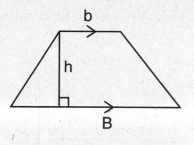

Figure 6.13 Trapezoid

The area of a trapezoid is the altitude times one-half the sum of the lengths of the two parallel sides called *bases*:

$$Area = h \times \frac{1}{2}(B + b)$$

In an **isosceles trapezoid**, the nonparallel sides (legs) have the same length and the bases angles have the same measure, as shown in Figure 6.14.

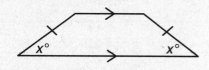

Figure 6.14 Isosceles trapezoid

➡ Example

To find the area of trapezoid *ABCD*, use the fact that the lengths of the sides of right triangle *AEB* form a (5-12-13) Pythagorean triple, where height *BE* = 12. The length of lower base *AD* = *AE* + *ED* = 5 + 27 = 32.

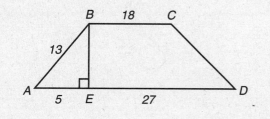

$$Area\ of\ trapezoid\ ABCD = Height \times \frac{(Sum\ of\ bases)}{2}$$

$$= (BE) \times \frac{(AD + BC)}{2}$$

$$= (12) \times \frac{(32 + 18)}{2}$$

$$= 300$$

➡ Example

To find the area of trapezoid *ABCD*, first find the length of base *CD* by drawing height *BH* to *CD*. Since parallel lines are everywhere equidistant, $BH = AD = 15$. The lengths of the sides of right triangle *BHC* form an (8-15-17) Pythagorean triple, where $CH = 8$. Thus, $CD = CH + HD = 8 + 10 = 18$.

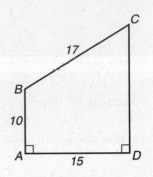

$$\text{Area of trapezoid } ABCD = \text{Height} \times \frac{(\text{Sum of bases})}{2}$$

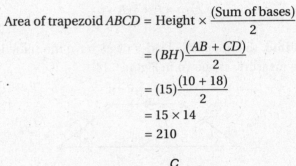

$$= (BH)\frac{(AB + CD)}{2}$$

$$= (15)\frac{(10 + 18)}{2}$$

$$= 15 \times 14$$

$$= 210$$

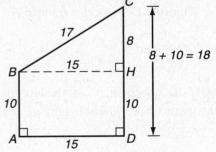

Multiple-Choice

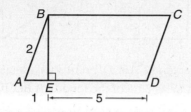

1. In the figure above, what is the area of parallelogram *ABCD*?

   (A) $4\sqrt{2}$
   (B) $4\sqrt{3}$
   (C) $6\sqrt{2}$
   (D) $6\sqrt{3}$

2. What is the area of a square with a diagonal of $\sqrt{2}$?

   (A) $\dfrac{1}{2}$
   (B) 1
   (C) $\sqrt{2}$
   (D) 2

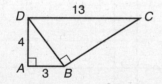

3. In the figure above, what is the area of quadrilateral *ABCD*?

   (A) 28
   (B) 32
   (C) 36
   (D) 42

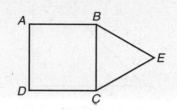

4. In the figure above, if the area of square *ABCD* is 64, what is the area of equilateral triangle *BEC*?

   (A) 8
   (B) $8\sqrt{3}$
   (C) $12\sqrt{3}$
   (D) $16\sqrt{3}$

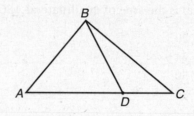

5. In the figure above, the ratio of *AD* to *DC* is 3 to 2. If the area of △*ABC* is 40, what is the area of △*BDC*?

   (A) 16
   (B) 24
   (C) 30
   (D) 36

Questions 6–7 are based on the diagram below.

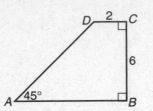

Note: Figure not drawn to scale.

6. What is the perimeter of quadrilateral *ABCD*?

(A) $16 + 3\sqrt{2}$
(B) $16 + 6\sqrt{2}$
(C) 28
(D) $22 + 6\sqrt{2}$

7. What is the area of quadrilateral *ABCD*?

(A) 20
(B) 24
(C) 30
(D) 36

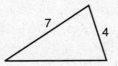

8. If the perimeter of the triangle above is 18, what is the area of the triangle?

(A) $2\sqrt{33}$
(B) $6\sqrt{5}$
(C) 14
(D) $9\sqrt{5}$

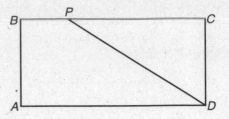

9. In rectangle *ABCD*, point *P* divides *BC* such that *BP* is 25% of the length of *BC*. If the area of quadrilateral *ABPD* is $\frac{3}{4}$, what is the area of rectangle *ABCD*?

(A) $\frac{15}{6}$

(B) $\frac{9}{8}$

(C) $\frac{6}{5}$

(D) $\frac{3}{2}$

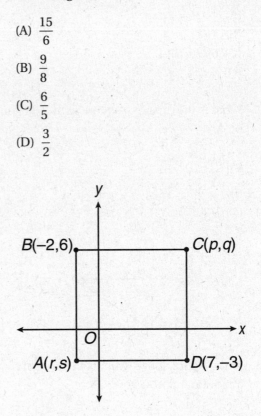

10. In the figure above, what is an equation of the line that contains diagonal *AC* of square *ABCD*?

(A) $y = 2x + 1$

(B) $y = \frac{1}{2}x - 2$

(C) $y = 2x - 8$

(D) $y = x - 1$

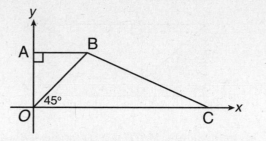

11. In the figure above, $\overline{OA} \perp \overline{AB}$, and $m\angle BOC =$ 45°. If the coordinates of point A are (0, 3) and the coordinates of point C are (7, 0), what is the number of square units in the area of quadrilateral $OABC$?

(A) 15
(B) 20
(C) 25
(D) 30

12. If one pair of opposite sides of a square are increased in length by 20% and the other pair of sides are increased in length by 50%, by what percent is the area of the rectangle that results greater than the area of the original square?

(A) 80%
(B) 77%
(C) 75%
(D) 70%

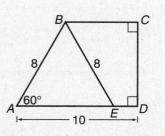

13. In the figure above, what is the area of quadrilateral $BCDE$?

(A) $8\sqrt{3}$
(B) $16\sqrt{3}$
(C) $8 + 4\sqrt{3}$
(D) $4 + 12\sqrt{3}$

Grid-In

1. Brand X paint costs $14 per gallon, and 1 gallon provides coverage of an area of at most 150 square feet. What is the minimum cost of the amount of brand X paint needed to cover the four walls of a rectangular room that is 12 feet wide, 16 feet long, and 8 feet high?

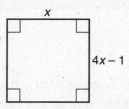

2. What is the area of the square above?

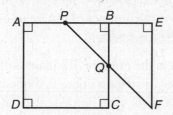

3. In the figure above, P and Q are the midpoints of sides AB and BC, respectively, of square $ABCD$. Line segment PB is extended by its own length to point E, and line segment PQ is extended to point F so that $FE \perp PE$. If the area of square $ABCD$ is 9, what is the area of quadrilateral $QBEF$?

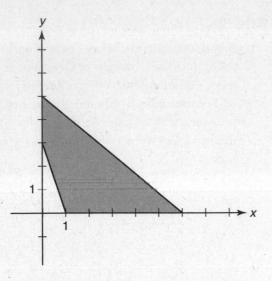

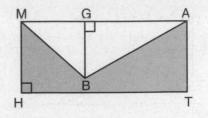

8. In the figure, *MATH* is a rectangle, *GB* = 4.8, *MH* = 6, and *HT* = 15. The area of the shaded region is how many times larger than the area of △*MBA*?

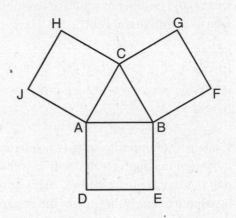

4. In the figure above, what is the number of square units in the area of the shaded region?

5. If the coordinates of the endpoints of a diagonal of a square are (−2, −3) and (5, 4), what is the number of square units in the area of the square?

6. What is the number of square units in the area of the region in the first quadrant of the *xy*-plane that is bounded by *y* = |*x*| + 2, the line *x* = 5, the positive *x*-axis, and the positive *y*-axis?

9. In the figure above, quadrilaterals *ABED*, *BFGC*, and *ACHJ* are squares. If the area of equilateral △*ABC* is 16√3 square inches, what is the number of inches in the perimeter of polygon *ADEBFGCHJA*?

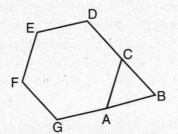

7. In the figure above, *ACDEFG* is a regular hexagon. Sides *DC* and *GA* are extended such that *A* is the midpoint of *BG* and *C* is the midpoint of *BD*. If the area of △*ABC* is 9√3 square centimeters, what is the number of centimeters in the perimeter of polygon *ABCDEFG*?

OVERVIEW

The SAT may also include questions that relate to basic relationships in a circle and its equation in the *xy*-coordinate plane.

CIRCLE TERMS

As illustrated in Figure 6.15,

- A **radius** of a circle is a segment whose endpoints are the center of the circle and a point on the circle. The plural of radius is radii.
- A **chord** of a circle is a segment whose endpoints are points on the circle. The longest chord of a circle is the **diameter,** which passes through the center of the circle. The radius length is one-half the length of the diameter of the circle.

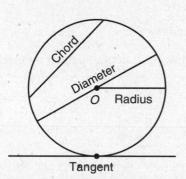
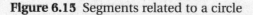

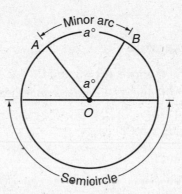
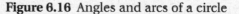

Figure 6.15 Segments related to a circle   **Figure 6.16** Angles and arcs of a circle

- A **tangent** of a circle is a line that intersects a circle in exactly one point.

ANGLES AND ARCS

Referring to Figure 6.16,

- An **arc** is a curved section of the circle. The number of degrees of arc in a circle is 360. A diameter divides a circle into two equal arcs, each of which is called a **semicircle**. An arc smaller than a semicircle is called a **minor arc** and is named by its endpoints. A **major arc** is an arc that is greater than a semicircle.
- An angle whose vertex is at the center and whose sides are radii is called a **central angle**. The measure of a central angle is equal to the degree measure of its intercepted arc and vice versa.

An **inscribed angle** is an angle whose vertex is on the circle and whose sides intercept an arc of a circle. Figure 6.17 illustrates that the measure of an inscribed angle is equal to one-half the measure of its intercepted arc. If $m\angle AC = x$, then the measure of inscribed angle ABC is $\frac{1}{2}x$.

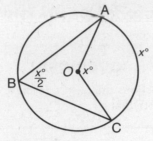

TIP

Inscribed angles that intercept the same arc of a circle are equal in measure.

Figure 6.17 Comparing measures of central and inscribed angles.

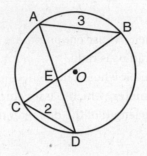

➡ Example

In circle O above, $AB = 3$ and $CD = 2$. Which of the following statements must be true?

I. $\angle ABC$ and $\angle DCB$ have equal measures

II. $\overline{AB} \parallel \overline{CD}$

III. $2(AE) = 3(CE)$

(A) I and II

(B) I and III

(C) I only

(D) III only

Solution

- Since inscribed angles BAD and DCB intercept the same arc, BD, they are equal in measure. Also, inscribed angles ABC and CDA intercept the same arc, AC, so their measures are also equal.

- Since triangles AEB and DEC agree in two pairs of angles, they are similar so the lengths of their corresponding sides have the same ratio. Since $\dfrac{AB}{CD} = \dfrac{3}{2}$, $\dfrac{AE}{CE} = \dfrac{3}{2}$ so $2(AE) = 3(CE)$, which makes statement III is true. There is no justification for concluding that $\overline{AB} \parallel \overline{CD}$ or that $\angle ABC$ and $\angle DCB$ have equal measures since they are *not* corresponding angles of similar triangles.

The correct choice is **(D)**.

CIRCLE FORMULAS

The distance around a circle is its circumference. The circumference, C, of a circle depends on its diameter, d:

$$C = \pi d = 2\pi r$$

The length, L, of an arc of a circle is a fractional part of the circumference:

$$L = \frac{n}{360} \times \overset{\text{circumference}}{2\pi r}$$

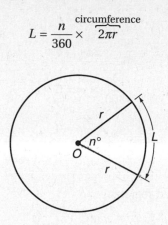

The area of a circle is the number of square units in the region it encloses. The area, A, of a circle depends on its radius:

$$A = \pi r^2$$

A sector of a circle is the region bounded by two radii and their intercepted arc. The area, A_s, of a sector of a circle is a fractional part of the area of the circle:

$$A_s = \frac{n}{360} \times \overset{\text{area of circle}}{\pi r^2}$$

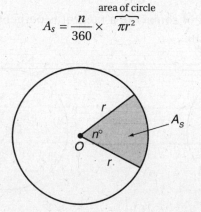

➡ Example

If in the accompanying figure the length of arc AB of circle O is 8π, what is the number of square units in the area of the shaded region?

(A) 16π

(B) 32π

(C) 64π

(D) 96π

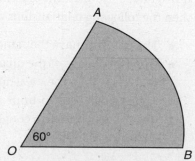

Solution

- Find the radius of circle O:

$$L = \frac{60}{360} \times 2\pi r = 8\pi$$

$$\frac{120}{360} \times r = 8$$

$$\frac{1}{3}r = 8$$

$$r = 8 \times 3 = 24$$

- Find the area of the shaded region:

$$A_s = \frac{60}{360} \times \pi(24)^2$$

$$= \frac{\pi}{6} \times 576$$

$$= 96\pi$$

The correct choice is **(D)**.

TANGENTS AND RADII

A radius drawn to the point of contact of a tangent is perpendicular to the tangent at that point, as shown in Figure 6.18.

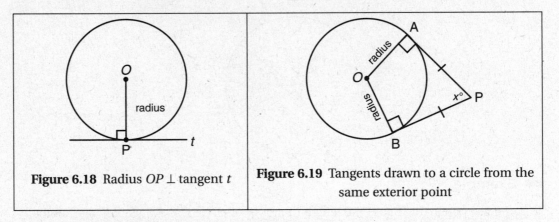

Figure 6.18 Radius $OP \perp$ tangent t

Figure 6.19 Tangents drawn to a circle from the same exterior point

If two tangents are drawn to a circle from the same exterior point P, as shown in Figure 6.19, then the following relationships are always true:

- The tangents have the same length so $PA = PB$.
- Opposite angles of the quadrilateral that is formed are supplementary. If the measure of $\angle P$ is x, then the measure of central angle AOB is $180 - x$.
- The segment drawn from point P to the center of the circle (not shown) divides the quadrilateral into two congruent right triangles and, as a result, bisects angles AOB and APB, as well as arc AB.

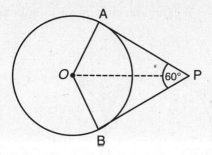

In the figure above, line segments *PA* and *PB* are tangent to circle *O* at points *A* and *B*, respectively, and the measure of ∠*APB* is 60°. Segment *OP* is drawn. If $OP = \dfrac{14}{\pi}$, what is the length of minor arc *AB*?

Solution

- Angle *AOB* measures 180° − 60° = 120°. Since *OP* divides quadrilateral *OAPB* into two congruent right triangles, it bisects ∠*AOB*:

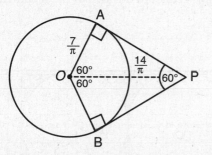

- The side opposite the 30° angle in a right triangle is one-half the length of the hypotenuse so $OA = \dfrac{7}{\pi}$.

- Since the measure of minor arc *AB* is 120°:

$$\text{Length of } AB = \frac{120}{360} \times 2\pi \times \frac{7}{\pi}$$

$$= \frac{1}{3} \times 14$$

$$= \frac{14}{3}$$

Grid-in **14/3**

FINDING AREAS OF SHADED REGIONS INDIRECTLY

To find the area of a shaded region, you may need to subtract the areas of figures that overlap.

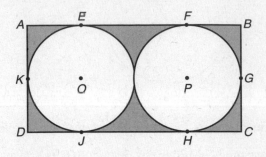

➡ **Example**

In the figure above, the radius of each circle is 1. If circles O and P touch the sides of rectangle $ABCD$ only at the lettered points, what is the area of the shaded region?

Solution

The area of the shaded regoin is équal to the area of the rectangle minus the sum of the areas of the two circles.

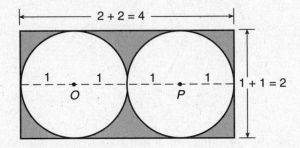

- The width of the rectangle is equal to the diameter of a circle, which is $1 + 1$ or 2.
- The length of the rectangle is equal to the sum of the two diameters, which is $2 + 2$ or 4.
- The area of the rectangle is length × width = $4 \times 2 = 8$, and the area of the each circle is $\pi r^2 = \pi 1^2 = \pi$.
- Thus,

$$\text{Area of shaded region} = 8 - (\pi + \pi) = \mathbf{8 - 2\pi}$$

DIAMETER AND CHORD RELATIONSHIPS

In a circle,

- If a line is drawn through the center of a circle and perpendicular to a chord, it bisects the chord and its arcs. See Figure 6.20.

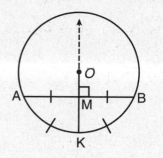

Figure 6.20 If $\overline{OK} \perp \overline{AB}$, then $AM = BM$ and $m\angle AK = m\angle BK$

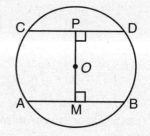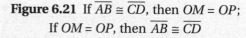

Figure 6.21 If $\overline{AB} \cong \overline{CD}$, then $OM = OP$; If $OM = OP$, then $\overline{AB} \cong \overline{CD}$

- Congruent chords are the same distance from the center of the circle, and chords that are the same distance from the center of a circle are congruent. See Figure 6.21.

➡ Example

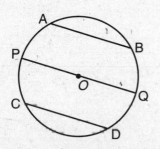

In the figure above, $\overline{PQ}$ is a diameter of circle O, $\overline{AB} \parallel \overline{PQ} \parallel \overline{CD}$, and $\overline{AB} \cong \overline{CD}$. The length of chord $\overline{AB}$ is $\frac{3}{4}$ of the length of diameter $\overline{PQ}$. If r represents the radius length of circle O, what is the distance between chords $\overline{AB}$ and $\overline{CD}$ in terms of r?

(A) $\frac{5}{4}r$

(B) $\frac{\sqrt{7}}{4}r$

(C) $\frac{3}{2}r$

(D) $\frac{\sqrt{7}}{2}r$

Solution

- It is given that $AB = \frac{3}{4} \times 2r = \frac{3}{2}r$. Draw $\overline{OH} \perp \overline{AB}$ and extend $\overline{OH}$ so it intersects $\overline{CD}$ at K. Since $\overline{AB} \parallel \overline{CD}$, $\overline{OK} \perp \overline{CD}$, HK represents the distance between the two chords.

When working in a circle in which the radius is not drawn, it may be necessary to draw your own radius in a way that helps solve the problem.

- Draw radius $\overline{OA}$ thereby forming right triangle OHA. Because $\overline{OH} \perp \overline{AB}$, it bisects AB so $AH = \dfrac{1}{2} \times \dfrac{3}{2}r = \dfrac{3}{4}r$:

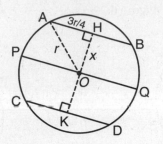

- Use the Pythagorean theorem to find the length of $\overline{OH}$:

$$x^2 + \left(\dfrac{3}{4}r\right)^2 = r^2$$

$$x^2 + \dfrac{9}{16}r^2 = r^2$$

$$x^2 = \dfrac{7}{16}r^2$$

$$\sqrt{x^2} = \sqrt{\dfrac{7}{16}r^2}$$

$$x = \dfrac{\sqrt{7}}{4}r$$

- Since $\overline{AB} \cong \overline{CD}$, $OH = OK = \dfrac{\sqrt{7}}{4}r$ and $HK = 2 \times \dfrac{\sqrt{7}}{4}r = \dfrac{\sqrt{7}}{2}r$.

The correct choice is **(D)**.

EQUATION OF A CIRCLE: CENTER-RADIUS FORM

The set of points (x, y) in the xy-coordinate plane that lie on a circle whose center is at (h, k) and has a radius of r, as shown in Figure 6.22, is described by the equation $(x - h)^2 + (y - k)^2 = r^2$.

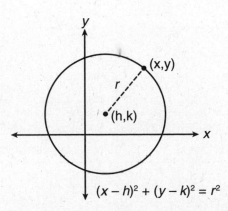

Figure 6.22 Equation of a circle in the xy-plane

The equation $(x - h)^2 + (y - k)^2 = r^2$ is referred to as the **center-radius form** of the equation of a circle. If the circle is centered at the origin, then $h = k = 0$ and the equation of the circle simplifies to $x^2 + y^2 = r^2$.

- If the center of a circle is at $(-2, 1)$ and its radius is 5, then an equation of the circle is $(x - h)^2 + (y - k)^2 = r^2$ where $h = -2$, $k = 1$, and $r = 5$. Making the substitutions gives $(x + 2)^2 + (y - 1)^2 = 25$.

- The center and radius of a circle can be read directly from the center-radius form of its equation. If the equation of a circle is given as

$$(x + 1)^2 + (y - 5)^2 = 64$$

then you know that $h = -1$, $k = 5$, and $r^2 = 64$ so $r = 8$. Hence, the center of the circle is $(-1, 5)$, and the radius is 8.

➡ Example

Which equation could represent the circle shown in the graph below that passes through the point $(0, -1)$?

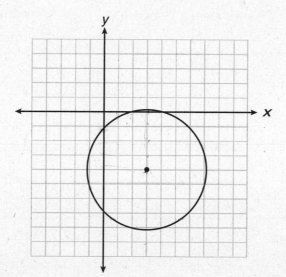

(A) $(x - 3)^2 + (y + 4)^2 = 16$

(B) $(x + 3)^2 + (y - 4)^2 = 18$

(C) $(x + 3)^2 + (y - 4)^2 = 16$

(D) $(x - 3)^2 + (y + 4)^2 = 18$

Solution

Counting unit boxes, the center of the circle is located at $(3, -4)$ so the equation of the circle has the form $(x - 3)^2 + (y + 4)^2 = r^2$. Therefore, you can eliminate choices (B) and (C). Since the radius is more than 4 units, $r^2 > 16$, which means the correct choice must be **(D)**.

➡ Example

What is the center and radius of a circle whose equation is $3x^2 + 3y^2 - 12x + 18y = 69$?

Solution

Rewrite the equation in center-radius form by completing the square for both x and y.

- Divide each term of the equation by 3, and then collect terms with same variable:

$$(x^2 - 4x) + (y^2 + 5y) = 23$$

- Complete the square for x and for y:

$$\left(x^2 - 4x + \underline{\underline{4}}\right) + \left(y^2 + 6y + \underline{\underline{9}}\right) = 23 + \underline{4 + 9}$$
$$(x - 2)^2 + (y + 3)^2 = 36$$

- The circle equation now has the center-radius form $(x - h)^2 + (y - k)^2 = r^2$ with center at $(h, k) = (2, -3)$ and radius $r = \sqrt{36} = 6$.

Multiple-Choice

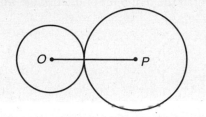

1. Circles O and P intersect at exactly one point, as shown in the figure above. If the radius of circle O is 2 and the radius of circle P is 6, what is the circumference of any circle that has OP as a diameter?

    (A) 4π
    (B) 8π
    (C) 12π
    (D) 16π

2. What is the area of a circle with a circumference of 10π?

    (A) $\sqrt{10\pi}$
    (B) 5π
    (C) 25π
    (D) 100π

3. Every time the pedals go through a 360° rotation on a certain bicycle, the tires rotate three times. If the tires are 24 inches in diameter, what is the minimum number of complete rotations of the pedals needed for the bicycle to travel at least 1 mile? [1 mile = 5,280 feet]

    (A) 24
    (B) 281
    (C) 561
    (D) 5,280

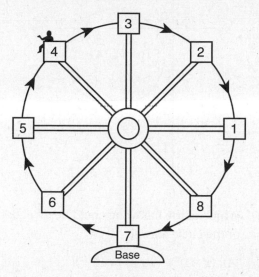

4. Kristine is riding in car 4 of the ferris wheel represented in the diagram above, which is $\dfrac{84}{\pi}$ meters from car 8. The ferris wheel is rotating in the direction indicated by the arrows. If each of the cars are equally spaced around the circular wheel, what is the best estimate of the number of meters in the distance through which Kristine's car will travel to reach the bottom of the ferris wheel before her car returns to the same position?

    (A) 42.0
    (B) 52.50
    (C) 64.75
    (D) 105.0

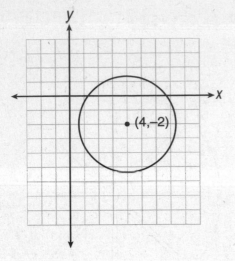

8. If a bicycle wheel has traveled $\dfrac{f}{\pi}$ feet after n complete revolutions, what is the length in feet of the diameter of the bicycle wheel?

(A) $\dfrac{f}{n\pi^2}$

(B) $\dfrac{\pi^2}{fn}$

(C) $\dfrac{nf}{\pi^2}$

(D) nf

5. Which of the following could be an equation of the circle above?

(A) $(x-4)^2 + (y+2)^2 = 17$
(B) $(x+4)^2 + (y-2)^2 = 17$
(C) $(x-4)^2 + (y+2)^2 = 13$
(D) $(x+4)^2 + (y-2)^2 = 13$

6. If the equation of a circle is $x^2 - 10x + y^2 + 6y = -9$, which of the following lines contains a diameter of the circle?

(A) $y = 2x - 7$
(B) $y = -2x + 7$
(C) $y = 2x + 13$
(D) $y = -2x + 13$

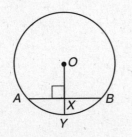

7. In the figure above, if the radius length of circle O is 10, $\overline{OY} \perp \overline{AB}$, and $AB = 16$, what is the length of segment XY?

(A) 2
(B) 3
(C) 4
(D) 6

$$x^2 + y^2 - 6x + 8y = 56$$

9. What is the area of a circle whose equation is given above?

(A) 25π
(B) 81π
(C) 162π
(D) $6,561\pi$

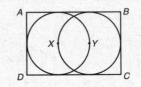

10. In the figure above, X and Y are the centers of two overlapping circles. If the area of each circle is 7, what is the area of rectangle $ABCD$?

(A) $14 - \dfrac{17}{\pi}$

(B) $7 + \dfrac{14}{\pi}$

(C) $\dfrac{28}{\pi}$

(D) $\dfrac{42}{\pi}$

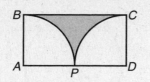

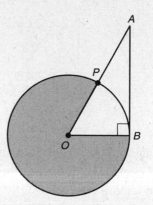

11. In rectangle *ABCD* above, arcs *BP* and *CP* are quarter circles with centers at points *A* and *D*, respectively. If the area of each quarter circle is π, what is the area of the shaded region?

(A) $4 - \dfrac{\pi}{2}$

(B) $4 - \pi$

(C) $8 - \pi$

(D) $8 - 2\pi$

13. The circle shown above has center *O* and a radius length of 12. If *P* is the midpoint of $\overline{OA}$ and $\overline{AB}$ is tangent to circle *O* at *B*, what is the area of the shaded region?

(A) 81π

(B) 96π

(C) 120π

(D) 128π

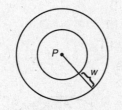

12. In the figure above, point *P* is the center of each circle. The circumference of the larger circle exceeds the circumference of the smaller circle by 12π. What is the width, *w*, of the region between the two circles?

(A) 4

(B) 6

(C) 8

(D) 9

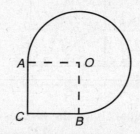

14. In the figure above, *OACB* is a square with area $4x^2$. If *OA* and *OB* are radii of a sector of a circle *O*, what is the perimeter, in terms of *x*, of the unbroken figure?

(A) $x(4 + 3\pi)$

(B) $x(3 + 4\pi)$

(C) $x(6 + 4\pi)$

(D) $4(x + 2\pi)$

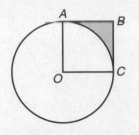

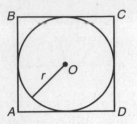

15. In the figure above, *OABC* is a square. If the area of circle *O* is 2π, what is the area of the shaded region?

   (A) $\dfrac{\pi}{2} - 1$

   (B) $2 - \dfrac{\pi}{2}$

   (C) $\pi - 2$

   (D) $\dfrac{\pi - 1}{2}$

17. In the figure above, if circle *O* is inscribed in square *ABCD* in such a way that each side of the square is tangent to the circle, which of the following statements must be true?

   I. $AB \times CD < \pi \times r \times r$

   II. Area $ABCD = 4r^2$

   III. $r < \dfrac{2(CD)}{\pi}$

   (A) I and II
   (B) I and III
   (C) II and III
   (D) II only

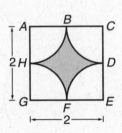

16. In the figure above, the vertices of square *ACEG* are the centers of four quarter circles of equal area. What is the best approximation for the area of the shaded region? (Use π = 3.14.)

   (A) 0.64
   (B) 0.79
   (C) 0.86
   (D) 1.57

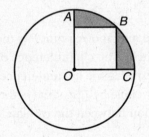

18. In the figure above, *OABC* is a square and *B* is a point on the circle with center *O*. If *AB* = 6, what is the area of the shaded region?

   (A) $9(\pi - 2)$
   (B) $9(\pi - 1)$
   (C) $12(\pi - 2)$
   (D) $18(\pi - 2)$

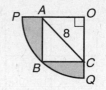

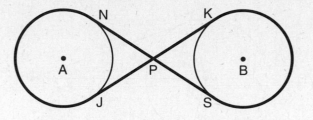

19. In the figure above, arc *PBQ* is one-quarter of a circle with center at *O*, and *OABC* is a rectangle. If *AOC* is an isosceles right triangle with *AC* = 8, what is the perimeter of the figure that encloses the shaded region?

(A) $24 - 4\pi$

(B) $24 - 4\sqrt{2} + 4\pi$

(C) $16 - 4\sqrt{2} + 4\pi$

(D) $16 + 4\pi$

20. The center of circle *Q* has coordinates $(3, -2)$ in the *xy*-plane. If an endpoint of a radius of circle *Q* has coordinates $R(7, 1)$, what is an equation of circle *Q*?

(A) $(x - 3)^2 + (y + 2)^2 = 5$

(B) $(x + 3)^2 + (y - 2)^2 = 25$

(C) $(x - 3)^2 + (y + 2)^2 = 25$

(D) $(x + 3)^2 + (y - 2)^2 = 5$

21. In the pully system illustrated in the figure above, a belt with negligible thickness is stretched tightly around two identical wheels represented by circles *A* and *B*. If the radius of each wheel is $\dfrac{12}{\pi}$ inches and the measure of $\angle NPJ$ is 60°, what is the length of the belt?

(A) $32 + \dfrac{48\sqrt{3}}{\pi}$

(B) $48 + \dfrac{32\sqrt{3}}{\pi}$

(C) $\left(32 + 48\sqrt{3}\right)\pi$

(D) $80\pi\sqrt{3}$

Grid-In

1.

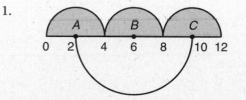

In the figure above, the sum of the areas of the three shaded semicircles with centers at *A*, *B*, and *C* is *X*, and the area of the larger semicircle below the line is *Y*. If $Y - X = k\pi$, what is the value of *k*?

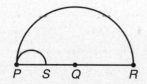

2. In the figure above, each arc is a semicircle. If *S* is the midpoint of *PQ* and *Q* is the midpoint of *PR*, what is the ratio of the area of semicircle *PS* to the area of semicircle *PR*?

3. What is the distance in the *xy*-plane from the point $(3, -6)$ to the center of the circle whose equation is $x(x + 4) + y(y - 12) = 9$?

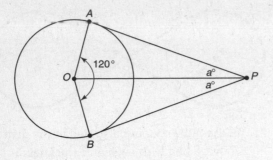

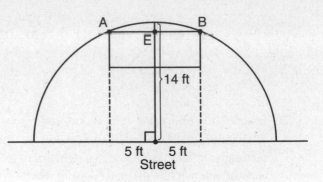

4. In the figure above, $\overline{PA}$ is tangent to circle O at point A, $\overline{PB}$ is tangent to circle O at point B. Angle AOB measures 120° and $OP = \dfrac{24}{\pi}$. What is the length of minor arc AB?

5. The diagram above shows a semicircular arch over a street that has a radius of 14 feet. A banner is attached to the arch at points A and B, such that $AE = EB = 5$ feet. How many feet above the ground are these points of attachment for the banner, correct to the *nearest tenth* of a foot?

OVERVIEW

The reference section at the beginning of each SAT Math section includes the volume formulas for these solids: rectangular box, right circular cylinder, sphere, right circular cone, and pyramid with a rectangular base.

VOLUME OF A RECTANGULAR SOLID AND CYLINDER

The SAT may include problems that make use of the formulas in Table 6.1. The **lateral area** of a right circular cylinder is the area of its curved surface that does not include the bases. The **surface area** (SA) of a solid is the sum of the areas of all of its surfaces.

Table 6.1 Formulas for a Rectangular Box and Circular Cylinder

Type of Solid	Diagram	Formulas
Rectangular box		• $V = l \times w \times h$ • $d = \sqrt{l^2 + w^2 + h^2}$ • $SA = 2[l \times w + l \times h + w \times h]$
Right circular cylinder		• $V = \pi r^2 h$ • $SA = \underbrace{2\pi r^2}_{\substack{\text{Area of}\\ \text{bases}}} + \underbrace{2\pi rh}_{\text{Lateral area}}$

➡ **Example**

An oil filter that has the shape of a right circular cylinder is 12 centimeters in height and has a volume of 108π cubic centimeters. What is the *diameter* of the base of the cylinder, in centimeters?

Solution

$$V = \pi r^2 h$$

$$\frac{108\pi}{12\pi} = \frac{12\pi(r^2)}{12\pi}$$

$$9 = r^2$$

$$3 = r$$

Since the radius of the base is 3 centimeters, the diameter is 6 centimeters.

➥ Example

A coffee shop makes coffee in a 5 gallon capacity urn and serves it in cylindrical-shaped mugs with an internal diameter of 3 inches. Coffee is poured into each mug at a height of about 4 inches. What is the largest number of full mugs of coffee that can be filled from the coffee urn if the urn is filled to capacity? (Note: There are 231 cubic inches in 1 gallon.)

Solution

- Five gallons is equivalent in volume to $231 \times 5 = 1{,}155$ cubic inches.
- Find the volume of each filled coffee mug:

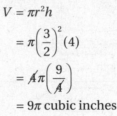

$$= 9\pi \text{ cubic inches}$$

- Find the number of filled coffee mugs by dividing the capacity of the urn by 9π:

$$\frac{1{,}155}{9\pi} \approx 40.85$$

Hence, **40** full coffee mugs can be filled from the coffee urn.

VOLUME OF A CONE, PYRAMID, AND SPHERE

The volume of a right circular cone and a pyramid are each calculated by multiplying one-third of the area of the base by the height of the figure. The volume of a sphere depends on its radius. See Table 6.2 where B represents the area of the base of the solid.

Table 6.2 Volume Formulas for a Cone, Pyramid, and Sphere

Type of Solid	Diagram	Formulas
Right circular cone	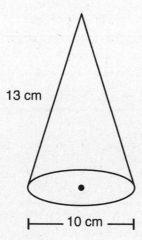	$V = \dfrac{1}{3}Bh$ $= \dfrac{1}{3}\pi r^2 h$
Pyramid with rectangle base		$V = \dfrac{1}{3}Bh$ $= \dfrac{1}{3}l \times w \times h$
Sphere		$V = \dfrac{4}{3}\pi r^3$

➥ Example

13 cm

10 cm

What is the volume, in cubic centimeters, of the right circular cone in the figure above, in terms of π?

Solution

Use the volume formula $V = \frac{1}{3}\pi r^2 h$ where $r = \frac{1}{2}(10) = 5$, and $h = 12$ since it is a leg in a 5-*12*-13 right triangle:

$$V = \frac{1}{3}\pi r^2 h$$

$$= \frac{1}{3}\pi \times 5^2 \times 12$$

$$= 100\pi \text{ cubic centimeters}$$

Multiple-Choice

1. A pyramid has a height of 12 centimeters and a square base. If the volume of the pyramid is 256 cubic centimeters, how many centimeters are in the length of one side of its base?

   (A) 8
   (B) 16
   (C) 32
   (D) $\dfrac{8}{\sqrt{3}}$

2. The volume of a rectangular box is 144 cubic inches. The height of the solid is 8 inches. Which measurements, in inches, could be the dimensions of the base?

   (A) 3.3×5.5
   (B) 2.5×7.2
   (C) 12.0×8.0
   (D) 9.0×9.0

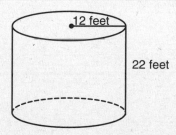

3. The cylindrical tank shown in the diagram above is to be painted. The tank is open at the top, and the bottom does *not* need to be painted. Only the outside needs to be painted. Each can of paint covers 500 square feet. What is the least number of cans of paint that must be purchased to complete the job?

   (A) 2
   (B) 3
   (C) 4
   (D) 5

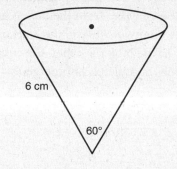

4. In the figure above of a right circular cone, what is the approximate number of cubic centimeters in the volume of the cone?

   (A) 49
   (B) 105
   (C) 210
   (D) 306

5. Sophie has a hard rubber ball whose circumference measures 13 inches. She wants to store it in a box. What is the number of cubic inches in the volume of the smallest cube-shaped box with integer dimensions that she can use?

   (A) 64
   (B) 81
   (C) 125
   (D) 216

6. The amount of light produced by a cylindrical fluorescent light bulb depends on its lateral area. A certain cylindrical fluorescent light bulb is 36 inches long, has a 1 inch diameter, and is manufactured to produce 0.283 watts of light per square inch. What is the best estimate for the total amount of light that it is able to produce?

   (A) 32 watts
   (B) 34 watts
   (C) 40 watts
   (D) 48 watts

7. A rectangular fish tank has a base 2 feet wide and 2 feet long. When the tank is partially filled with water, a solid cube with an edge length of 1 foot is placed in the tank. If no overflow of water from the tank is assumed, by how many *inches* will the level of the water in the tank rise when the cube becomes completely submerged?

   (A) $\dfrac{1}{6}$

   (B) $\dfrac{1}{2}$

   (C) 2

   (D) 3

8. The volume of a cylinder of radius r is $\dfrac{1}{4}$ of the volume of a rectangular box with a square base of side length x. If the cylinder and the box have equal heights, what is r in terms of x?

   (A) $\dfrac{x^2}{2\pi}$

   (B) $\dfrac{x}{2\sqrt{\pi}}$

   (C) $\dfrac{\sqrt{2}x}{\pi}$

   (D) $\dfrac{\pi}{\sqrt{2}x}$

9. The height of sand in a cylinder-shaped can drops 3 inches when 1 cubic foot of sand is poured out. What is the diameter, in *inches*, of the cylinder?

   (A) $\dfrac{2}{\sqrt{\pi}}$

   (B) $\dfrac{4}{\sqrt{\pi}}$

   (C) $\dfrac{16}{\pi}$

   (D) $\dfrac{48}{\sqrt{\pi}}$

10. The height h of a cylinder equals the circumference of the cylinder. In terms of h, what is the volume of the cylinder?

    (A) $\dfrac{h^3}{4\pi}$

    (B) $\dfrac{h^2}{2\pi}$

    (C) $\dfrac{h^3}{2}$

    (D) $h^2 + 4\pi$

11. As shown in the figure above, a worker uses a cylindrical roller to help pave a road. The roller has a radius of 9 inches and a width of 42 inches. To the *nearest square inch*, what is the area the roller covers in one complete rotation?

   (A)   2,374
   (B)   2,375
   (C) 10,682
   (D) 10,688

12. The density of lead is approximately 0.41 pounds per cubic inch. What is the approximate mass, in pounds, of a lead ball that has a 5 inch diameter?

   (A)   26.8
   (B)   78.5
   (C)   80.4
   (D)   214.7

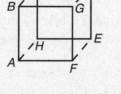

13. In the figure above, if the edge length of the cube is 4, what is the shortest distance from *A* to *D*?

   (A) $4\sqrt{2}$
   (B) $4\sqrt{3}$
   (C) 8
   (D) $4\sqrt{2} + 4$

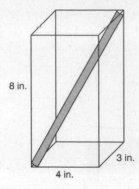

16. If a pyramid with a square base with side length s and a right cone with radius r have equal heights and equal volumes, then which equation must be true?

(A) $s = \sqrt{\pi r}$

(B) $s = \dfrac{\sqrt{r}}{\pi}$

(C) $s = \pi\sqrt{r}$

(D) $s = r\sqrt{\pi}$

14. A cylindrical tube with negligible thickness is placed into a rectangular box that is 3 inches by 4 inches by 8 inches, as shown in the accompanying diagram. If the tube fits exactly into the box diagonally from the bottom left corner to the top right back corner, what is the best approximation of the number of inches in the length of the tube?

(A) 3.9

(B) 5.5

(C) 7.8

(D) 9.4

17. An ice cube has a surface area of 150 square centimeters. If the ice cube melts at a constant rate of 13.0 cubic centimeters per minute, the number of minutes that elapse before the ice cube is completely melted is closest to which of the following?

(A) 10

(B) 11

(C) 12

(D) 14

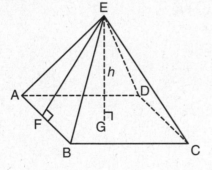

18. A hot water tank with a capacity of 85.0 gallons of water is being designed to have the shape of a right circular cylinder with a diameter 1.8 feet. Assuming that there are 7.48 gallons in 1 cubic foot, how high in feet will the tank need to be?

(A) 4.50

(B) 4.75

(C) 5.00

(D) 5.25

15. In the pyramid shown in the diagram above G is the center of square base $ABCD$, $\overline{EF} \perp \overline{AB}$, and h the height of the pyramid. Which statements must be true?

  I. $EA = EC$

  II. $\triangle BEC$ is isosceles

  III. $EF = EG$

(A) I and II only

(B) I and III only

(C) I only

(D) II only

For Questions 19–20 use the figure below.

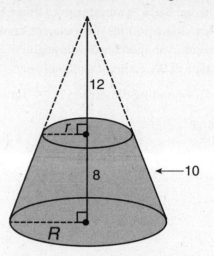

A lamp shade with a circular base is an example of a solid shape called a *frustrum*. In the figure above, the shaded region represents a frustrum of a right cone in which the portion of the original cone that lies 12 inches below its vertex has been cut off by a slicing plane (not shown) parallel to the base.

19. If the height and slant height of the frustrum are 8 inches and 10 inches, respectively, what is the number of inches in the radius length, *R*, of the original cone?

(A)   9
(B) 12
(C) 15
(D) 18

20. What is the volume, in cubic inches, of the frustrum?

(A) 324π
(B) 812π
(C) 1,089π
(D) 1,176π

Grid-In

1. The dimensions of a rectangular box are integers greater than 1. If the area of one side of this box is 12 and the area of another side is 15, what is the volume of the box?

2. The Parkside Packing Company needs a rectangular shipping box. The box must have a length of 1 foot, a width of 8 inches, and a volume of *at least* 700 cubic inches. What is the least number of inches in height of the box such that the height is a whole number?

3. A planned building was going to be 100 feet long, 75 feet deep, and 30 feet high. The owner decides to increase the volume of the building by 10% without changing the dimensions of the depth and the height. What will be the new length of this building in feet?

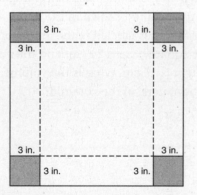

Note: Figure is not drawn to scale.

4. A box is constructed by cutting 3-inch squares from the corners of a square sheet of cardboard, as shown in the accompanying diagram, and then folding the sides up. If the volume of the box is 75 cubic inches, find the number of square inches in the area of the *original* sheet of cardboard.

5. Two spheres that are tangent to each other have volumes of 36π cubic centimeters and 972π cubic centimeters. What is the greatest possible distance, in centimeters, between a point on one sphere and a second point on the other sphere?

6. A sealed cylindrical can holds three tennis balls each with a diameter of 2.5 inches. If the can is designed to have the smallest possible volume, find the number of cubic inches of unoccupied space inside the can correct to the *nearest tenth of a cubic inch*.

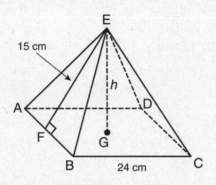

7. The *slant height* of a pyramid is the perpendicular distance from the vertex of the pyramid to the base of a triangular side. In the pyramid shown in the figure above, the height is labeled h, the length of a side of the square base is 24 cm, and the slant height is 15 cm. What is the volume, in cubic centimeters, of the pyramid?

8. A bookend that weights 0.24 pounds is shaped like a pyramid with a square base. How many pounds does a larger, similar pyramid-shaped bookend weigh if it is made of the same material and each corresponding dimension is $2\frac{1}{2}$ times as large?

OVERVIEW

This section reviews some basic trigonometric facts and relationships that you need to know for the new SAT. The Greek letter θ (theta) is sometimes used to represent the unknown measure of an angle.

MEASURING ANGLES IN RADIANS

Degrees are a measure of rotation where $1° = \dfrac{1}{360}$ th of a complete rotation. A *radian* is a unit of angle measure that express the measure of an angle as a real number. One **radian** is the measure of a central angle of a circle that intercepts an arc whose length is the same as a radius of that circle. In Figure 6.23, central angle θ intercepts an arc that is 6π inches in length in a circle whose radius measures 12 inches. The measure of angle θ, in radians, is $\dfrac{6\pi}{12} = \dfrac{\pi}{2}$ radians.

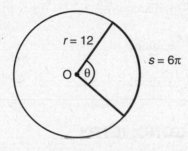

Figure 6.23 Radian measure: $\theta = \dfrac{s}{r}$

In general, $\theta = \dfrac{s}{r}$ or $s = r \cdot \theta$ where s represents the length of the arc intercepted by a central angle that measures θ radians in a circle with radius r.

> **MATH REFERENCE FACT**
>
> The number of radians of arc in a semicircle is π and the number of radians of arc in a circle is 2π.

Converting Between Radians and Degrees

Since the degree measure of arc of a circle is 360°,

$$2\pi \text{ radians} = 360° \text{ so } \pi \text{ radians} = \frac{360°}{2} = 180°$$

This relationship provides a way of converting between radian and degree measures:

- To change from degrees to radians, multiply the number of degrees by $\frac{\pi}{180°}$. For example,

$$60° = \overset{3}{\cancel{60°}} \times \frac{\pi}{\cancel{180°}} = \frac{\pi}{3} \text{ radians}$$

- To change from radians to degrees, multiply the number of radians by $\frac{180°}{\pi}$. For example,

$$\frac{7}{12}\pi \text{ radians} = \frac{7\cancel{\pi}}{\cancel{12}} \times \frac{\overset{15°}{\cancel{180°}}}{\cancel{\pi}} = 105°$$

If you memorize the radian equivalents of 30°, 45°, and 90°, then you can use these values to quickly figure out the radian equivalents of their multiples. For example,

- $60° = 2 \times 30° = 2 \times \frac{\pi}{6} = \frac{\pi}{3}$ radians

- $150° = 5 \times 30° = 5 \times \frac{\pi}{6} = \frac{5\pi}{6}$ radians

- $270° = 3 \times 90° = 3 \times \frac{\pi}{2} = \frac{3\pi}{2}$ radians

- $315° = 7 \times 45° = 7 \times \frac{\pi}{4} = \frac{7\pi}{4}$ radians

RIGHT TRIANGLE TRIGONOMETRY

The three basic trigonometric functions of **sin**e, **cos**ine, and **tan**gent are defined in Table 6.3.

Table 6.3 The Three Basic Trigonometric Functions

Basic Three Trigonometric Functions	Quotient Relationships	
$\sin A = \dfrac{\text{leg opposite } \angle A}{\text{hypotenuse}} = \dfrac{a}{c}$		
$\cos A = \dfrac{\text{leg adjacent } \angle A}{\text{hypotenuse}} = \dfrac{b}{c}$	$\dfrac{\sin A}{\cos A} = \tan A$	
$\tan A = \dfrac{\text{leg opposite } \angle A}{\text{leg adjacent } \angle A} = \dfrac{a}{b}$		

COFUNCTION RELATIONSHIPS

Two angles are complementary if their measures add up to $90° \left(= \dfrac{\pi}{2} \text{ radians} \right)$. The sine and cosine functions have equal values when their angles are complementary. For example, $\sin 50° = \cos 40°$ and $\cos \dfrac{\pi}{3} = \sin \dfrac{\pi}{6}$. In general,

- If x is an acute angle measured in degrees, then

$$\sin x = \cos (90 - x) \text{ or, equivalently, } \cos x = \sin (90 - x)$$

- If x is an acute angle measured in radians, then

$$\sin x = \cos \left(\dfrac{\pi}{2} - x \right) \text{ or, equivalently, } \cos \dfrac{\pi}{2} = \sin \left(\dfrac{\pi}{2} - x \right)$$

INDIRECT MEASUREMENT

Trigonometric functions are particularly useful when it is necessary to calculate the measure of a side or an angle of a right triangle that may be difficult, if not impossible, to measure directly.

 **Example**

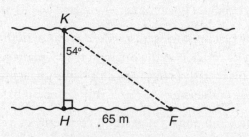

To determine the distance across a river, as shown in the figure above, a surveyor marked two points on one riverbank, H and F, 65 meters apart. She also marked one point, K, on the opposite bank such that $\overline{KH} \perp \overline{HF}$. If $\angle K = 54°$ and x is the width of the river, which of the following equations could be used to find x?

(A) $\tan 54° = \dfrac{x}{65}$

(B) $\sin 36° = \dfrac{x}{65}$

(C) $\tan 36° = \dfrac{x}{65}$

(D) $\sin 54° = \dfrac{x}{65}$

Solution

Represent the width of the river, *KH*, by *x*.

Because the problem involves the sides opposite and adjacent to the given angle, use the tangent ratio:

$$\tan\angle K = \frac{\text{opposite side} \,(HF)}{\text{adjacent side} \,(KH)}$$

$$\tan 54° = \frac{65}{x}$$

Since this is not one of the answer choices, consider the other acute angle of the right triangle at *F*, which measures 90° − 54° = 36°:

$$\tan\angle F = \frac{KH}{HF}$$

$$\tan 36° = \frac{x}{65}$$

The correct choice is (**C**).

ANGLES IN STANDARD POSITION

Trigonometric functions of angles greater than $\frac{\pi}{2}$ radians (= 90°) or less than 0 radians (= 0°) can be given meaning by placing angles in *standard position* in the *xy*-plane. An angle is in **standard position** when its vertex is at the origin in the *xy*-plane and one of its sides, called the **initial side**, remains fixed on the *x*-axis. The side of the angle that rotates about the origin is called the **terminal side** of the angle.

- If the terminal side of an angle rotates in a counterclockwise direction, as shown in Figure 6.24a, the angle is *positive*.
- If the terminal side of an angle rotates in a clockwise direction, as shown in Figure 6.24b, the angle is *negative*.

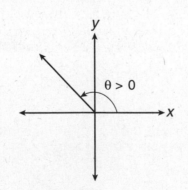

Figure 6.24a Positive angle

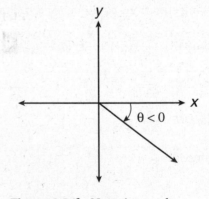

Figure 6.24b Negative angle

DEFINING TRIGONOMETRIC FUNCTIONS USING COORDINATES

If $P(x, y)$ is any point on the terminal side of an angle in standard position and r is the distance of point P from the origin, then trigonometric functions can be defined in terms of x, y, and r as shown in Table 6.5. If a perpendicular is drawn from $P(x, y)$ to the x-axis, a right triangle is formed in which x, y, and r are related by the Pythagorean equation $x^2 + y^2 = r^2$.

Table 6.5 Coordinate Definitions of the Basic Trigonometric Functions

Coordinate Definitions	Standard Position of Angle θ In the xy-plane
If $P(x, y)$ is on the terminal side of an angle, θ in standard position then: ■ $\sin \theta = \dfrac{y}{r}$ ■ $\cos \theta = \dfrac{x}{r}$ ■ $\tan \theta = \dfrac{y}{x}$ where $x^2 + y^2 = r^2$	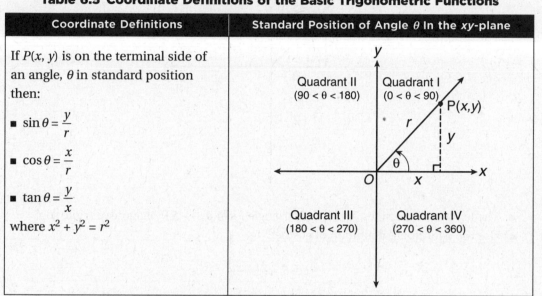

TIP

If $P(x, y)$ is a point on the terminal side of an angle θ in standard position, then the x-coordinate of point P is $r\cos\theta$ and the y-coordinate is $r\sin\theta$ where r, the distance from the origin to point P, is $\sqrt{x^2 + y^2}$.

SIGNS OF TRIGONOMETRIC FUNCTIONS IN THE FOUR QUADRANTS

The signs of trigonometric functions of angle θ depend on the signs of x and y in the particular quadrant in which the terminal side of θ lies, as shown in Table 6.6. For example, in Quadrant II, $x < 0$ and $y > 0$ so $\tan\theta = \dfrac{y}{x} = \dfrac{+}{-} = -$. Quadrant I is the only quadrant in which all of the trigonometric functions are positive at the same time.

Table 6.6 Quadrants in Which Trigonometric Functions Are Positive

Quadrant I	Quadrant II	Quadrant III	Quadrant IV
All are +	Sine is +	Tangent is +	Cosine is +

➡ **Example**

If $P(-3, 4)$ is a point on the terminal side of angle θ, what is the value of $\cos\theta$?

(A) $-\dfrac{3}{4}$

(B) $-\dfrac{3}{5}$

(C) $\dfrac{3}{4}$

(D) $\dfrac{4}{5}$

TIP

The first letter of each word of the phrase "*A*ll *S*tudents *T*ake *C*alculus" can help you remember the quadrants in which a trigonometric function is positive.

Solution

Since $x = -3$ and $y = +4$, the terminal side of the angle θ lies in Quadrant II:

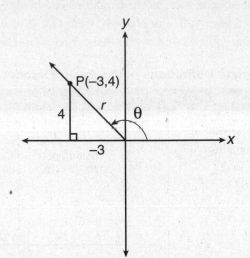

- The lengths of the sides of the right triangle form a 3-4-5 Pythagorean triple so $r = 5$.
- Use the coordinate definition of cosine:

$$\cos\theta = \frac{x}{r} = -\frac{3}{5}$$

The correct choice is **(B)**.

➡ Example

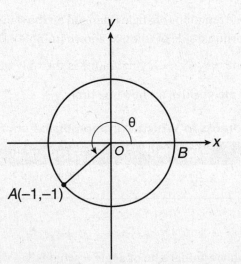

In the xy-plane above, O is the center of the circle, and the measure of angle θ is $k\pi$ radians. If $0 \le \theta \le 2\pi$, what is the value of k?

Solution

Since $(-1, -1) = (x, y)$, $\frac{y}{x} = 1$ so $\triangle AOC$ is a 45°-45° right triangle with $m\angle AOC = 45$. Hence, θ measures 225° or, equivalently, $\frac{5}{4}\pi$ radians.

Grid-in **5/4**

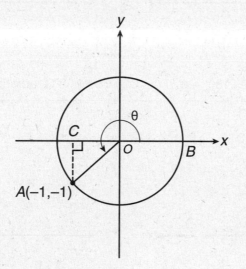

Angles Greater Than 2π Radians

An angle of 410° exceeds one complete rotation by 410° − 360° = 50° so its terminal side will lie in Quadrant I and form an angle of 50° with the positive x-axis. A trigonometric function of an angle greater than 2π radians or less than 0 radians can be written as the same function of an angle between 0 and 2π radians by subtracting or adding a multiple of 2π:

- $\sin 410° = \sin(410° - 360°) = \sin 50°$

- $\cos 870° = \cos(870° - 720°) = \cos 150°$

- $\tan \dfrac{9\pi}{4} = \tan\left(\dfrac{9\pi}{4} - 2\pi\right) = \tan \dfrac{\pi}{4}$

- $\cos\left(-\dfrac{\pi}{3}\right) = \cos\left(-\dfrac{\pi}{3} + 2\pi\right) = \cos \dfrac{5\pi}{3}$

"Reducing" Angles of Trigonometric Functions

The **reference angle**, denoted by θ', is the acute angle formed by the terminal side of the angle in standard position and the x-axis as illustrated in Figure 6.25.

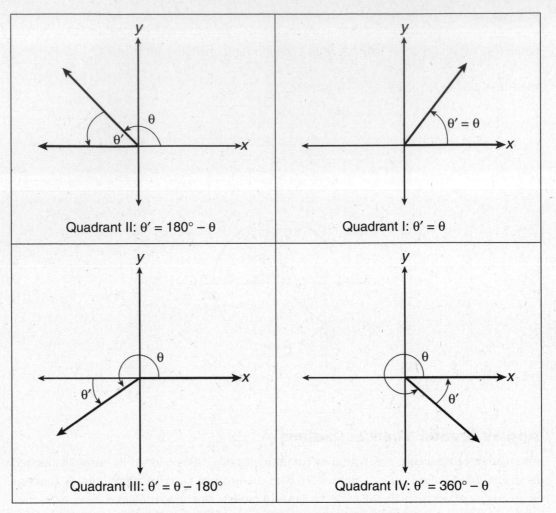

Figure 6.25 Reference angle θ' in the four quadrants

The trigonometric function of any angle θ can be expressed as either plus or minus the same trigonometric function of its *reference angle*. To express cos 135° as a function of its reference angle,

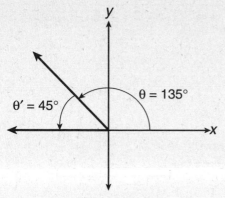

- Locate the reference angle, θ', as shown in the accompanying figure, and find its measure:

$$\theta' = 180 - 135 = 45$$

- Determine the sign of the cosine function in the quadrant in which θ' is located. Since cosine is negative in Quadrant II,

$$\cos 135° = -\cos 45°$$

Similarly, because sine is positive in Quadrant II while tangent is negative,

$$\sin 135° = \sin 45° \quad \text{and} \quad \tan 135° = -\tan 45°$$

You should draw your own diagram and confirm that

- $\sin 310° = -\sin 50°$

- $\cos 670° = \cos (670 - 360)° = \cos 50°$

- $\tan 880° = \tan (880 - 720)° = -\tan 20°$

- $\tan \dfrac{5\pi}{4} = \tan \dfrac{\pi}{4}$.

- $\sin (-140°) = \sin 220° = -\sin 40°$

- $\cos \dfrac{10\pi}{3} = \cos \left(\dfrac{10\pi}{3} - 2\pi \right) = -\cos \dfrac{\pi}{3}$

Multiple-Choice

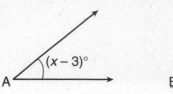 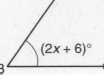

1. If in the figure above $\dfrac{\sin A}{\cos B} = 1$, then $x =$

   (A) 6

   (B) 26

   (C) 29

   (D) 59

2. By law, a wheelchair service ramp may be inclined no more than 4.76°. If the base of a ramp begins 15 feet from the base of a public building, which equation could be used to determine the maximum height, h, of the ramp where it reaches the building's entrance?

   (A) $h = 15 \sin 4.76°$

   (B) $h = \dfrac{15}{\sin 4.76°}$

   (C) $h = \dfrac{\tan 4.76°}{15}$

   (D) $h = 15 \tan 4.76°$

3. What is the number of radians through which the minute hand of a clock turns in 24 minutes?

   (A) 0.2π

   (B) 0.4π

   (C) 0.6π

   (D) 0.8π

4. If $x = 1.75$ radians, then the value of $\cos x$ is closest in value to which of the following?

   (A) $-\cos 1.39$

   (B) $\cos 4.89$

   (C) $\cos 4.53$

   (D) $-\cos 0.18$

5. If $\sin \dfrac{2}{9}\pi = \cos x$, then $x =$

   (A) $\dfrac{7}{9}\pi$

   (B) $\dfrac{5}{18}\pi$

   (C) $\dfrac{\pi}{3}$

   (D) $\dfrac{13}{18}\pi$

6. The bottom of a pendulum traces an arc 3 feet in length when the pendulum swings through an angle of $\dfrac{1}{2}$ radian. What is the number of feet in the length of the pendulum?

   (A) 1.5

   (B) 6

   (C) $\dfrac{1.5}{\pi}$

   (D) 6π

7. What is the radian measure of the smaller angle formed by the hands of a clock at 7 o'clock?

   (A) $\dfrac{\pi}{2}$

   (B) $\dfrac{2\pi}{3}$

   (C) $\dfrac{5\pi}{6}$

   (D) $\dfrac{7\pi}{6}$

8. A wheel has a radius of 18 inches. The distance, in inches, the wheel travels when it rotates through an angle of $\frac{2}{5}\pi$ radians is closest to which value?

(A) 45
(B) 23
(C) 13
(D) 11

9. A wedge-shaped piece is cut from a circular pizza. The radius of the pizza is 14 inches and the angle of the pointed end of the pizza measures 0.35 radians. The number of inches in the length of the rounded edge of the crust is closest to which value?

(A) 4.0
(B) 4.9
(C) 5.7
(D) 7.5

   I. $x = y$
   II. $2(x + y) = \pi$
   III. $\cos x = \sin y$

10. If $0 < x, y < \frac{\pi}{2}$, and $\sin x = \cos y$, then which of the statements above must be true?

(A) I and II only
(B) II and III only
(C) II only
(D) III only

11. If $\cos\theta = -\frac{3}{4}$ and $\tan\theta$ is negative, the value of $\sin\theta$ is

(A) $-\frac{4}{5}$

(B) $-\frac{\sqrt{7}}{4}$

(C) $\frac{4\sqrt{7}}{7}$

(D) $\frac{\sqrt{7}}{4}$

12. If $\cos A = \frac{4}{5}$ and $\angle A$ is *not* in Quadrant I, what is the value of $\sin A$?

(A) -0.6
(B) -0.2
(C) 0.6
(D) 0.75

13. If $\sin A = b$, what is the value of the product $\sin A \cdot \cos A \cdot \tan A$ in terms of b?

(A) 1

(B) $\frac{1}{b}$

(C) b

(D) b^2

14. The equatorial diameter of the earth is approximately 8,000 miles. A communications satellite makes a circular orbit around the earth at a distance of 1,600 miles from the earth. If the satellite completes one orbit every 5 hours, how many miles does the satellite travel in 1 hour?

(A) $1,120\pi$
(B) $1,940\pi$
(C) $2,240\pi$
(D) $2,560\pi$

15. A rod 6 inches long is pivoted at one end. If the free end of the rod rotates in a machine at the rate of 165 revolutions per minute, what is the total distance, in inches, traveled by the end of the rod in one *second*?

(A) 14.5π
(B) 16.5π
(C) 29π
(D) 33π

Grid-In

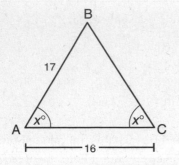

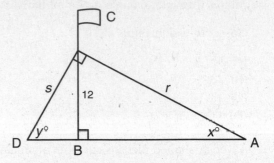

For **Questions 4 and 5** refer to the diagram.

1. In the figure above, what is the value of sin A – cos A?

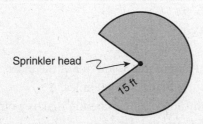

2. A lawn sprinkler sprays water in a circular pattern at a distance of 15 feet from the sprinkler head which rotates through an angle of $\frac{5\pi}{3}$ radians, as shown by the shaded area in the diagram above. What is the area of the lawn, to the *nearest square foot*, that receives water from this sprinkler?

A flagpole that stands on level ground. Two cables, r and s, are attached to the pole at a point 12 feet above the ground and form a right angle with each other. Cable r is attached to the ground at a point that makes tan $x = 0.75$.

4. What is the value of cos x?

5. What is the sum of the lengths of cables r and s?

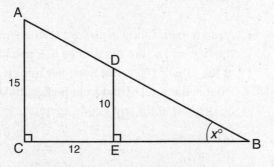

Note: Figure not drawn to scale.

3. In the figure above, angles *ACB* and *DEB* are right angles, $AC = 15$, $CE = 12$, and $DE = 10$. What is the value of cos x?

OVERVIEW

A **unit circle** is a circle with a radius of 1. The coordinates of a point on the unit circle centered at the origin in the *xy*-plane can be expressed in terms of cosine and sine.

THE UNIT CIRCLE

If the terminal side of angle θ intersects a unit circle in the *xy*-plane at $P(x, y)$, as shown in Figure 6.26, then

- $P(x, y) = P(\cos\theta, \sin\theta)$ since $\cos\theta = \dfrac{x}{1} = x$ and $\sin\theta = \dfrac{y}{1} = y$.

- $\sin^2\theta + \cos^2\theta = 1$ because $x^2 + y^2 = 1$.

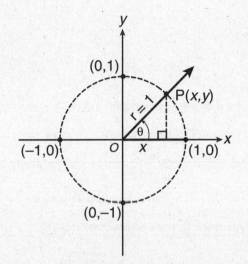

Figure 6.26 The unit circle

RANGE OF VALUES OF SINE AND COSINE

Since the *y*-values of the sine and cosine functions are coordinates of points on the unit circle, $-1 \le \sin\theta \le 1$ and $-1 \le \cos\theta \le 1$ where

- $\sin\dfrac{\pi}{2} = 1$ $\qquad$ and $\quad \sin\dfrac{3\pi}{2} = -1$.

- $\cos 0 = \cos 2\pi = 1$ $\quad$ and $\quad \cos\pi = -1$.

MATH REFERENCE FACT

The maximum and minimum values of $\sin\theta$ and $\cos\theta$ occur when the terminal side of angle θ coincides with a coordinate axis.

➡ Example

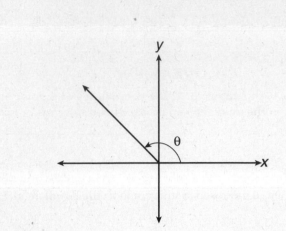

In the figure above, if $\cos\theta = -0.36$, what is the value of $\sin\theta$ to the *nearest hundredth*?

(A) 0.64
(B) 0.80
(C) 0.93
(D) −0.93

Solution

Use the relationship $\sin^2\theta + \cos^2\theta = 1$ to solve for $\sin\theta$:

$$\sin^2\theta + \cos^2\theta = 1$$
$$\sin^2\theta + (0.36)^2 = 1$$
$$\sin^2\theta = 1 - 0.1296$$
$$\sin\theta = \pm\sqrt{0.8704} \approx \pm 0.93$$

Since sine is positive in Quadrant II, the correct choice is **(C)**.

➡ Example

If $\sin w = a$ and $\dfrac{\pi}{2} < w < \pi$, what is $\tan w$ in terms of a?

(A) $\dfrac{a}{\sqrt{1-a^2}}$

(B) $\dfrac{-a}{\sqrt{1-a^2}}$

(C) $\dfrac{1}{1-a}$

(D) $\dfrac{-a}{1-a}$

Solution

The terminal side of angle w in standard position intersects the unit circle in Quadrant II at a point P whose x-coordinate is $\cos w$ and y-coordinate is $\sin w$:

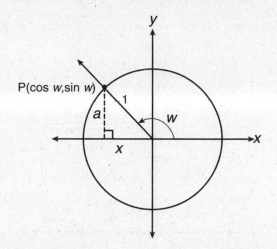

- Form a right triangle by dropping a perpendicular to the x-axis as shown in the accompanying figure. The lengths of the legs of the right triangle correspond to the coordinates of point P.
- Use the Pythagorean theorem to express x ($= \cos w$) in terms of a:

$$x^2 + a^2 = 1$$
$$x^2 = 1 - a^2$$
$$x = -\sqrt{1-a^2} = \cos w \; \leftarrow x \text{ is negative in Quadrant II}$$

- Find $\tan w$:

$$\tan w = \frac{\sin w}{\cos w}$$
$$= \frac{a}{-\sqrt{1-a^2}}$$
$$= \frac{-a}{\sqrt{1-a^2}}$$

The correct choice is **(B)**.

➥ Example

Point P is on the unit circle with center at O and point A is the point at which the unit circle intersects the positive x-axis. If angle AOP measures $\dfrac{7\pi}{6}$ radians, what are the coordinates of point P?

Solution

If $\theta = \dfrac{7\pi}{6}$, then θ is in Quadrant III, and the reference angle is $\dfrac{\pi}{6}(= 30°)$. The right triangle that contains the reference angle is a 30-60 right triangle with a hypotenuse of 1. Since x and y are both negative in Quadrant III, $x = -\dfrac{\sqrt{3}}{2}$ and $y = -\dfrac{1}{2}$ so the coordinates of point P are $\left(-\dfrac{\sqrt{3}}{2}, -\dfrac{1}{2}\right)$.

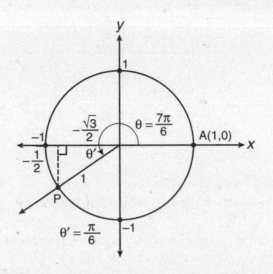

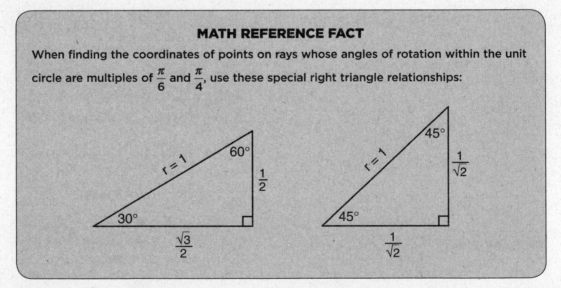

MATH REFERENCE FACT

When finding the coordinates of points on rays whose angles of rotation within the unit circle are multiples of $\dfrac{\pi}{6}$ and $\dfrac{\pi}{4}$, use these special right triangle relationships:

➡ **Example**

If $\cos x = a$, $\cos w = -a$, and $-\dfrac{\pi}{2} < x < 0$, which of the following is a possible value of w?

(A) $\pi - x$

(B) $x - \pi$

(C) $2\pi - x$

(D) $x + 2\pi$

Solution

Since $-\dfrac{\pi}{2} < x < 0$ and $\cos x = a$, locate point $P(a, y)$ in Quadrant IV with reference angle x. We need to find the angle, w, that the ray through $P(-a, y)$ makes with the positive x-axis. Locate $P(-a, y)$ in Quadrant II on the ray opposite OP. Since the reference angle is also x, the angle of rotation, w, is $\pi - x$ as shown in the accompanying figure.

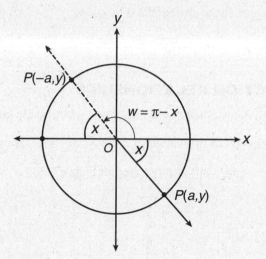

The correct choice is **(A)**.

PERIODIC FUNCTIONS

The sine and cosine functions are *periodic* functions. A function is **periodic** if its values repeat at regular intervals. The sine and cosine functions each have a period of 2π radians since each time 2π is added to an angle, or subtracted from an angle, we go around the unit circle and return to the same point. For example, $\sin(2\pi + x) = \sin x$ and $\cos(x - 2\pi) = \cos x$.

SYMMETRY IN THE UNIT CIRCLE

In Figure 6.27, P' is the reflection of P in the x-axis.

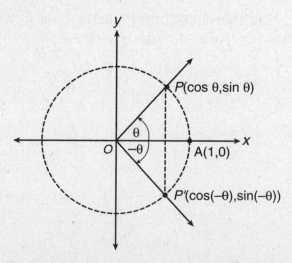

Figure 6.27 Symmetry in the unit circle

- The angle ray OP' makes with the terminal side is equal in measure but clockwise in rotation compared to the angle ray OP makes with the terminal side. Thus, the coordinates of point P' are $(\cos(-\theta), \sin(-\theta))$.
- Because P' is the reflection of P in the x-axis, it has the same x-coordinate as point P but the opposite y-coordinate:

$$\cos(-\theta) = \cos\theta \quad \text{and} \quad \sin(-\theta) = -\sin\theta$$

- Since $\tan\theta$ is the quotient of $\sin\theta$ divided by $\cos\theta$,

$$\tan(-\theta) = -\tan\theta$$

GENERAL REDUCTION RELATIONSHIPS

For SAT test purposes, to determine whether a general relationship such as $\sin(x + \pi) = -\sin x$ is true or false, use a convenient test value for x. If $x = 30°\left(= \dfrac{\pi}{6}\right)$, then $x + \pi$ is a Quadrant III angle and the reference angle is 30° as illustrated in Figure 6.28.

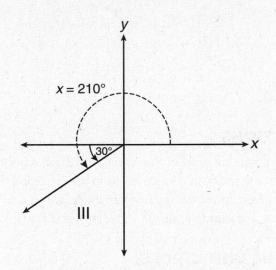

Figure 6.28 Illustrating $\sin 210° = -\sin 30°$

Sine is negative in Quadrant III so $\sin(30° + 180°) = \sin 210° = -\sin 30°$. You can then make the generalization that $\sin(x + \pi) = -\sin x$. Similarly, $\sin(x - \pi) = -\sin x$.

Multiple-Choice

1. The path traveled by a roller coaster is modeled by the equation $y = 27 \sin 13x + 30$ where y is measured in meters. What is the number of meters in the maximum altitude of the roller coaster?

   (A) 13
   (B) 27
   (C) 30
   (D) 57

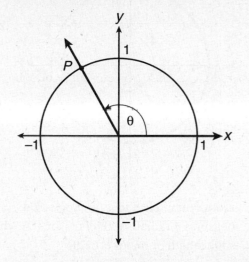

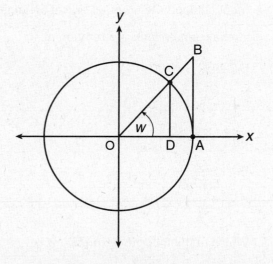

2. The unit circle above has radius $\overline{OC}$, angle AOB measures w radians, $\overline{BA}$ is tangent to circle O at A, and $\overline{CD}$ is perpendicular to the x-axis. The length of which line segment represents $\sin w$?

   (A) $\overline{OD}$
   (B) $\overline{CD}$
   (C) $\overline{AB}$
   (D) $\overline{OB}$

3. If x is an acute angle, which expression is *not* equivalent to $\cos x$?

   (A) $-\cos(-x)$
   (B) $\sin\left(\dfrac{\pi}{2} - x\right)$
   (C) $-\cos(x + \pi)$
   (D) $\cos(x - 2\pi)$

4. In the figure above, θ is an angle in standard position and its terminal side passes through the point $P\left(-\dfrac{1}{2}, \dfrac{\sqrt{3}}{2}\right)$ on the unit circle. What is a possible value of θ?

   (A) $\dfrac{2}{3}\pi$
   (B) $\dfrac{5}{6}\pi$
   (C) $\dfrac{7}{6}\pi$
   (D) $\dfrac{4}{3}\pi$

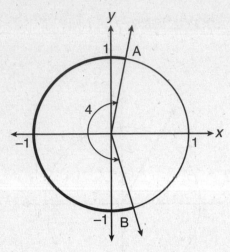

5. In the unit circle above, an angle that measures 4 radians intercepts arc AB. What is the length of major arc AB?

(A) $\dfrac{\pi}{2}$

(B) 4

(C) $\dfrac{\pi+2}{4}$

(D) $\dfrac{4}{\pi}$

6. If θ is an angle in standard position and its terminal side passes through the point $\left(\dfrac{\sqrt{3}}{2},-\dfrac{1}{2}\right)$ on the unit circle, then a possible value of θ is

(A) $\dfrac{7\pi}{6}$

(B) $\dfrac{4\pi}{3}$

(C) $\dfrac{5\pi}{3}$

(D) $\dfrac{11\pi}{6}$

7. What are the coordinates of the image of the point $(1, 0)$ on the terminal side of an angle after a clockwise rotation of $\dfrac{\pi}{6}$ radians?

(A) $\left(\dfrac{\sqrt{3}}{2},-\dfrac{1}{2}\right)$

(B) $\left(\dfrac{1}{2},-\dfrac{\sqrt{3}}{2}\right)$

(C) $\left(-\dfrac{\sqrt{3}}{2},1\right)$

(D) $\left(\dfrac{1}{2},-\dfrac{1}{2}\right)$

8. What are the coordinates of the image of the point $(1, 0)$ on the terminal side of an angle after a counterclockwise rotation of $\dfrac{3}{4}\pi$ radians?

(A) $\left(\dfrac{\sqrt{2}}{2},-\dfrac{\sqrt{2}}{2}\right)$

(B) $\left(-\dfrac{\sqrt{2}}{2},\dfrac{\sqrt{2}}{2}\right)$

(C) $\left(-\sqrt{2},1\right)$

(D) $\left(-\dfrac{1}{2},\dfrac{1}{2}\right)$

9. Which of the following expressions is equivalent to $\dfrac{\sin^2 x}{1+\cos x}$?

(A) $1-\sin x$

(B) $1-\cos x$

(C) $\sin x+\cos x$

(D) $\sin x-\cos x$

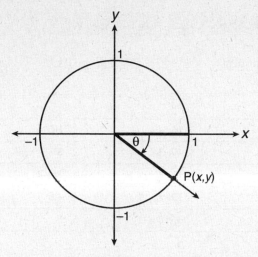

11. If x is a positive acute angle and $\cos x = a$, an expression for $\tan x$ in terms of a is

(A) $\dfrac{1-a}{a}$

(B) $\sqrt{1-a^2}$

(C) $\dfrac{\sqrt{1-a^2}}{a}$

(D) $\dfrac{1}{1-a}$

10. In the unit circle above, the ordered pair (x, y) represents a point P where the terminal side intersects the unit circle, as shown in the accompanying figure. If $\theta = -\dfrac{\pi}{3}$ radians, what is the value of y?

(A) $-\dfrac{\sqrt{3}}{2}$

(B) $-\dfrac{\sqrt{2}}{2}$

(C) $-\sqrt{3}$

(D) $-\dfrac{1}{2}$

NOTE: See pages 456–477 for worked out solutions.

Lesson 6-1

1. B	6. D	11. C	16. C	1. 115
2. B	7. C	12. A	17. D	2. 35
3. B	8. B	13. A	18. C	3. 25/2
4. D	9. D	14. B	19. C	4. 45
5. A	10. C	15. A	**GRID-IN**	5. 5

Lesson 6-2

1. D	6. B	11. A	2. 1/9	7. 42
2. B	7. C	12. A	3. 27/8	8. 1.5
3. C	8. B	13. B	4. 13.5	9. 72
4. D	9. C	**GRID-IN**	5. 49	
5. A	10. D	1. 42	6. 45/2	

Lesson 6-3

1. B	6. B	11. D	16. C	21. A	4. 8
2. C	7. C	12. B	17. C	**GRID-IN**	5. 13.1
3. B	8. A	13. C	18. D	1. 2	
4. B	9. B	14. A	19. D	2. 2/32	
5. C	10. D	15. B	20. C	3. 13	

Lesson 6-4

1. A	6. A	11. B	16. D	**GRID-IN**	5. 24
2. B	7. A	12. A	17. A	1. 60	6. 12.3
3. C	8. B	13. B	18. A	2. 8	7. 1728
4. A	9. D	14. D	19. C	3. 110	8. 3.75
5. C	10. A	15. A	20. D	4. 121	

Lesson 6-5

			GRID-IN	5. 35
1. C	6. B	11. D	1. 7/17	
2. D	7. C	12. A	2. 589	
3. D	8. B	13. D	3. .923	
4. A	9. B	14. C	4. .8	
5. B	10. B	15. D		

Lesson 6-6

1. D	3. A	5. B	7. A	9. B	11. C
2. B	4. A	6. D	8. B	10. A	

SOLUTIONS FOR TUNE-UP EXERCISES

LESSON 3-1
Multiple-Choice

1. **(B)** To divide powers with the *same* base, keep the base and *subtract* the exponents. If $5 = a^x$, then

$$\frac{5}{a} = \frac{a^x}{a}$$
$$= a^{x-1}$$

2. **(A)** If $y = 25 - x^2$, the smallest possible value of y is obtained by subtracting the largest possible value of x^2 from 25. Since $1 \le x \le 5$, the largest possible value of x^2 is $5^2 = 25$. When $x^2 = 25$, then $y = 25 - 25 = 0$.

3. **(D)** Given $y = wx^2$ and y is not 0. Since the values of x and w are each doubled, replace w with $2w$ and x with $2x$ in the original equation:

$$y_{new} = (2w)(2x)^2$$
$$= (2w)(4x^2)$$
$$= 8(wx^2)$$
$$= 8y$$

Hence, the original value of y is multiplied by 8.

4. **(D)** Since $\dfrac{x^{23}}{x^m} = x^{15}$,

$x^m = \dfrac{x^{23}}{x^{15}} = x^{23-15} = x^8$, so $m = 8$. If

$(x^4)^n = x^{20}$, then $4n = 20$, so $n = \dfrac{20}{4} = 5$.

Thus, $mn = 8 \times 5 = 40$.

5. **(C)** If $2 = p^3$, then

$$(2)^3 = (p^3)^3$$
$$8 = p^{3 \times 3}$$
$$= p^9$$

Hence, $8p = p^9 \times p = p^{10}$.

6. **(A)** If $10^{k-3} = m$, then $\dfrac{10^k}{10^3} = m$ so

$10^k = 10^3 m = 1{,}000m$.

7. **(B)** If w is a positive number and $w^2 = 2$, then $w = \sqrt{2}$, so

$$w^3 = w^2 \cdot w = 2\sqrt{2}$$

8. **(B)** If $x = \sqrt{6}$ and $y^2 = 12$, then $y = \sqrt{12}$, so

$$\frac{4}{xy} = \frac{4}{\sqrt{6}\sqrt{12}} = \frac{4}{\sqrt{72}}$$
$$= \frac{4}{\sqrt{36}\sqrt{2}}$$
$$= \frac{4}{6\sqrt{2}}$$
$$= \frac{2}{3\sqrt{2}} \cdot \frac{\sqrt{2}}{\sqrt{2}}$$
$$= \frac{2\sqrt{2}}{3 \cdot 2}$$
$$= \frac{\sqrt{2}}{3}$$

9. **(C)** Test each answer choice in turn by replacing r with x^9 and w with x^5. Only choice (C) is true:
   - (A) $rw - 1 = x^9 \cdot x^5 - 1 = x^{9+5} - 1$
     $= x^{14} - 1 \ne x^{13}$
   - (B) $r + w - 1 = x^9 + x^5 - 1 \ne x^{13}$
   - (C) $\dfrac{r^2}{w} = \dfrac{(x^9)^2}{x^5} = \dfrac{x^{18}}{x^5} = x^{18-5} = x^{13}$
   - (D) $r^2 - w = (x^9)^2 - x^5 = x^{18} - x^5 \ne x^{13}$

10. **(A)** The value of the card is reduced $\$175 - \$172.25 = \$2.75$ each time it is used. If the starting value is $\$175$, then after the card is used n times, it is worth $175 - 2.75n$.

11. **(A)** Three times the sum of a number x and four $[3(x + 4)]$ is equal to five times the number decreased by 2 $[5x - 2]$. Hence $3(x + 4) = 5x - 2$ represents the correct translation.

12. **(C)** Owen gets paid $\$280$ plus 5% $(= 0.05)$ commission on all sales. If he sells d dollars worth of equipment, his commission is $0.05d$. In one week, he earns $280 + 0.05d$, and in w weeks, assuming the same amount of sales, he earns $(280 + 0.05d)$ times w or $(280 + 0.05d)w$.

13. **(C)** In one day there are 24 hours so in one week there are $24 \times 7 = 168$ hours. Hence, in d days there are $24d$ hours and in w weeks there are $168w$ hours so in w weeks and d days there are $168w + 24d$ hours.

14. **(C)** The expression $2(x - 5)$ is two times or twice the difference between x and 5. Each of the other choices can be translated algebraically as $2x - 5$.

15. **(A)** If k pencils cost c cents, then each pencil costs $\dfrac{c}{k}$ cents so p pencils cost
$$p\left(\dfrac{c}{k}\right) = \dfrac{pc}{k}.$$

16. **(B)** If it takes h hours to paint a rectangular wall that is x feet wide and y feet long, then it take $60h$ minutes to paint an area of xy square feet. Hence, it takes $\dfrac{60h}{xy}$ to paint 1 square foot.

17. **(B)** If Carol earns x dollars a week for 3 weeks and y dollars for 4 weeks, she earns for the 7 weeks a total of $3x + 4y$ dollars, which is an average of $\dfrac{3x + 4y}{7}$ dollars per week.

18. **(C)** If Tim bought a skateboard and two helmets for a total of d dollars and the skateboard costs s dollars, then the two helmets cost $d - s$ dollars so each helmet costs $\dfrac{d - s}{2}$ dollars.

19. **(A)** The inequality symbol that represents "at most" is $\leq$. If the width of a rectangle is represented by x, then $2x - 3$ represents a length that is three feet less than twice the width. If the area of the rectangle is at most 30 square feet, then $x(2x - 3) \leq 30$.

20. **(D)** If one-third of a debt of $\$4,920,000,000,000 = 4.92 \times 10^{12}$ was paid off, then
$$\dfrac{2}{3} \times 4.92 \times 10^{12} = \dfrac{9.84}{3} \times 10^{12} = 3.28 \times 10^{12}$$
was not paid off.

21. **(A)** Use the laws of integer exponents to simplify the expression:
$$\dfrac{\left(b^{2n+1}\right)^3}{b^n \cdot b^{5n+1}} = \dfrac{b^{3(2n+1)}}{b^{n+5n+1}}$$
$$= \dfrac{b^{6n+3}}{b^{6n+1}}$$
$$= b^{(6n+3)-(6n+1)}$$
$$= b^2$$

Grid-In

1. **(2)** Given $2^4 \times 4^2 = 16^x$, find the value of x by expressing each side of the equation as a power of the same base:
$$2^4 \times \left(2^2\right)^2 = \left(2^4\right)^x$$
$$2^4 \times 2^4 = 2^{4x}$$
$$2^{4+4} = 2^{4x}$$
$$2^8 = 2^{4x}$$
$$8 = 4x, \text{ so } x = 2$$

2. **(707)** Since $\dfrac{a^6}{b} = 11$, $a^6 = 11b$. Thus,
$$a^7 = a \times a^6 = 7777$$
$$a \times (11b) = 7777$$
$$11ab = 7777$$
$$\dfrac{11ab}{11} = \dfrac{7777}{11}$$
$$ab = 707$$

3. **(32)** If $y = 2^{2p-1}$ and $z = p - 2$, then when $p = 2.5$, $z = 2.5 - 2 = \dfrac{1}{2}$ so
$$\dfrac{y}{z} = \dfrac{2^{2(2.5)-1}}{\dfrac{1}{2}}$$
$$= 2 \cdot 2^4$$
$$= (2)(16)$$
$$= 32$$

4. **(4)** The fraction will have its largest value when $k - p$ has its greatest value and m has its least value. The largest value of $k - p$ is $21 - 9$ or 12. The inequality $2 < m < 6$ means that m is greater than 2 but less than 6. Since m is an integer, the least value of m is 3.

Hence, the largest possible value of $\dfrac{k - p}{m}$

is $\dfrac{12}{3}$ or 4.

LESSON 3-2
Multiple-Choice

1. **(A)** If $\dfrac{m - n}{n} = \dfrac{4}{9}$, then $\dfrac{m}{n} - \dfrac{n}{n} = \dfrac{4}{9}$ so

$\dfrac{m}{n} - 1 = \dfrac{4}{9}$ and $\dfrac{m}{n} = \dfrac{4}{9} + 1 = \dfrac{13}{9}$. Hence,

$\dfrac{n}{m} = \dfrac{9}{13}$.

2. **(D)** If $\dfrac{p + 4}{p - 4} = 13$, then

$$13(p - 4) = p + 4$$
$$13p - 52 = p + 4$$
$$13p - p = 4 + 52$$
$$12p = 56$$
$$p = \dfrac{56}{12} = \dfrac{14}{3}$$

3. **(C)** If $4x + 7 = 12$, then $8x + 14 = 24$ and $(8x + 14) - 11 = 24 - 11$ so $8x + 3 = 13$.

4. **(D)** If $\dfrac{6}{x} = \dfrac{4}{x - 9}$, then $6(x - 9) = 4x$ and $6x - 54 = 4x$ so $2x = 54$ and $x = 27$. Hence,

$\dfrac{x}{18} = \dfrac{27}{18} = \dfrac{3}{2}$.

5. **(B)** If $3j - (k + 5) = 16 - 4k$, then $3j - k - 5 = 16 - 4k$ and $3j + 4k - k = 16 + 5$ so $3j + 3k = 21$. Then $3(j + k) = 21$ so $(j + k) = \dfrac{21}{3} = 7$.

6. **(A)** If $\dfrac{1}{2}(10p + 2) = p + 7$, then $5p + 1 = p + 7$ and $4p = 6$.

7. **(A)** If $0.25y + 0.36 = 0.33y - 1.48$, then $0.25y - 0.33y = -0.36 - 1.48$ so $-0.08y = -1.84$ and $y = \dfrac{-1.84}{-0.08} = 23$ so $\dfrac{y}{10} = \dfrac{23}{10} = 2.30$.

8. **(B)** If $\dfrac{4}{7}k = 36$, then

$$\dfrac{1}{7}k = \dfrac{1}{4}(36) = 9$$

Since $\dfrac{1}{7}k = 9$, then

$$\dfrac{3}{7}k = 3(9) = 27$$

9. **(D)**
$$\dfrac{1}{2}x + \dfrac{1}{4}x + \dfrac{1}{8}x = 14$$
$$\dfrac{4}{8}x + \dfrac{2}{8}x + \dfrac{1}{8}x = 14$$
$$\dfrac{7}{8}x = 14$$
$$\dfrac{8}{7}\left(\dfrac{7}{8}x\right) = \dfrac{8}{7}(14)$$
$$x = 16$$

10. **(C)** If $\dfrac{2}{x} = 2$, then $x = 1$ since $\dfrac{2}{1} = 2$.

Hence,

$$x + 2 = 1 + 2 = 3$$

11. **(A)** If $\dfrac{y - 2}{2} = y + 2$, then $y - 2 = 2(y + 2)$.

Eliminate the parentheses, and then collect all the terms involving y on the same side of the equation:

$$y - 2 = 2(y + 2)$$
$$= 2y + 4$$
$$y - 2y = 4 + 2$$
$$-y = 6, \text{ so } y = -6$$

12. **(D)** If $\dfrac{2y}{7} = \dfrac{y + 3}{4}$, set the cross-products equal and then solve the resulting equation:

$$\dfrac{2y}{7} = \dfrac{y + 3}{4}$$
$$4(2y) = 7(y + 3)$$
$$8y = 7y + 21$$
$$8y - 7y = 21$$
$$y = 21$$

13. **(D)** If $\dfrac{y}{3} = 4$, then $y = 3(4) = 12$, so

$$3y = 3(12) = 36$$

14. **(C)** According to the conditions of the problem, $5k = k + 5$ so $5k - k = 5$.

Since $4k = 5$, $k = \dfrac{5}{4}$.

15. **(B)** If $\dfrac{1}{2}r + 1 = s + 1$, then $\dfrac{1}{2}r = s$ so $r = 2s$.

Substituting $2s$ for r in $\dfrac{8r + 7}{4s} = 11$:

$$\dfrac{8(2s) + 7}{4s} = 11$$

$$\dfrac{16s + 7}{4s} = 11$$

$$\cancel{4s}\left(\dfrac{16s + 7}{\cancel{4s}}\right) = 4s(11)$$

$$16s + 7 = 44s$$

$$\dfrac{7}{28} = \dfrac{28s}{28}$$

$$\dfrac{1}{4} = s$$

Hence, $r = 2s = 2\left(\dfrac{1}{4}\right) = \dfrac{1}{2}$ so $r + s = \dfrac{1}{2} + \dfrac{1}{4} = \dfrac{3}{4}$.

16. **(A)** If $m + 1 = \dfrac{5(m - 1)}{3}$, then

$$3(m + 1) = 5(m - 1)$$

$$3m + 3 = 5m - 5$$

$$\dfrac{8}{2} = \dfrac{2m}{2}$$

$$4 = m$$

Hence, $\dfrac{1}{m} = \dfrac{1}{4}$.

17. **(B)** If $\dfrac{5(p - 1) + 6}{8} = \dfrac{7 - (3 - 2p)}{12}$, then

$$\dfrac{5p + 1}{8} = \dfrac{4 + 2p}{12}$$

$$12(5p + 1) = 8(4 + 2p)$$

$$60p + 12 = 16p + 32$$

$$44p = 20$$

$$p = \dfrac{20}{44} = \dfrac{5}{11}$$

18. **(B)** Given that $h = 61.4 + 2.3F$ where h and F are measured in centimeters, change a height of 5 feet 8 inches into centimeters:

$h = 5$ feet 8 inches $= 68$ inches

$$= 68 \text{ inches} \times 2.54 \, \dfrac{\text{cm}}{\text{inch}}$$

$$= 172.72 \text{ centimeters}$$

Substitute 172.72 for h in the formula and solve for F:

$$172.72 = 61.4 + 2.3F$$

$$172.72 - 61.4 = 2.3F$$

$$\dfrac{111.32}{2.3} = \dfrac{2.3F}{2.3}$$

$$48.4 = F$$

19. **(C)** If x represents the sum of the emails received by Ryan and Taylor, then Sara received $1.25x$ emails. Since a total of 882 emails were received,

$$x + 1.25x = 882$$

$$2.25x = 882$$

$$x = \dfrac{882}{2.25} = 392$$

and $\quad 1.25x = 1.25(392) = 490$

20. **(B)** It is given that Emily received 163 votes. If x represents the number of votes Alexis received and Emily received 25% more votes than Alexis, then $163 = 1.25x$ so

$$x = \dfrac{163}{1.25} = 130.4 \approx 130$$

21. **(A)** Let x represent the number of freshman students. Then $x + 60$ represents the number of sophomores, $2x - 50$ represents the number of juniors, and $3x$ represents the number of seniors. Since there is a total of 1,424 students:

$$x + (x + 60) + (2x - 50) + 3x = 1,424$$

$$7x + 10 = 1,424$$

$$7x = 1,414$$

$$x = \dfrac{1,414}{7} = 202$$

Grid-In

1. **(5/4)** If $2w - 1 = 2$, then $2w = 3$ and $w = \dfrac{3}{2}$. Then

$$w^2 - 1 = \left(\frac{3}{2}\right)^2 - 1$$

$$= \frac{9}{4} - 1$$

$$= \frac{5}{4}$$

2. **(4)** If $11 - 3x$ is 4 less than 13, then $11 - 3x = 9$ so $3x = 2$ and $6x = 4$.

3. **(7/10)** If $2x + 1 = 8$, then $2x = 7$ so $x = \dfrac{7}{2}$. If $15 - 3y = 0$, then $3y = 15$ so $y = \dfrac{15}{3} = 5$. Hence, $\dfrac{x}{y} = \dfrac{7}{2} \div 5 = \dfrac{7}{10}$.

4. **(42)** If the losing team scored x points, then $x + (x + 12) = 72$ and $2x = 60$ so $x = \dfrac{60}{2} = 30$. Hence, the winning team scored $30 + 12 = 42$ points.

5. **(36)** Let x represent the number of texts received. Then

$$(x + x + 15)0.13 = 7.41$$

$$2x + 15 = \frac{7.41}{0.13}$$

$$2x + 15 = 57$$

$$2x = 42$$

$$x = \frac{42}{2} = 21$$

Hence, $x + 15 = 21 + 15 = 36$.

6. **(7/2)** If $\dfrac{7}{12}x - \dfrac{1}{3}x = \dfrac{1}{2} + \dfrac{3}{8}$, then

$$\frac{7}{12}x - \frac{4}{12}x = \frac{4}{8} + \frac{3}{8}$$

$$\frac{3}{12}x = \frac{7}{8}$$

$$\frac{1}{4}x = \frac{7}{8}$$

$$x = \frac{7}{8} \times 4 = \frac{7}{2}$$

7. **(15/7)** If $\dfrac{5(y - 2)}{y} - \dfrac{1}{3} = 0$, then

$$\frac{5y - 10}{y} = \frac{1}{3}$$

$$3(5y - 10) = y$$

$$15y - 30 = y$$

$$14y = 30$$

$$y = \frac{30}{14} = \frac{15}{7}$$

8. **(17.5)** If supply equals demand, then

$$141 - 0.16p = 64 + 0.28p$$

$$141 - 64 = 0.16p + 0.28p$$

$$77 = 0.44p$$

$$\frac{77}{0.44} = p$$

$$175 = p$$

Since \$175 represents the price of a lot of 10 memory chips, the price per memory chip is $\dfrac{\$175}{10} = \$17.5$.

9. **(6)** If $\dfrac{7(x + 9)}{4} - 1 = 41$, then $\dfrac{7(x + 9)}{4} = 42$, and $7(x + 9) = (4)(42)$ so $x + 9 = \dfrac{(4)(42)}{7} = 24$. Hence,

$$(x + 9) - 18 = 24 - 18$$

$$x - 9 = 6$$

10. **(6,000)** If x represents the total sales, then

$$185 + 0.04x = 275 + 0.025x$$

$$0.04x - 0.025x = 275 - 185$$

$$0.015x = 90$$

$$x = \frac{90}{0.015} = 6,000$$

11. **(41)** If 7 quarters plus n nickels is equivalent to 380 pennies, then $7(25) + 5n = 380$ so $5n = 205$ and $n = \dfrac{205}{5} = 41$.

12. **(4)** If x represents the number of years it takes for the two trees to grow to the same height, then

$$36 + 15x = 60 + 9x$$

where 3 feet = 36 inches and 5 feet = 60 inches. Then $6x = 24$ so $x = 4$.

13. **(19/6)** If $\dfrac{9n - (5n - 3)}{8} = \dfrac{2(n-1) - (3 - 7n)}{12}$, then removing parentheses gives

$$\frac{9n - 5n + 3}{8} = \frac{2n - 2 - 3 + 7n}{12}$$

$$\frac{4n + 3}{8} = \frac{9n - 5}{12}$$

$$8(9n - 5) = 12(4n + 3)$$

$$72n - 40 = 48n + 36$$

$$n = 76$$

$$n = 76 = 19 \text{ or } 19/6$$

14. **(23.5)** If x represents the number of cats, then $22 - x$ represents the number of dogs. Hence,

$$2.35x + 5.5(22 - x) = 89.50$$

$$2.35x + 121 - 5.5x = 89.50$$

$$-3.15x + 121 = 89.50$$

$$-3.15x = -31.5$$

$$x = \frac{-31.5}{-3.15} = 10$$

Hence, $\$2.35 \times 10 = \$23.5$ was spent caring for all of the cats.

15. **(3.59)** Form and solve the equation $p - r = 0.93$:

$$(2.89 + .10x) - (2.24 + 0.06x) = 0.93$$

$$0.65 + 0.04x = 0.93$$

$$0.04x = 0.28$$

$$x = \frac{0.28}{0.04} = 7$$

If $x = 7$, then $p = 2.89 + 0.10(7) = 3.59$.

LESSON 3-3
Multiple-Choice

1. **(C)** If $V = \dfrac{1}{3}Bh$, then $Bh = 3V$ so $h = \dfrac{3V}{B}$.

2. **(A)** If $F = \dfrac{kmM}{r^2}$, then $kmM = Fr^2$ so $m = \dfrac{Fr^2}{kM}$.

3. **(B)** If $P = 2(L + W)$, then $P = 2L + 2W$ so $2W = P - 2L$ and $W = \dfrac{P - 2L}{2}$.

4. **(A)** If $A = \dfrac{1}{2}h(x + y)$, then $2A = h(x + y) = hx + hy$ so $hy = 2A - hx$ and $y = \dfrac{2A - hx}{h}$.

5. **(A)** If $s = \dfrac{2x + t}{r}$, then $2x + t = rs$ so $2x = rs - t$ and $x = \dfrac{rs - t}{2}$.

6. **(B)** If $x = x_0 + \dfrac{1}{2}(v + v_0)t$, then $(v + v_0)t = 2(x - x_0)$ so $vt + v_0 t = 2(x - x_0)$ and $vt = 2(x - x_0) - v_0 t$. Dividing each term of the last equation by t gives

$$v = \frac{2(x - x_0)}{t} - \frac{v_0 t}{t} = \frac{2(x - x_0)}{t} - v_0$$

7. **(B)** For the given equation, $2s - 3t = 3t - s$, finding s in terms of t means solving the equation for s by treating t as a constant. Work toward isolating s by first adding $3t$ on each side of the equation:

$$2s = 3t + 3t - s$$
$$= 6t - s$$

Next, add s on each side of the equation:

$$2s + s = 6t$$
$$3s = 6t$$
$$s = \frac{6t}{3} = 2t$$

8. **(C)** If $xy + z = y$, then $xy = y - z$, so $x = \dfrac{y - z}{y}$.

9. **(D)** If $b(x + 2y) = 60$, then $bx + 2by = 60$.
Since it is also given that $by = 15$:

$$bx + 2by = 60$$
$$bx + 2(15) = 60$$
$$bx + 30 = 60$$
$$bx = 60 - 30 = 30$$

10. **(D)** If $\dfrac{a-b}{b} = \dfrac{2}{3}$, then

$$\frac{a}{b} - \frac{b}{b} = \frac{2}{3}$$
$$\frac{a}{b} - 1 = \frac{2}{3}$$
$$\frac{a}{b} = 1 + \frac{2}{3}$$
$$= \frac{5}{3}$$

The value of $\dfrac{a}{b}$ is $\dfrac{5}{3}$.

11. **(C)** Since it is given that $s + 3s$ is 2 more
than $t + 3t$, $s + 3s = (t + 3t) + 2$ or,
equivalently, $4s = 4t + 2$, so $4s - 4t = 2$. Dividing
each member of the equation by 4 gives
$\dfrac{4s}{4} - \dfrac{4t}{4} = \dfrac{2}{4}$, which simplifies to $s - t = \dfrac{1}{2}$.

12. **(D)** Since $\dfrac{1}{p+q} = r = \dfrac{r}{1}$, eliminate the
fractions by cross-multiplying:

$$r(p + q) = 1(1)$$
$$rp + rq = 1$$
$$rp = 1 - rq$$
$$p = \frac{1-rq}{r}$$

13. **(B)** Since $\dfrac{a+b+c}{3} = \dfrac{a+b}{2}$, eliminate the
fractions by cross-multiplying:

$$2(a + b + c) = 3(a + b)$$
$$2a + 2b + 2c = 3a + 3b$$
$$2c = (3a - 2a) + (3b - 2b)$$
$$= \quad a \quad + \quad b$$
$$c = \frac{a+b}{2}$$

14. **(A)** The value in cents of n nickels plus d
dimes is $5n + 10d$, which you are told is
equal to c cents. Hence, $5n + 10d = c$ or
$5n = c - 10d$, so

$$n = \frac{c}{5} - \frac{10d}{5} = \frac{c}{5} - 2d$$

15. **(D)** Since $a = 2c$ and $b = 5d$, replace a in
the equation $\dfrac{c}{d} - \dfrac{a}{b} = x$, with $2c$ and replace
b with $5d$:

$$\frac{c}{d} - \frac{2c}{5d} = x$$
$$\frac{5c}{5d} - \frac{2c}{5d} = x$$
$$\frac{5c - 2c}{5d} = x$$
$$\frac{3c}{5d} = x$$

To find the value of $\dfrac{c}{d}$ in terms of x, multiply
both sides of the equation $\dfrac{3c}{5d} = x$ by the
reciprocal of $\dfrac{3}{5}$:

$$\frac{3}{5}\left(\frac{3c}{5d}\right) = \frac{5}{3}x$$
$$\frac{c}{d} = \frac{5}{3}x$$

16. **(A)** If $\dfrac{4}{t-1} = \dfrac{2}{w-1}$, then

$$2(t-1) = 4(w-1)$$
$$2t - 2 = 4w - 4$$
$$2t = 4w - 2$$
$$t = \dfrac{4w-2}{2}$$
$$= \dfrac{\cancel{2}(2w-1)}{\cancel{2}}$$
$$= 2w - 1$$

Grid-In

1. **(8)** Since $(4 \times b)^2 = 4^2 \times b^2 = 16 \times b^2$,

$$16 \times a^2 \times 64 = 16 \times b^2$$
$$\cancel{16} \times a^2 \times 64 = \cancel{16} \times b^2$$
$$a^2 \times 64 = b^2$$
$$\sqrt{a^2 \times 64} = \sqrt{b^2}$$
$$a \times 8 = b$$

Hence, b is 8 times as great as a.

2. **(5)** Rearrange the terms of $3a + 3b - c = 40$ to get $(3a - c) + 3b = 40$. Because $3a - c = 5b$, $5b + 3b = 40$; so, $8b = 40$ and $b = \dfrac{40}{8} = 5$.

3. **(18)** If $a = 2x + 3$ and $b = 4x - 7$, when $3b = 5a$, x must satisfy the equation $3(4x - 7) = 5(2x + 3)$. Removing parentheses makes $12x - 21 = 10x + 15$. Collecting like terms gives $12x - 10x = 15 + 21$ or $2x = 36$ so $x = \dfrac{36}{2} = 18$.

4. **(5)** Multiplying both sides of $\dfrac{x}{8} = \dfrac{y}{5} = \dfrac{31}{40}$ by 40 produces the eqivalent equation $5x + 8y = 31$. Substitute consecutive positive integer values for x until you find one that makes y have a positive integer value:

x	$5x$	$y = \dfrac{31-5x}{8}$
1	5	$y = \dfrac{31-5}{8} = \dfrac{26}{8}$
2	10	$y = \dfrac{31-10}{8} = \dfrac{21}{8}$
3	15	$y = \dfrac{31-15}{8} = \dfrac{16}{8} = 2$ Stop!

Hence, $x + y = 3 + 2 = 5$.

5. **(7/17)** If $\dfrac{1}{8}x + \dfrac{1}{8}y = y - 2x$, then

$$\dfrac{x+y}{8} = y - 2x, \text{ so}$$

$$x + y = 8(y - 2x)$$
$$x + y = 8y - 16x$$
$$17x = 7y$$
$$\dfrac{17x}{17y} = \dfrac{7y}{17y}$$
$$\dfrac{x}{y} = \dfrac{7}{17}$$

LESSON 3-4
Multiple-Choice

1. **(D)** Write each term of the polynomial numerator separately over the monomial denominator. Then divide powers of the same base by subtracting their exponents:

$$\dfrac{20b^3 - 8b}{4b} = \dfrac{20b^3}{4b} - \dfrac{8b}{4b}$$
$$= 5b^{3-1} - 2$$
$$= 5b^2 - 2$$

2. **(A)** To solve $\dfrac{3}{w} - \dfrac{4}{3} = \dfrac{5w}{10w^2}$, simplify the right hand side:

$$\dfrac{3}{w} - \dfrac{4}{3} = \dfrac{\overset{1}{\cancel{5w}}}{\underset{2w}{\cancel{10w^2}}}$$
$$\dfrac{3}{w} - \dfrac{1}{2w} = \dfrac{4}{3}$$
$$\dfrac{6-1}{2w} = \dfrac{4}{3}$$
$$\dfrac{5}{2w} = \dfrac{4}{3} \leftarrow \text{cross-multiply}$$
$$8w = 15$$
$$w = \dfrac{15}{8}$$

3. **(C)** If $P = 4x - z + 3y$ and $Q = -x + 4z + 3y$, then

$$2P - Q = 2(4x - z + 3y) - (-x + 4z + 3y)$$
$$= 8x - 2z + 6y + x - 4z - 3y$$
$$= (8x + x) + (-2z - 4z) + (6y - 3y)$$
$$= 9x - 6z + 3y$$

4. **(D)** If $(x - y)^2 = 50$ and $xy = 7$, then

$$(x - y) = x^2 - 2xy + y^2 = 50$$
$$x^2 - 2(7) + y^2 = 50$$
$$x^2 - 14 + y^2 = 50$$
$$x^2 + y^2 = 50 + 14$$
$$= 64$$

5. **(B)** Solution 1: Do the algebra. If $p = \dfrac{a}{a - b}$, then

$$1 - p = 1 - \frac{a}{a - b}$$
$$= \frac{a - b}{a - b} - \frac{a}{a - b}$$
$$= \frac{a - b - a}{a - b}$$
$$= \frac{-b}{a - b}$$

Since $\dfrac{-b}{a - b}$ is not one of the answer choices, eliminate the negative sign in the numerator by multiplying the numerator and the denominator by -1:

$$1 - p = \left(\frac{-1}{-1}\right)\frac{-b}{a - b}$$
$$= \frac{(-1)(-b)}{(-1)(a - b)} = \frac{b}{-a + b} = \frac{b}{b - a}$$

Solution 2: Substitute numbers for the letters. Let $a = 3$ and $b = 2$; then

$$p = \frac{b}{a - b} = \frac{3}{3 - 2} = 3$$

so $1 - p = 1 - 3 = -2$. When you plug in 3 for a and 2 for b in each of the answer choices, you find that only choice (B) produces -2.

6. **(B)** If $(a - b)^2 + (a + b)^2 = 24$, then

$$\left(a^2 - 2ab + b^2\right) + \left(a^2 + 2ab + b^2\right) = 24$$
$$2a^2 + 2b^2 = 24$$
$$\frac{2a^2}{2} + \frac{2b^2}{2} = \frac{24}{2}$$
$$a^2 + b^2 = 12$$

7. **(B)** The given equation is

$$\frac{2}{p} - \frac{1}{2p} = \frac{p^2 + 1}{p^2 + 1}$$

Combine terms on the left side and reduce the fraction on the right side to 1:

$$\frac{4}{2p} - \frac{1}{2p} = 1$$
$$\frac{3}{2p} = 1$$
$$2p = 3, \text{ so}$$
$$p = \frac{3}{2} \text{ and } \frac{1}{p}$$

8. **(C)** Rewrite $s + 2 = t$ as $s = t - 2$. Then $rs = (t + 2)(t - 2) = t^2 - 4$.

9. **(C)** Examine each choice in turn:
   - (A) $(x + y)^2 = x^2 + y^2$ is false since $(x + y)^2 = x^2 + 2xy + y^2$.
   - (B) $x^2 + x^2 = x^4$ is false since $x^2 + x^2 = 1x^2 + 1x^2 = 2x^2$.
   - (C) $\dfrac{2^{x+2}}{2^x} = 4$ is true since

$$\frac{2^{x+2}}{2^x} = 2^{(x+2)-x}$$
$$= 2^2$$
$$= 4$$

   - (D) $(3x)^2 = 6x^2$ is false since $(3x)^2 = 3^2 \cdot x^2 = 9x^2$.

10. **(D)** You are given that $(p - q)^2 = 25$ and $pq = 14$. Use the formula for the square of a binomial to expand $(p - q)^2$:

$$(p - q)^2 = p^2 - 2pq + q^2 = 25$$
$$p^2 - 2(14) + q^2 = 25$$
$$p^2 + q^2 = 25 + 28 = 53$$

Now use the formula for $(p + q)^2$:

$$(p + q)^2 = p^2 + 2pq + q^2$$
$$= 53 + 2(14)$$
$$= 53 + 28 = 81$$

11. **(B)** If $\dfrac{a}{2} - \dfrac{b}{3} = 1$, then multiplying each term of the equation by 4 gives $2a - \dfrac{4b}{3} = 4$, so

$2a = \dfrac{4b}{3} + 4$. Substituting in $2a + 3b$ gives

$\left(\dfrac{4b}{3} + 4\right) + 3b = \left(\dfrac{4b}{3} + 4\right) + \dfrac{9b}{3} = \dfrac{13b}{3} + 4.$

12. **(D)** Use FOIL to multiply the left side of the given equation:

$$(x+5)(x+p) = x^2 + 2x + k$$
$$x^2 + 5x + px + 5p = x^2 + 2x + k$$

Since $5x + px$ must be equal to $2x$, $p = -3$. Hence,

$$k = 5p = 5(-3) = -15$$

13. **(D)** Multiply on each side of the given equation, $(x-2)(x+2) = x(x-p)$. The result is

$$x^2 - 4 = x^2 - xp$$

so $4 = xp$ and $p = \dfrac{4}{x}$.

14. **(B)** Clear the given equation of fractions by multiplying each term by 30:

$$\overset{15}{\cancel{30}}\left(\dfrac{m}{\cancel{2}}\right) - \overset{6}{\cancel{30}}\left[\dfrac{3(m-4)}{\cancel{5}}\right] = \overset{5}{\cancel{30}}\left[\dfrac{5(3-m)}{\cancel{6}}\right]$$
$$15m - 18(m-4) = 25(3-m)$$
$$15m - 18m + 72 = 75 - 25m$$
$$-3m + 72 = 75 - 25m$$
$$22m = 3$$
$$m = \dfrac{3}{22}$$

15. **(D)** Use the formula for the square of a binomial to expand the left side of the given equation:

$$\left(k + \dfrac{1}{k}\right)^2 = 16$$
$$k^2 + 2(k)\left(\dfrac{1}{k}\right) + \left(\dfrac{1}{k}\right)^2 = 16$$
$$k^2 + 2 + \dfrac{1}{k^2} = 16$$
$$k^2 + \dfrac{1}{k^2} = 16 - 2 = 14$$

16. **(D)** Change the right side of the given equation, $\dfrac{a}{b} = 1 - \dfrac{x}{y}$, into a single fraction:

$$\dfrac{a}{b} = 1 - \dfrac{x}{y}$$
$$= \dfrac{y}{y} - \dfrac{x}{y}$$
$$= \dfrac{y-x}{y}$$

Since $\dfrac{b}{a}$ is the reciprocal of $\dfrac{a}{b}$,

$$\dfrac{b}{a} = \dfrac{y}{y-x}$$

17. **(A)** Clear the given equation of fractions by multiplying each term by $12r$:

$$\overset{1}{\cancel{12r}}\left(\dfrac{11+s}{\cancel{12r}}\right) = \overset{2r}{\cancel{12r}}\left(\dfrac{1}{\cancel{6}}\right) + \overset{3}{\cancel{12r}}\left(\dfrac{1-3s}{\cancel{4r}}\right)$$
$$11 + s = 2r + 3(1-3s)$$
$$11 + s = 2r + 3 - 9s$$
$$8 + 10s = 2r$$
$$\dfrac{8 + 10s}{2} = r$$
$$\dfrac{\cancel{2}(4+5s)}{\cancel{2}} = r$$
$$4 + 5s = r$$

Grid-In

1. **(19)** Since

$$(3y-1)(2y+k) = ay^2 + by - 5$$

and the product of the last terms of the two binomial factors is equal to the constant term, $(-1)(k) = -5$, so $k = 5$. Now multiply the two binomials together:

$$(3y-1)(2y+5) = (3y)(2y) + (3y)(5) + (-1)(2y) + (-1)(5)$$
$$= 6y^2 \quad + 15y \quad - 2y \quad - 5$$
$$= 6y^2 \quad + 13y \quad - 5$$

Since

$$(3y-1)(2y+5) = 6y^2 + 13y - 5 = ay^2 + by - 5$$

equating the coefficients makes $a = 6$ and $b = 13$, so

$$a + b = 6 + 13 = 19$$

2. **(20)**

$$4x^2 + 20x + r = (2x + s)^2$$
$$= (2x + s)(2x + s)$$
$$= (2x)(2x) + (2x)(s) + (s)(2x) + (s)(s)$$
$$= 4x^2 \quad + 2sx \quad + 2sx \quad + s^2$$
$$= 4x^2 \quad + 4sx \quad + s^2$$

Since the coefficients of x on each side of the equation must be the same, $20 = 4s$, so $s = 5$. Comparing the last terms of the polynomials on the two sides of the equation makes $r = s^2 = 5^2 = 25$. Hence,

$$r - s = 25 - 5 = 20$$

3. **(9/4)** Add $\dfrac{1}{8}$ to each side of the equation and then clear the equation of its fractions by multiplying each term by $4y$:

$$\frac{5}{8} + \frac{1}{8} = \frac{-(11 - 7y)}{4y} + \frac{1}{2y} + \left(-\frac{1}{8} + \frac{1}{8}\right)$$

$$\overset{y}{\cancel{4}}y\left(\frac{3}{\cancel{4}}\right) = \overset{1}{\cancel{4y}}\left[\frac{-(11 - 7y)}{\cancel{4y}}\right] + \overset{2}{\cancel{4y}}\left(\frac{1}{\cancel{2y}}\right)$$

$$3y = -(11 - 7y) + 2$$
$$3y = -11 + 7y + 2$$
$$-4y = -9$$
$$y = \frac{-9}{-4} = \frac{9}{4}$$

LESSON 3-5
Multiple-Choice

1. **(B)** If $q = \dfrac{d}{d + n}$, then $q(d + n) = d$ so

$$qd + nq = d$$
$$nq = d - qd$$
$$nq = d(1 - q)$$
$$\frac{nq}{1 - q} = d$$

2. **(A)** Combine the fractions and then simplify:

$$\frac{a}{a^2 - b^2} + \frac{b}{a^2 - b^2} = \frac{a + b}{a^2 - b^2}$$

$$= \frac{\overset{1}{\cancel{(a + b)}}}{\cancel{(a + b)}(a - b)}$$

$$= \frac{1}{a - b}$$

3. **(A)** To solve $ax + x^2 = y^2 - ay$ for a in terms of x and y, isolate a on the left side of the equation:

$$ax + x^2 = y^2 - ay$$
$$ax + ay = y^2 - x^2$$
$$a(x + y) = (y - x)(y + x)$$

$$\frac{a(x + y)}{(x + y)} = \frac{(y - x)\overset{1}{\cancel{(y + x)}}}{\cancel{(x + y)}}$$

$$a = y - x$$

4. **(C)** If $\dfrac{xy}{x + y} = 1$, then multiplying both sides of the equation by $x + y$ gives $xy = x + y$, so $xy - x = y$. Hence, $x(y - 1) = y$, so

$$x = \frac{y}{y - 1}$$

5. **(B)** Factor the numerator and the denominator of the second fraction:

$$\frac{4x}{x - 1} + \frac{4\overset{1}{\cancel{(x + 1)}}}{(x - 1)\cancel{(x + 1)}} = \frac{4x}{x - 1} + \frac{4}{x - 1}$$

$$= \frac{4x + 4}{x - 1}$$

$$= \frac{4(x + 1)}{x - 1}$$

6. **(A)** Simplify each fraction, then add:

$$h = \frac{x^2 - 1}{x + 1} + \frac{x^2 - 1}{x - 1}$$

$$= \frac{(x + 1)(x - 1)}{x + 1} + \frac{(x + 1)(x - 1)}{x - 1}$$

$$= (x - 1) + (x + 1)$$

$$= 2x$$

$$\frac{h}{2} = x$$

7. **(A)** Collect the terms involving a on one side of the given equation and the terms involving b on the opposite side of the equation:

$$ax^2 - bx = ay^2 + by$$
$$ax^2 - ay^2 = bx + by$$

Factor each side of the equation:

$$ax^2 - ay^2 = bx + by$$
$$a(x^2 - y^2) = b(x + y)$$

Divide each side of the equation by b and $x^2 - y^2$:

$$\frac{a}{b} = \frac{x + y}{x^2 - y^2}$$
$$= \frac{x + y}{(x + y)(x - y)}$$
$$= \frac{1}{x - y}$$

8. **(C)** Solve the given equation, $\frac{a^2 - b^2}{9} = a + b$, for $a^2 - b^2$. Then solve the equation that results for $a - b$:

$$\frac{a^2 - b^2}{9} = a + b$$
$$a^2 - b^2 = 9(a + b)$$
$$(a + b)(a - b) = 9(a + b)$$
$$a - b = \frac{9(a + b)}{a + b} = 9$$

9. **(D)** Factor out the GCF from both the numerator and denominator of $\frac{8r + 8s}{15x - 15y}$:

$$\frac{8r + 8s}{15x - 15y} = \frac{8(r + s)}{15(x - y)}$$
$$= \frac{8}{15} \times \frac{r + s}{x - y}$$
$$= \frac{\overset{2}{\cancel{8}}}{\underset{5}{\cancel{15}}} \times \frac{\cancel{3}}{\cancel{4}}$$
$$= \frac{2}{5}$$

10. **(A)** Factor the numerator of the given fraction, $x^4 - 1$, as the difference of two squares:

$$\frac{x^4 - 1}{x^2 + 1} = \frac{(x^2 + 1)\overset{1}{\cancel{}}(x^2 - 1)}{\cancel{x^2 + 1}}$$
$$= x^2 - 1$$
$$= (k + 1) - 1$$
$$= k$$

11. **(D)** Write p in factored form. If $p = x(3x + 5) - 8$, then $p = 3x^2 + 5x - 28 = (3x - 7)(x + 4)$ so p is divisible by $3x - 7$ and $x + 4$.

12. **(C)** It is given that $x^2 + 16x + 64$ and $4x^2 + 37x + k$ have a factor in common. Since $x^2 + 16x + 64 = (x + 8)^2$, $x + 8$ must be a factor of $4x^2 + 37x + k$:

$$4x^2 + 37x + k = (?x + ?)(x + 8)$$
$$= (4x + ?)(x + 8) \leftarrow \text{product of}$$
$$\text{first terms is } 4x^2$$
$$= (4x + 5)(x + 8) \leftarrow \text{since}$$
$$32x + \boxed{5}x = 37x$$

Since k represents the product of the last terms of $(4x + 5)(x + 8)$, $k = (8)(5) = 40$.

LESSON 3-6
Multiple-Choice

1. **(C)** A fraction is NOT defined when its *denominator* has a value of 0. Thus, the fraction $\frac{x - 2}{(x + 7)(x - 3)}$ is NOT defined if $(x + 7)(x - 3) = 0$, which occurs if $x = -7$ or $x = 3$.

2. **(D)** Multiplying both sides of the given equation, $\frac{a^2}{2} = 2a$, by 2 gives $a^2 = 4a$. To apply the zero-product rule, one side of the equation must be 0. After $4a$ is subtracted from both sides of the equation, $a^2 - 4a = 0$, which can be factored as $a(a - 4) = 0$. Thus, either $a = 0$ or $a - 4 = 0$. Hence, a equals 0 or 4.

3. **(A)** Since $0^2 = 0$, $(s-3)^2 = 0$ means that $s - 3 = 0$, so $s = 3$. Hence,

$$(s + 3)(s + 5) = (3 + 3)(3 + 5)$$
$$= (6)(8)$$
$$= 48$$

4. **(D)** If $k = 7 + \dfrac{8}{k}$, then $\left(k - \dfrac{8}{k}\right)^2 = 7^2$ so

$k^2 - 8 - 8 + \dfrac{64}{k^2} = 49$ and $k^2 + \dfrac{64}{k^2} = 49 + 16 = 65$.

5. **(C)** The given equation is $\dfrac{18 - 3w}{w + 6} = \dfrac{w^2}{w + 6}$. If $w \neq -6$, then $18 - 3w = w^2$ so $w^2 + 3w - 18 = 0$. Hence, $(w + 6)(w - 3) = 0$ so $w = -6$ or $w = 3$. But $w \neq -6$ so the only possible solution is $w = 3$.

6. **(B)** For equation (1), $2x^2 + 7x - 4 = 0$ so $(2x - 1)(x + 4) = 0$ and $x = \dfrac{1}{2}$ or $x = -4$. For equation (2), $(y - 1)^2 = 9$ so $y - 1 = 3$ or $y - 1 = -3$. Hence, the two solutions are $y = 4$ and $y = -2$. Thus, $f = \dfrac{1}{2}$ and $g = -2$ so

$$f \times g = \dfrac{1}{2} \times (-2) = -1.$$

7. **(C)** If $x^3 - 20x = x^2$, then $x^3 - x^2 - 20x = x(x^2 - x - 20) = x(x - 5)(x + 4)$ so $x = -4, 0$, and 5. Hence, $(-4 + 0 + 5) + (-4)(0)(5) = 1 + 0 = 1$.

8. **(D)** If $\dfrac{x^2}{3} = x$, then $x^2 = 3x$, so $x^2 - 3x = 0$. Factoring $x^2 - 3x = 0$ gives $x(x - 3) = 0$. Thus, $x = 0$ or $x = 3$.

9. **(D)** If $(x + 1)(x - 3) = 0$, then $x + 1 = 0$ or $x - 3 = 0$. Hence, $x = -1$ or $x = 3$. The sum of these roots is $-1 + 3$ or 2, and their product is $(-1) \times (3) = -3$. Since

$$2 - (-3) = 2 + 3 = 5$$

the sum of the roots of the equation exceeds the product of its roots by 5.

10. **(B)** If $x^2 - 63x - 64 = 0$, then

$$(x - 64)(x + 1) = 0$$

so $x = 64$ or $x = -1$. If p and n are integers such that $p^n = x$, then either $p^n = 64$ or

$p^n = -1$. Examine each answer choice in turn until you find a number that cannot be the value of p in either $p^n = 64$ or $p^n = -1$.

- (A) If $p = -8$, then $(-8)^n = 64$, so $n = 2$.
- (B) If $p = -4$, then there is no integer value of n for which $(-4)^n = 64$ or $(-4)^n = -1$.

11. **(A)** Isolate variable r by dividing both sides of the equation by r^t:

$$r^t = 6.25r^{t+2}$$
$$1 = \dfrac{6.25r^{t+2}}{r^t}$$
$$= 6.25r^{(t+2)-t}$$
$$= 6.25r^2$$

Using a calculator, divide both sides of the equation by 6.25. Since $\dfrac{1}{6.25} = 0.16 = r^2$,

$$r = \sqrt{0.16} = 0.4 \text{ or } \dfrac{2}{5}$$

12. **(C)** The given equation is:

$$\dfrac{x}{2x - 1} = \dfrac{2x + 1}{x + 2}$$

Cross-multiplying gives $4x^2 - 1 = x^2 + 2x$ so $3x^2 - 2x - 1 = 0$. Then $(3x + 1)(x - 1) = 0$ so $x = -\dfrac{1}{3}$ or $x = 1$. The sum of the two solutions is $\dfrac{2}{3}$.

13. **(B)** The given equation is

$$\dfrac{1}{(t - 2)^2} = 6 + \dfrac{1}{(t - 2)}$$

Eliminate the fractions by multiplying each term of the equation by $(t - 2)^2$:

$$(t - 2)^2 \left[\dfrac{1}{(t - 2)^2}\right] = (t - 2)^2 (6) + (t - 2)^2 \left[\dfrac{1}{(t - 2)}\right]$$
$$6(t - 2)^2 + (t - 2) - 1 = 0 \quad \leftarrow \text{Let } x = (t - 2)$$
$$6x^2 + x - 1 = 0$$
$$(3x - 1)(2x + 1) = 0$$
$$x = \dfrac{1}{3} \quad or \quad x = -\dfrac{1}{2}$$

Since $x = t - 2$ or, equivalently, $t = x + 2$:

- If $x = \dfrac{1}{3}$, then $t = \dfrac{7}{3}$
- If $x = -\dfrac{1}{2}$, then $t = \dfrac{3}{2}$
- The product of the two roots is $\dfrac{7}{3} \times \dfrac{3}{2} = \dfrac{7}{2}$.

Grid-In

1. **(2)** If $(4p + 1)^2 = 81$ and $p > 0$, the expression inside the parentheses is either 9 or –9. Since $p > 0$, let $4p + 1 = 9$; then $4p = 8$ and $p = 2$. A possible value of p is 2.

2. **(2)** If $(x - 1)(x - 3) = -1$, then
$$x^2 - 4x + 3 = -1$$
so $x^2 - 4x + 4 = 0$. Factoring this equation gives $(x - 2)(x - 2) = 0$. Hence, a possible value of x is 2.

3. **(13)** The roots of the equation $(x - 5)(x + 2) = 0$ are the values of x that make the equation a true statement: $x = 5$ or $x = -2$. The sum of the roots is $5 + (-2)$ or 3, and the product of the roots is $(5)(-2)$ or –10. Hence, the sum, 3, exceeds the product, –10, by $3 - (-10)$ or 13.

4. **(16/3)** If $(3k + 14)k = 5$, then $3k^2 + 14k - 5 = 0$ so $(3k - 1)(k + 5) = 0$ and $k = \dfrac{1}{3}$ or $k = -5$.
Lastly, subtract the smaller root from the larger: $\dfrac{1}{3} - (-5) = \dfrac{1}{3} + 5 = \dfrac{1}{3} + \dfrac{15}{3} = \dfrac{16}{3}$.

5. **(4)** If $x^4 + 16 = 10x^2$, then $x^4 - 10x^2 + 16 = 0$ so $(x^2 - 2)(x^2 - 8) = 0$ and $x^2 = 2$ or $x^2 = 8$ so $x = -\sqrt{2}, \sqrt{2}, -\sqrt{8}, \sqrt{8}$. Since the two roots, p and q, must be different and their product positive, it must be the case that either $(-\sqrt{2})(-\sqrt{8}) = \sqrt{16} = 4$ or $(\sqrt{2})(\sqrt{8}) = \sqrt{16} = 4$.

6. **(4/5)** In general, if $x^2 = y^2$, then there are two possibilities: either $x = y$ or $x = -y$. If $(2a - 5)^2 = (4 - 3a)^2$, then
   - The first possibility is $2a - 5 = 4 - 3a$ so $5a = 9$ and $a = \dfrac{9}{5}$.

- The second possibility is $2a - 5 = -(4 - 3a)$ $= -4 + 3a$ so $a = -1$.

The sum of the roots is $\dfrac{9}{5} + (-1) = \dfrac{4}{5}$.

LESSON 3-7
Multiple-Choice

1. **(A)** First solve the equation that contains one variable. Since $3x + 15 = 0$, then $3x = -15$, so
$$x = \dfrac{-15}{3} = -5$$

Substituting –5 for x in the other equation, $2x - 3y = 11$, gives $2(-5) - 3y = 11$ or $-10 - 3y = 11$. Adding 10 to both sides of the equation makes $-3y = 21$, so
$$y = \dfrac{-21}{3} = -7$$

2. **(B)** If $2a = 3b$, then $4a = 6b$. Substituting $6b$ for $4a$ in $4a + b = 21$ gives $6b + b = 21$ or $7b = 21$, so
$$b = \dfrac{21}{7} = 3$$

3. **(B)** Add corresponding sides of the two given equations:
$$\begin{array}{r} 2p + q = 11 \\ + \ p + 2q = 13 \\ \hline 3p + 3q = 24 \end{array}$$

Dividing each member of $3p + 3q = 24$ by 3 gives $p + q = 8$.

4. **(A)** The given system of equations is $2(x + y) = 3y + 5$ and $3x + 2y = -3$.
   - Solve the first equation for y:
$$2x + 2y = 3y + 5$$
$$2x = 5 + 3y - 2y$$
$$x = \dfrac{5 + y}{2}$$

■ Substitute $\dfrac{5+y}{2}$ for x in the second equation:

$$3\left(\dfrac{5+y}{2}\right) + 2y = -3$$

5. **(D)** Eliminate y by adding corresponding sides of the two equations:

$$
\begin{aligned}
x - y &= 3 \\
+\, x + y &= 5 \\
\hline
2x + 0 = 8,\ \text{so}\ x &= \dfrac{8}{2} = 4
\end{aligned}
$$

Since $x = 4$ and $x - y = 3$, then $4 - y = 3$, so $y = 1$.

6. **(C)** Subtract corresponding sides of the two given equations:

$$
\begin{array}{ccc}
5x + y = 19 & & 5x + y = 19 \\
-(x - 3y = 7) & \rightarrow & +\ -x + 3y = -7 \\
\hline
& & 4x + 4y = 12
\end{array}
$$

Dividing each member of the equation $4x + 4y = 12$ by 4 gives $x + y = 3$.

7. **(D)** Subtract corresponding sides of the two given equations:

$$
\begin{array}{ccc}
x + 3 = 5y & & x + 3 = 5y \\
-(x - 9 = 2y) & \rightarrow & +\ -x + 9 = -2y \\
\hline
& & 0 + 12 = 3y
\end{array}
$$

$$\dfrac{12}{3} = y\ \text{or}\ y = 4$$

Since $y = 4$ and $x + 3 = 5y$, then $x + 3 = 5(4) = 20$, so

$$x = 20 - 3 = 17$$

8. **(C)** Eliminate y by adding corresponding sides of the given equations:

$$
\begin{aligned}
\dfrac{1}{x} + \dfrac{1}{y} &= \dfrac{1}{4} \\
+\ \dfrac{1}{x} + \dfrac{1}{y} &= \dfrac{3}{4} \\
\hline
\dfrac{2}{x} + 0 &= \dfrac{4}{4} = 1
\end{aligned}
$$

Since $\dfrac{2}{x} = 1$, then $x = 2$.

9. **(A)** In the equation $\dfrac{a}{b} = \dfrac{2}{5}$, cross-multiplying gives $5a = 2b$. Since $5a + 3b = 35$ and $5a = 2b$, then

$$
\begin{aligned}
2b + 3b &= 35 \\
5b &= 35 \\
b &= 7
\end{aligned}
$$

Since $5a = 2b = 2(7) = 14$,

$$a = \dfrac{14}{5}$$

10. **(B)** If $\dfrac{x}{y} = 6$ and $x = 36$, then $\dfrac{36}{y} = 6$, so $y = 6$. Since $\dfrac{y}{w} = 4$ and $y = 6$,

$$
\begin{aligned}
\dfrac{6}{w} &= 4 \\
4w &= 6 \\
w &= \dfrac{6}{4} = \dfrac{3}{2}
\end{aligned}
$$

11. **(B)** Add corresponding sides of the given equations:

$$
\begin{aligned}
4r + 7s &= 23 \\
+\ \ r - 2s &= 17 \\
\hline
5r + 5s &= 40
\end{aligned}
$$

Dividing each member of $5r + 5s = 40$ by 5 gives $r + s = 8$. Since $r + s = 8$, then

$$3r + 3s = 3(8) = 24$$

12. **(C)** If $\dfrac{p-q}{2} = 3$ and $rp - rq = 12$, then

$$p - q = 2(3) = 6$$

and

$$r(p - q) = 12$$

so $r(6) = 12$ or $6r = 12$. Hence,

$$r = \dfrac{12}{6} = 2$$

13. **(C)** If $(a + b)^2 = 9$, then

$$a^2 + 2ab + b^2 = 9$$

If $(a - b)^2 = 49$, then

$$a^2 - 2ab + b^2 = 49$$

Add corresponding sides of the two equations:

$$\begin{array}{r} a^2 + 2ab + b^2 = 9 \\ + \ a^2 - 2ab + b^2 = 49 \\ \hline 2a^2 + 0 \ + 2b^2 = 58 \end{array}$$

Dividing each member of $2a^2 + 2b^2 = 58$ by 2 gives

$$a^2 + b^2 = 29$$

14. **(A)** The given system of equations is $3x - y = 8 - x$ and $6x + 4y = 2y - 9$.
- Solve the first equation for y: $3x - y = 8 - x$ $\Rightarrow y = 4x - 8$. Collect like terms with y in the second equation. Then substitute $4x - 8$ for y. Thus, $6x + 4y = 2y - 9$ becomes $6x + 2y = -9$ so

$$6x + 2(4x - 8) = -9$$
$$6x + 8x - 16 = -9$$
$$14x = 7$$
$$x = \frac{7}{14} = \frac{1}{2}$$

- Substitute $\frac{1}{2}$ for x in the first equation:

$$3\left(\frac{1}{2}\right) - y = 8 - \frac{1}{2}$$
$$\frac{3}{2} - y = \frac{15}{2}$$
$$y = \frac{3}{2} - \frac{15}{2}$$
$$= -\frac{12}{2}$$
$$= -6$$

- Hence, $xy = \left(\frac{1}{2}\right)(-6) = -3$.

15. **(B)** Eliminate y by subtracting corresponding sides of the given equations:

$$\begin{array}{r} 3x + y = c \\ - \ x + y = b \\ \hline 2x + 0 = c - b \end{array}$$

so $x = \dfrac{c - b}{2}$.

16. **(D)** If $a + b = 11$ and $a - b = 7$, then adding corresponding sides of the two equations gives $2a = 18$, so

$$a = \frac{18}{2} = 9$$

If $a + b = 11$ and $a = 9$, then $9 + b = 11$, so

$$b = 11 - 9 = 2$$

Hence,

$$ab = (9)(2) = 18$$

17. **(D)** For the given system of three equations,

$$x - z = 7$$
$$x + y = 3$$
$$z - y = 6$$

add the equations two at a time to eliminate the variables y and z.
- Eliminate y by adding corresponding sides of the second and third equations. The result is $x + z = 9$.
- Eliminate z by adding $x + z = 9$ to the first equation. The result is $2x = 16$.

Hence, $x = \dfrac{16}{2} = 8$.

18. **(C)** If $a = 4c$, $c = re$, and $a = 5e$, then substituting the second equation into the first equation gives $a = 4re$. Since $a = 4re = 5e$ and $e \neq 0$, the coefficients of e must be equal, so $4r = 5$ and $r = \dfrac{5}{4}$.

19. **(D)** If a player's earnings, x, are 0.005 million dollars more than that of a teammates' earnings, y, then $x = y + 0.005$. If the two players earn a total of 3.95 million dollars, then $x + y = 3.95$. Choice (D) includes these two equations.

20. **(D)** Let x represent the required number of servings of cereal and y the required number of servings of milk.
- If 35 grams of protein are needed, then $5x + 8y = 35$. If the servings must be 470 calories, then $90x + 80y = 470$.
- Solve the system of two equations simultaneously. Multiply each member of the first equation by 18. Then subtract the second equation from it:

$$18(5x) + 18(8y) = 18(35) \rightarrow \begin{array}{r} 90x + 144y = 630 \\ 90x + 80y = 470 \\ \hline 64y = 160 \\ y = \dfrac{160}{64} = 2.5 \end{array}$$

- Substitute 2.5 for y in the first equation to find x:

$$5x + 8(2.5) = 35$$
$$5x + 20 = 35$$
$$5x = 15$$
$$x = \frac{15}{5} = 3$$

Hence, $2\frac{1}{2}$ servings of milk and 3 servings of cereal are needed.

Grid-In

1. **(3)** Since

$$5 \text{ sips} + 4 \text{ gulps} = 1 \text{ glass}$$

and

$$13 \text{ sips} + 7 \text{ gulps} = 2 \text{ glasses}$$

then

$$13\,\text{sips} + 7\,\text{gulps} = \overbrace{2(5\,\text{sips} + 4\,\text{gulps})}^{1\,\text{glass}}$$
$$= 10\,\text{sips} + 8\,\text{gulps}$$
$$13 - 10\,\text{sips} = 8 - 7\,\text{gulps}$$
$$3\,\text{sips} = 1\,\text{gulp}$$

2. **(26)** If x represents the number of minutes jogging and y represents the number of minutes pedaling, then

$$x + y = 60 \qquad \leftarrow \text{total time exercising}$$
$$9x + 6.5y = 475 \qquad \leftarrow \text{total calories burned}$$

Solving for x in the first equation gives $x = 60 - y$. Eliminate x in the second equation by substituting $60 - y$ for x:

$$9(60 - y) + 6.5y = 475$$
$$540 - 9y + 6.5y = 475$$
$$-2.5y = -65$$
$$\frac{-2.5y}{-2.5} = \frac{-65}{-2.5}$$
$$y = 26$$

3. **(1.45)** If $p = $ the cost of a pen and $n = $ the cost of a notebook then

$$\begin{array}{cc} & \overset{2\times}{\longrightarrow} \\ 1p + 2n = 3.50 & 2p + 4n = 7.00 \\ 2p + 3n = 5.55 & -2p + 3n = 5.55 \\ \hline & n = 1.45 \end{array}$$

The charge for one notebook is \$1.45.

4. **(206)** Remove the fractions in the two equations by multiplying each member of the first equation by 6 and multiplying each member of the second equation by 8:

$$6\left(\frac{1}{2}r\right) - 6\left(\frac{1}{3}s\right) = 6(8) \rightarrow 3r - 2s = 48$$
$$8\left(\frac{5}{8}r\right) - 8\left(\frac{1}{4}s\right) = 8(29) \rightarrow 5r - 2s = 232$$
$$-2r + 0s = -184$$
$$r = \frac{-184}{-2} = 92$$

- Eliminate s by subtracting corresponding sides of the two equations:

$$\begin{array}{r} 3r - 2s = 48 \\ 5r - 2s = 232 \\ \hline -2r + 0s = -184 \\ r = \dfrac{-184}{-2} = 92 \end{array}$$

- Substitute 92 for r in the first equation to find s:

$$\frac{1}{2}(92) - \frac{1}{3}s = 8$$

$$46 - \frac{1}{3}s = 8$$

$$-\frac{1}{3}s = -38$$

$$s = (-3)(-38) = 114$$

Hence, $r + s = 92 + 114 = 206$.

5. **(20)** If x represents the number of apples sold during the first week and y the number of oranges sold during the first week, then

$x + y = 108$ ←total sold during first week
$5x + 3y = 452$ ←total sold during second week

Solving for x in the first equation gives $x = 108 - y$, and then substituting this for x in the second equation gives $5(108 - y) + 3y = 452$. After simplifying, $540 - 2y = 452$ so $-2y = -88$ and $y = \dfrac{-88}{-2} = 44$. Since $y = 44$, $x = 108 - 44 = 64$. Hence, $64 - 44 = 20$ more apples than oranges were sold in the first week.

6. **(8)** If x represents the cost of a bag of popcorn and y the cost of a drink, then

$2x + 3y = 18.25$ ← amount spent by Jacob
$4x + 2y = 27.50$ ← amount spent by Zachary

Eliminate x by multiplying the first equation by 2 and then subtracting corresponding sides of the two equations:

$$4x + 6y = 36.50$$
$$\underline{4x + 2y = 27.50}$$
$$4y = 9.00$$
$$y = \frac{9.00}{4} = 2.25$$

Find x by substituting 2.25 for y in the first equation:

$$2x + 3(2.25) = 18.25$$
$$2x + 6.75 \quad = 18.25$$
$$2x = 11.50$$
$$x = \frac{11.50}{2} = 5.75$$

Hence, the cost of one bag of popcorn and one drink is $x + y = \$5.75 + \$2.25 = \$8$.

7. **(.21)** If x represents the charge per minute and y represents the charge per mile, then

$2 + 14x + 10y = 16.94$ ← Ariel
$2 + 10x + 6y = 11.30$ ← Victoria

Simplify each equation by dividing each member of the equation by 2:

$1 + 7x + 5y = 8.47$ $\rightarrow$ $7x + 5y = 7.47$
$1 + 5x + 3y = 5.65$ $\rightarrow$ $5x + 3y = 4.65$

Eliminate y by multiplying the first equation by -3, multiplying the second equation by 5, and then add corresponding sides of the two equations:

$(-3)(7x + 5y) = -3(7.47) \rightarrow -21x - 15y = -22.41$
$5(5x + 3y) = 5(4.65) \quad \rightarrow \underline{25x + 15y \quad = 23.25}$
$$4x \qquad\quad = 0.84$$
$$x = \frac{0.84}{4} = 0.21$$

8. **(9.68)** If $x = 0.21$, then $1 + 7(0.21) + 5y = 8.47$ so $5y = 6$ and $y = \dfrac{6}{5} = 1.20$. Hence, a ride that takes 8 minutes and travels 5 miles costs $2 + 8(.21) + 5(1.20) = 2 + 1.68 + 6 = 9.68$.

LESSON 3-8
Multiple-Choice

1. **(A)** Solution 1: Since $4 + 3p < p + 1$, then

$$3p - p < 1 - 4 \quad \text{or} \quad 2p < -3$$

so $p < -\dfrac{3}{2}$. Hence, the largest integer value for p is -2.

Solution 2: Plug each of the answer choices for p into $4 + 3p < p + 1$ until you find one that makes the inequality a true statement. Since choice (A) gives

$$4 + 3(-2) < (-2) + 1$$

there is no need to continue.

2. **(D)** Solution 1: Solve $-3 < 2x + 5 < 9$ by first subtracting 5 from each member. The result is $-8 < 2x < 4$. Now divide each member of this inequality by 2, obtaining $-4 < x < 2$. Examine each of the answer choices until you find one (D) that is not between -4 and 2. Since x is less than 2, 2 is not a possible value of x.

   Solution 2: Plug each of the answer choices for x into $-3 < 2x + 5 < 9$ until you find one (D) that does not make the inequality a true statement.

3. **(B)** If each package contains 8 hot dogs, then p packages contains $8p$ hot dogs. Hence, the inequality $8p \geq 78$ could be used to determine the number of packages needed if each of the 78 guests is served at least one hot dog.

4. **(A)** If Peter starts Kindergarten being able to spell 10 words and can learn 2 new words every day, then after d days he is able to spell a total of $2d + 10$ words. Hence, the number of days it takes Peter to be able to spell *at least* 85 words is represented by $2d + 10 \geq 85$.

5. **(A)** If $7 - 5x \leq -3(x - 5)$, then

$$7 - 5x \leq -3x + 15$$
$$3x - 5x \leq 15 - 7$$
$$\frac{-2x}{-2} \geq \frac{8}{-2} \quad \leftarrow \text{reverse inequality}$$
$$x \geq -4$$

Since -5 is less than -4, it is not a solution of the inequality.

6. **(A)** If x represents the number of minutes Tamara could use her phone each month, then

$$19 + 0.07x \leq 29.50$$
$$0.07x \leq 10.50$$
$$x \leq \frac{10.50}{0.07}$$
$$x \leq 150$$

The maximum number of minutes is 150.

7. **(B)** If $3(2m - 1) \leq 4m + 7$, then $6m - 3 \leq 4m + 7$ and $2m \leq 10$ so $m \leq 5$.

8. **(A)** If x represents the maximum number of songs Emma can buy with $50.00 to spend, then

$$0.49x + 13.95 \leq 50$$
$$0.49x \leq 36.05$$
$$x \leq \frac{36.05}{0.49}$$
$$x \leq 73.57$$

Since x must be an integer, $x = 73$.

9. **(D)** If 200 ninth graders attended the picnic at $0.75 each and x guests attended at $1.25 per guest, then the total revenue is $(0.75)(200) + 1.25x$, which must be at least $250.00 in order to cover the cost of the permit. Hence, $(0.75)(200) + 1.25x \geq 250.00$.

10. **(D)** If $2(x - 4) \geq \frac{1}{2}(5 - 3x)$, then $4(x - 4) \geq (5 - 3x)$ so $4x - 16 \geq 5 - 3x$ and $7x \geq 21$. Hence, $x \geq 3$. Since the smallest integer value of x is 3, the smallest integer value of x^2 is 9.

11. **(B)** If x represents the number of hours Edith tutors, then $11 - x$ represents the number of hours show works as a library assistant. Then $20x + 15(11 - x) \geq 185$.

12. **(A)** Guy is paid $185 + 0.03d$ and Jim is paid $275 + 0.025d$. Hence,

$$185 + 0.03d > 275 + 0.025d$$
$$0.005d > 90$$
$$\frac{0.005d}{0.005} > \frac{90}{0.005}$$
$$d > 18,000$$

13. **(C)** The sum of the fixed cost of admission plus the cost of r rides is $4.50 + 0.79r \le 16.00$:

$$4.50 + 0.79r \le 16.00$$
$$0.79r \le 16.00 - 4.50$$
$$\frac{0.79r}{0.79} \le \frac{11.50}{0.79}$$
$$r \le 14.55$$

Since r must be a whole number, $r = 14$.

14. **(D)** If $b + 3 > 0$, then $b > -3$. Since $1 > 2b - 9$, then $10 > 2b$, so $5 > b$ or $b < 5$. Since b is an integer, b may be equal to any of these seven integers: $-2, -1, 0, 1, 2, 3,$ or 4.

Grid-In

1. **(4)** If $2y - 3 < 7$, then $2y < 10$, so $y < 5$. Since

$$y + 5 > 8 \quad \text{and} \quad 2y - 3 < 7$$

then $y > 3$ and at the same time $y < 5$. The integer for which the question asks must be 4.

2. **(6)** When 2 times an integer x is increased by 5, the result is always greater than 16 and less than 29, so $16 < 2x + 5 < 29$. Subtracting 5 from each member of this inequality gives $11 < 2x < 24$. Then

$$\frac{11}{2} < \frac{2x}{2} < \frac{24}{2}$$

so $5\frac{1}{2} < x < 12$. According to this inequality, x is greater than $5\frac{1}{2}$, so the least integer value of x is 6.

3. **(.76)** If $2 < 20x - 13 < 3$, adding 13 to each member of the combined inequality makes $15 < 20x < 16$ or $\frac{15}{20} < x < \frac{16}{20}$, which can also be written as $0.75 < x < 0.80$. Hence, one possible value for x is 0.76. Grid in as .76.

4. **(6)** Canceling identical terms on either side of the given inequality,

$$\frac{1}{7} + \frac{1}{8} - \frac{1}{9} + \frac{1}{10} < \frac{1}{8} - \frac{1}{9} + \frac{1}{10} + \frac{1}{n}, \text{ results in}$$

$$\frac{1}{7} < \frac{1}{n} \text{ or, equivalently, } n < 7. \text{ Hence, the}$$

greatest possible integer value for n is 6.

5. **(15)** If x represents the maximum number of games Chelsea can play, then $20 + 15 + 0.65x \le 45$ so $0.65x \le 10$ and $x \le \frac{10}{0.65}$ ($\cong 15.38$). Since x must be an integer, $x = 15$.

6. **(40)** If x represents the number of picture frames that must be sold to make a profit, then

$$\underset{\text{Revenue}}{13x} - \underset{\text{Expenses}}{(75 + 6x)} \ge 200$$
$$7x - 75 \ge 200$$
$$\frac{7x}{7} \ge \frac{275}{7}$$
$$x \ge 39.29$$

Since x must be an integer, $x = 40$.

7. **(8)** If x represents the maximum number of printers that can be shipped each day, then $15 - x$ represents the corresponding number of monitors that can be shipped on the same day. Hence,

$$125x + 225(15 - x) \ge 2,500$$
$$125x + 3,375 - 225x \ge 2,500$$
$$\frac{-100x}{-100} \ge \frac{-875}{-100}$$
$$x \le 8.75$$

Since x must be an integer, $x = 8$.

8. **(6.1)** The given inequality is
$$-\frac{5}{3} < \frac{1}{2} - \frac{1}{3}x < -\frac{3}{2}.$$
To solve for x:
- Eliminate the fractions by multiplying each member of the inequality by 6:

$$6\left(-\frac{5}{3}\right) < 6\left(\frac{1}{2} - \frac{1}{3}x\right) < 6\left(-\frac{3}{2}\right)$$

$$-10 < -2x + 3 < -9$$

- Multiply each member of inequality by -1 and, at the same time, reverse the direction of the inequality signs:

$10 > 2x - 3 > 9$ or, equivalently, $9 < 2x - 3 < 10$

- Isolate x:

$$9 + 3 < 2x < 10 + 3$$
$$12 < 2x < 13$$
$$\frac{12}{2} < \frac{2x}{2} < \frac{13}{2}$$
$$6 < x < 6.5$$

Hence, x can be any number between 6 and 6.5, say 6.1.

LESSON 3-9
Multiple-Choice

1. **(C)** If $|n - 1| < 4$, then $-4 < n - 1 < 4$, so $-4 + 1 < n < 4 + 1$ and $-3 < n < 5$. There are seven integers between -3 and 5: $-2, -1, 0, 1, 2, 3,$ and 4.

2. **(A)** If $|x| \leq 2$ and $|y| \leq 1$, then $-2 \leq x \leq 2$ and $-1 \leq y \leq 1$. The least possible value of $x - y$ is $-2 - 1 = -3$.

3. **(B)** If $\left|\frac{1}{2}x\right| \geq \frac{1}{2}$, then either $\frac{1}{2}x \geq \frac{1}{2}$ so $x \geq 1$ or $\frac{1}{2}x \leq -\frac{1}{2}$ so $x \leq -1$.

4. **(D)** If $\frac{1}{2}|x| = 1$, then $|x| = 2$, so $x = \pm 2$. If $x = -2$, then $|y| = x + 1 = -1$, which is impossible. If $x = 2$, then $|y| = x + 1 = 3$, so $y = \pm 3$ and $y^2 = (\pm 3)^2 = 9$.

5. **(A)** Since the temperature, F, can range from 7° below 79° to 7° above 79°, the positive difference between F and 79° is always less than or equal to 7°, which is expressed by the inequality $|F - 79| \leq 7$.

6. **(D)** If $\frac{|a + 3|}{2} = 1$, then $|a + 3| = 2$, so $a + 3 = 2$ or $a + 3 = -2$. Hence, $a = -1$ or $a = -5$. If $2|b + 1| = 6$, then $|b + 1| = 3$, so $b + 1 = 3$ or $b + 1 = -3$. Hence, $b = 2$ or $b = -4$. Then $|a + b|$ could equal the following:
- $|-1 + 2| = 1$
- $|-1 + (-4)| = 5$
- $|-5 + 2| = 3$
- $|-5 + (-4)| = 9$
Thus, $|a + b|$ could not equal 7.

7. **(D)** To find the value of x that satisfied $|1 + x| = |1 - x|$, if any, plug the answer choices into the equation to see if one works. If $x = 0$, $|1 + 0| = |1 - 0| = 1$. ✓

8. **(A)** The midpoint of the interval $-1 < x < 3$ is $\frac{-1 + 3}{2} = 1$ so the equivalent absolute value inequality has the form $|x - 1| < k$. Since the difference between 1 and both -1 and 3 is 2, $k = 2$ so $-1 < x < 3$ is equivalent to $|x - 1| < 2$.

9. **(A)** The midpoint of the interval $45 < t < 85$ is $\frac{45 + 85}{2} = \frac{130}{2} = 65$ so the equivalent absolute value inequality has the form $|t - 65| < k$. Since the difference between 65 and both 45 and 85 is 20, $k = 20$ so $45 < t < 85$ is equivalent to $|t - 65| < 20$.

10. **(A)** If $|1.5C - 24| \leq 30$, then $-30 \leq 1.5C - 24 \leq 30$. Adding 24 to each member of the inequality gives $24 - 30 \leq (1.5C - 24) + 24 \leq 30 + 24$ or $-6 \leq 1.5C \leq 54$. Isolate C by dividing each member of the inequality by 1.5:

$$\frac{-6}{1.5} \leq \frac{1.5C}{1.5} \leq \frac{54}{1.5}$$
$$-4 \leq C \leq 36$$

Hence, the lowest temperature is -4.

Grid-In

$$|t - 7| = 4$$
$$|9 - t| = 2$$

1. **(11)** If $|t - 7| = 4$, then $t - 7 = 4$ or $t - 7 = -4$. Then $t = 11$ or $t = 3$. If $|9 - t| = 2$, then $9 - t = 2$ or $9 - t = -2$. Then $t = 7$ or $t = 11$.

2. **(1/2)** If $|-3y + 2| < 1$, then $-1 < -3y + 2 < 1$, so

$$-1 - 2 < -3y < 1 - 2$$
$$\frac{-3}{-3} > \frac{-3y}{-3} > \frac{-1}{-3}$$
$$1 > y > \frac{1}{3}$$

Hence, y can take on any value between $\frac{1}{3}$ and 1 such as $\frac{1}{2}$. Grid-in 1/2.

3. **(16)** It is given that $|x - 16| \leq 4$ and $|y - 6| \leq 2$. The greatest possible value of $x - y$ occurs when x takes on its greatest value and, at the same time, y takes on its smallest value.

- If $|x - 16| \leq 4$, then $-4 \leq x - 16 \leq 4$. Adding 16 to each member of the inequality makes $12 \leq x \leq 20$. Hence, the greatest value of x is 20.

- If $|y - 6| \leq 2$, then $-2 \leq y - 6 \leq 2$. Adding 6 to each member of the inequality makes $4 \leq y \leq 8$. Hence, the smallest value of y is 4.

- The greatest value of $x - y$ is $20 - (4) = 16$.

4. **(652)** It is given that $|d - x| \leq 0.05d$ and $x = 620$. Hence,

$$|d - 620| \leq 0.05d$$
$$-0.05d \leq d - 620 \leq 0.05d$$
$$620 - 0.05d \leq d \leq 620 + 0.05d$$

The maximum depth, d, must satisfy the inequality $d \leq 620 + 0.05d$ so $0.95d \leq 620$ and $d \leq \dfrac{620}{0.95}$. Since $d \leq 652.63$, to the *nearest foot*, the maximum depth is 652.

LESSON 3-10
Multiple-Choice

1. **(A)** Since average speed is measured by $\dfrac{\Delta \text{distance}}{\Delta \text{time}}$, the interval over which the average speed was the greatest is the interval segment having the steepest slope. Of the choices given, the interval from the first hour to the second hour has the steepest slope and, therefore, the greatest average speed: $\dfrac{110 - 40}{2 - 1} = 70 \dfrac{\text{miles}}{\text{hour}}$.

2. **(D)** A horizontal segment has a slope of 0 (segment a), a segment that slopes down from left to right has a negative slope (segment c), and a segment that slopes up from left to right has a positive slope with the steeper line having the greater slope ($d < b$). Hence, the correct ordering of slopes is $c < a < d < b$.

3. **(A)** The midpoint of the segment with endpoints $(-2, 4)$ and $(8, 4)$ is $\left(\dfrac{-2 + 8}{2}, 4\right) = (3, 4)$. The perpendicular bisector of this segment is the vertical line through $(3, 4)$ so its equation must be $x = 3$.

4. **(D)** Since points O, A, and B lie on the same line,

$$\text{Slope } OA = \text{Slope } OB$$
$$\frac{\text{Change in } y}{\text{Change in } x} = \frac{t - 0}{2 - 0} = \frac{5 - 0}{r - 0}$$
$$\frac{t}{2} = \frac{5}{r}$$
$$r \times t = 2 \times 5$$
$$r = \frac{10}{t}$$

5. **(C)** If $2(x + 2y) = 0$, then solving for y gives $y = -\dfrac{1}{2}x$ so the slope of the line is $-\dfrac{1}{2}$.

6. **(B)** It is given that for $A(-1, 0)$ and $P(4, 12)$, $AP = BP$. Points A, B, and P may or may not lie on the same line.

- If Points A, B, and P lie on the same line, then P is the midpoint of $\overline{AB}$. If the coordinates of B are represented by (x, y), then $\left(\dfrac{-1 + x}{2}, \dfrac{0 + y}{2}\right) = (4, 12)$ so $\dfrac{-1 + x}{2} = 4$ and $x = 9$. Also, $\dfrac{y}{2} = 12$ so $y = 24$. Hence, the coordinates of B could be $(9, 24)$, which means Statement II is true.

- Statement I is not true since $\left(\dfrac{3}{2}, 6\right)$ represents the midpoint of $\overline{AP}$.

- Use the distance formula to test if the coordinates of B could be $(-8, 7)$:

$$AP = \sqrt{(\Delta x)^2 + (\Delta y)^2}$$
$$= \sqrt{(4 - (-1))^2 + (12 - 0)^2}$$
$$= \sqrt{5^2 + 12^2}$$
$$= \sqrt{169}$$
$$= 13$$

$$PB = \sqrt{(\Delta x)^2 + (\Delta y)^2}$$
$$= \sqrt{(-8-4)^2 + (7-12)^2}$$
$$= \sqrt{(-12)^2 + (-5)^2}$$
$$= \sqrt{169}$$
$$= 13$$

Since $AP = PB$, Statement III is true. Hence, only Statements II and III are true.

7. **(B)** Parallel lines have the same slope. Find the slope of the given line, $\frac{1}{2}y - \frac{2}{3}x = 6$, by writing it in $y = mx + b$ slope-intercept form where m, the coefficient of x, is the slope of the line. Clear the equation of its fraction by multiplying each term by 6, which gives $3y - 4x = 36$ so $y = \frac{4x}{3} + 12$. Hence, the slope of the given line is $\frac{4}{3}$. Find the line in the set of answer choices whose slope is $\frac{4}{3}$. If $y = 4\left(\frac{x-1}{3}\right)$, as in choice (B), then $y = \frac{4}{3}x - \frac{4}{3}$ which has a slope of $\frac{4}{3}$.

8. **(C)** Perpendicular lines have slopes that are negative reciprocals. Find the slope of the given line, $y = -2(x + 1)$, by writing it as $y = -2x - 2$. Since the slope of the given line is -2, the slope of a line that is perpendicular to it must be $\frac{1}{2}$. Find the line in the set of answer choices whose slope is $\frac{1}{2}$. If $\frac{x-1}{6} = \frac{y}{3}$, as in choice (C), then $6y = 3x - 3$. Dividing each member of the equation by 6 gives $y = \frac{1}{2}x - \frac{1}{2}$, which has a slope of $\frac{1}{2}$.

9. **(C)** Test each point in the set of answer choices until you find the point that makes the slope m of the line containing that point and $(4, -2)$ equal to $\frac{3}{2}$. Let $(x_A, y_A) = (4, -2)$. Choice (C) works since, if $(x_B, y_B) = (6, 1)$, then

$$m = \frac{y_B - y_A}{x_B - x_A} = \frac{1 - (-2)}{6 - 4} = \frac{1 + 2}{2} = \frac{3}{2}$$

The coordinates of another point on the line are $(6, 1)$.

10. **(D)** Since it is given that point $E(5, h)$ is on the line that contains $A(0, 1)$ and $B(-2, -1)$, slope of $\overline{EA}$ = slope of $\overline{AB}$

$$= \frac{-1 - 1}{-2 - 0} = \frac{-2}{-2} = 1$$

Hence,

$$\text{slope of } \overline{EA} = \frac{h - 1}{5 - 0} = 1$$
$$h - 1 = 5$$
$$h = 6$$

11. **(B)** If the slope of line ℓ is m, then

$$m = \frac{(h + m) - 0}{0 - h} = \frac{h + m}{-h}$$

Hence, $-hm = h + m$ or $-h = m + mh$. Factoring out m from the right side of the equation gives $-h = m(1 + h)$ so

$$m = \frac{-h}{1 + h}$$

12. **(A)** The slope of a line through $(1, 1)$ and $(0, 2)$ is $\frac{2 - 1}{0 - 1} = -1$ and the slope of a line through $(1, 1)$ and $(0, 3)$ is $\frac{3 - 1}{0 - 1} = -2$. Hence, the slope of a line through $(1, 1)$ and a point between $(0, 2)$ and $(0, 3)$ must have a value between -1 and -2 such as $-\frac{3}{2}$.

13. **(D)** If $y + 2x = b$, then $y = -2x + b$, so the slope of this line is -2. The slope of a line perpendicular to this line is $\frac{1}{2}$, the negative reciprocal of -2. If this line passes through the origin, its y-intercept is 0, so its equation is $y = \frac{1}{2}x$. Because the point of intersection of the two lines is $(k + 2, 2k)$, the coordinates of this point must satisfy both equations.

Find the value of k by substituting the coordinates of this point into the equation $y = \frac{1}{2}x$:

$$2k = \frac{1}{2}(k + 2)$$
$$4k = k + 2$$
$$3k = 2$$
$$\frac{3k}{3} = \frac{2}{3}$$
$$k = \frac{2}{3}$$

14. **(A)** Let $y = mx + b$ represent the equation of the desired line.
    - If $y - 4x = 0$, then $y = 4x$, so the slope of this line is 4. Since parallel lines have equal slopes, $m = 4$.
    - If $y + 3 = -x + 1$, then $y = -x - 2$, so its y-intercept is -2. Since the desired line has the same y-intercept, $b = -2$.
    - The equation of the desired line is $y = 4x - 2$.

15. **(D)** Since $ABCD$ is a square, C has the same x-coordinate as D and the same y-coordinate as B. Hence, $p = 7$ and $q = 6$. Point A has the same x-coordinate as B and the same y-coordinate as point D. Hence, $r = -2$ and $s = -3$.
    - To find an equation of the line that contains diagonal $\overline{AC}$, first find the slope of $\overline{AC}$:

$$m = \frac{\Delta y}{\Delta x} = \frac{6 - (-3)}{7 - (-2)} = \frac{9}{9} = 1$$

Hence, the equation of $\overline{AC}$ has the form $y = x + b$.
Note: Instead of using the slope formula, you could reason that since the diagonal of a square forms two right triangles in which the vertical and horizontal sides always have the same length, their ratio is always 1. Since $\overline{AC}$ rises from left to right, its slope is positive. Therefore, the slope of $\overline{AC}$ is 1 (and the slope of diagonal $\overline{BD}$ is -1).

- To find b, substitute the coordinates of $C(7,6)$ into $y = x + b$, which makes $6 = 7 + b$, so $b = -1$.
- An equation of $\overline{AC}$ is $y = x - 1$.

16. **(B)** The general form of an equation of line p is $y = mx + b$. It is given that the slope of the line is $\frac{3}{2}$ and the line contains the point $(6, 2)$. Substitute $m = \frac{3}{2}$, $x = 6$, and $y = 2$ in the equation of the line and solve for b:

$$y = mx + b$$
$$2 = \frac{3}{2}(6) + b$$
$$2 = 9 + b$$
$$-7 = b$$

17. **(B)** The graph that shows a line where each value of y is three more than half of x will have as its equation $y = \frac{1}{2}x + 3$. The slope of the line is $\frac{1}{2}$ and its y-intercept is 3. You can eliminate Graph (1) since its y-intercept is 0 rather than 3. You can also eliminate Graph (4) as it has a negative rather than a positive slope. Since the x-intercept of $y = \frac{1}{2}x + 3$ is -6, the correct choice is Graph (2).

18. **(B)** Pick two ordered pairs from the table, say $(8, 70)$ and $(15, 113.75)$, calculate the average rate of change:

$$\frac{\Delta d}{\Delta h} = \frac{113.75 - 70}{15 - 8} = \frac{43.75}{7} = 6.25$$

Hence, the equation has the form $d = 6.25h + b$. To find the constant b, substitute 8 for h and 70 for d:

$$70 = 6.25(8) + b$$
$$70 = 50 + b$$
$$20 = b$$

The equation is $d = 6.25h + 20$.

19. **(C)** For the lines $y = ax + b$ and $y = bx + a$, if a and b are nonzero constants and $a + b = 0$, then the lines have different slopes, which are not necessarily negative reciprocals, so they are neither parallel nor perpendicular to each other. To find the x-intercepts of the lines, set $y = 0$ and solve for x:

- If $y = ax + b$, then $0 = ax + b$ and $x = \dfrac{-b}{a}$.

  Since it is given that $a + b = 0$, $a = -b$ so
  $$x = \frac{-b}{a} = \frac{-b}{-b} = 1.$$

- If $y = bx + a$, then $0 = bx + a$ and
  $x = \dfrac{-a}{b} = \dfrac{-(-b)}{b} = \dfrac{b}{b} = 1.$ Since it is given

  that $a + b = 0$, $a = -b$ so $x = \dfrac{-b}{a} = \dfrac{-b}{-b} = 1.$

Hence, the lines have the same x-intercept.

20. **(C)** Using the points $(-2, 0)$ and $(0, 3)$, the

slope of the line is $\dfrac{\text{rise}}{\text{run}} = \dfrac{3}{2}$. A line that is

perpendicular to this line has a slope of $-\dfrac{2}{3}$.

Solve each equation in the answer choices for y in terms of x until you find the equation

in which the coefficient of x is $-\dfrac{2}{3}$ as in

choice (C): $3y = -2x + 4$ so $y = -\dfrac{2}{3}x + \dfrac{4}{3}$.

Grid-In

1. **(3/4)** Since the line that passes through

points $(7, 3k)$ and $(0, k)$ has a slope of $\dfrac{3}{14}$,

$$\frac{3k - k}{7 - 0} = \frac{3}{14}$$
$$\frac{2k}{7} = \frac{3}{14}$$
$$28k = 21$$
$$k = \frac{21}{28} = \frac{3}{4}$$

Grid in as 3/4.

2. **(3/2)** Since the slope of line ℓ_1 is $\dfrac{5}{6}$, then

$$\frac{y_1 - 0}{3 - 0} = \frac{5}{6} \quad \text{or} \quad y_1 = \frac{15}{6} = \frac{5}{2}$$

The slope of line ℓ_2 is $\dfrac{1}{3}$, so

$$\frac{y_2 - 0}{3 - 0} = \frac{1}{3} \quad \text{or} \quad y_2 = \frac{3}{3} = 1$$

Since points A and B have the same x-coordinates, they lie on the same vertical line, so

$$\text{Distance from } A \text{ to } B = y_1 - y_2$$
$$= \frac{5}{2} - 1 \text{ or } \frac{3}{2}$$

Grid in as 3/2.

3. **(15/2)** First, find the slope of line q. Since line q passes through $(0, 0)$ and $(-1, -2)$, its

slope is $\dfrac{-2 - 0}{-1 - 0} = 2$. Since line p passes

through $(-1, -2)$ and $(-20, k)$, its slope is

$\dfrac{k - (-2)}{-20 - (-1)} = \dfrac{k + 2}{-19}$. The slopes of

perpendicular lines are negative reciprocals.

Since the negative reciprocal of 2 is $-\dfrac{1}{2}$:

$$\frac{k + 2}{-19} = -\frac{1}{2}$$
$$2(k + 2) = 19$$
$$2k + 4 = 19$$
$$2k = 15$$
$$k = \frac{15}{2}$$

4. **(5)** Find the value of the y-intercept "b" in the equation of the line that is perpendicular to $\overline{AB}$ at point M.

- The slope of $\overline{AB} =$

$$\frac{\Delta y}{\Delta x} = \frac{12 - (-4)}{6 - (-2)} = \frac{12 + 4}{6 + 2} = \frac{16}{8} = 2$$

Hence, the slope of a line perpendicular to $\overline{AB}$ is the negative reciprocal of 2, which is $-\dfrac{1}{2}$.

- The equation of the desired line has the form $y = -\dfrac{1}{2}x + b$. To find b, substitute the coordinates of $M(2, 4)$ into the equation. Since $4 = -\dfrac{1}{2}(2) + b$, $4 = -1 + b$, so $b = 5$.

- Since $b = 5$, 5 is the y-coordinate of the point at which the line that is perpendicular to $\overline{AB}$ at its midpoint crosses the y-axis.

5. **(10)** Since points $O(0, 0)$, $A(c, 40)$, and $B(5, 2c)$ lie on the same line:

$$\text{slope of } \overline{OA} = \text{slope of } \overline{OB}$$
$$\frac{40 - 0}{c - 0} = \frac{2c - 0}{5 - 0}$$
$$\frac{40}{c} = \frac{2c}{5}$$
$$2c^2 = 200$$
$$c^2 = 100$$
$$c = \pm 10$$

Hence, a possible value of c is 10.

LESSON 3-11

Multiple-Choice

1. **(C)** A system of linear equations has no solution when the lines have the same slope and, as a result, are parallel. Solve each equation for y in terms of x:

- If $4x + 6y = 12$, then $6y = -4x + 12$ and $y = -\dfrac{4}{6}x + \dfrac{12}{6}$ or $y = -\dfrac{2}{3}x + 2$ so the slope of the line is $-\dfrac{2}{3}$.

- If $y = 8 - kx$, then $y = -kx + 8$ so the slope of the line is $-k$.

- Setting the slopes equal gives $k = \dfrac{2}{3}$.

2. **(B)** If the two linear equations that Sara correctly solves have no solution, then the lines are parallel. If one equation is $\dfrac{y}{6} - \dfrac{x}{4} = 1$, then multiplying each term of the equation by 12 gives $2y - 3x = 12$ and

$y = \dfrac{3}{2}x + 6$ so the slope of the line is $\dfrac{3}{2}$. The other equation in the system must also have a slope of $\dfrac{3}{2}$ as in the equation in choice (B): $y = \dfrac{3}{2}x$.

3. **(C)** If a system of two linear equations has an infinite number of solutions, then the two equations can be made to look exactly the same. Write and then compare the two equations in $y = mx + b$ slope-intercept form. If one of the two equations is $3(x + y) = 6 - x$, then $3x + 3y = 6 - x$ so $3y = -4x + 6$ and $y = -\dfrac{4}{3}x + 2$. This looks exactly like the equation in choice (C).

4. **(D)** The graph of the inequality $y \le 2x$ includes all the points on and below the line $y = 2x$:

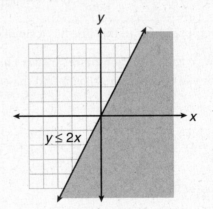

The graph includes all of the points in Quadrant IV.

5. **(C)** To find the value of k that makes the given system of equations have no solution, write each equation in $y = mx + b$ slope-intercept form:

$\dfrac{1}{2}x - \dfrac{5}{6}y = 5 \rightarrow 3x - 5y = 30$ so $y = \dfrac{3}{5}x - 6$

$-2x + ky = 3 \rightarrow ky = 2x + 3$ so $y = \dfrac{2}{k}x + \dfrac{3}{k}$

Set the slopes equal and solve for k:

$$\frac{3}{5} = \frac{2}{k}$$
$$3k = 10$$
$$k = \frac{10}{3}$$

6. **(D)** If a line has a slope of $\frac{1}{2}$ and contains the point (0, 7), then its equation is $y = \frac{1}{2}x + 7$. If another line passes through the points (0, 0) and (–1, 3), then its slope is $\frac{3-0}{-1-0} = -3$. Since it passes through the origin, its equation is $y = -3x$.

- If the lines $y = \frac{1}{2}x + 7$ and $y = -3x$ intersect, then $\frac{1}{2}x + 7 = -3x$ so $x + 14 = -6x$ and $7x = -14$, which means $x = -2$ and $y = -3x = -3(-2) = 6$.
- Since the point of intersection is (–2, 6), $r = -2$ and $s = 6$ so $r + s = -2 + 6 = 4$.

7. **(B)** The graph of the system $y + x > 2$ or $y > -x + 2$ and $y \le 3x - 2$ must have a broken line with a negative slope that corresponds to $y > -x + 2$ and a solid line with a positive slope that corresponds to $y \le 3x - 2$. This eliminates Graphs A and C. Graph D shows the solution set of $y > -x + 2$ as the region *below* the boundary line, which is not correct as it must lie *above* the boundary line as in Graph B because of the > relation. To be sure, choose a point in the solution region of Graph B, say (2, 1), and check that it satisfies the two inequalities:
- $y + x > 2$: $1 + 2 > 2$. ✓
- $y \le 3x - 2$: $1 \le 3(2) - 2$ and $1 \le 4$. ✓

8. **(D)** The graph of the system $2y < x$ or $y < 2x$ and $y \ge 3x + 1$, must have a broken line with a positive slope that corresponds to $y < 2x$ and a solid line with a positive slope that corresponds to $y \ge 3x + 1$ The solution set for $y < 2x$ is the region below the broken line and the solution set for $y \ge 3x + 1$ is the region above (to the left) of the solid line. These two regions overlap in the region labeled Z.

9. **(B)** Write each of the given equations in $y = mx + b$ slope-intercept form:

$3x + 5 = 2y \rightarrow y = \frac{3}{2}x + \frac{5}{2}$

$\frac{x}{3} + \frac{y}{2} = \frac{2}{3} \rightarrow 3y = -2x + 4$ so $y = \frac{-2}{3}x + \frac{4}{3}$

Since the slopes of the lines are negative reciprocals, the lines intersect at right angles.

10. **(C)** If the system $y < 2x + 4$ and $y \ge -x + 1$ are graphed on the same set of axes, then Quadrant III will not contain any solutions to the system.

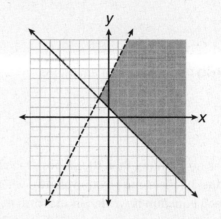

Grid-In

1. **(15/2)** If the system of equations $6x + py = 21$ and $qx + 5y = 7$ have infinitely many solutions, then the two equations can be made to look exactly alike. Multiply each member of the second equation by 3 so that its constant term matches the constant term of the first equation:

$$6x + py = 21$$
$$3qx + 15y = 21$$

The two equations will look exactly alike when $q = 2$ and $p = 15$ so $\frac{p}{q} = \frac{15}{2}$.

2. **(7/6)** If the system of equations

$$(k - 1)x + \frac{1}{3}y = 4$$
$$k(x + 2y) = 7$$

has no solution, then the lines have the same slope. Find the value of k that makes the slopes equal by first writing each equation in $y = mx + b$ slope-intercept form in order to determine their slopes:

$(k - 1)x + \frac{1}{3}y = 4 \rightarrow y = -3(k - 1)x + 12$

$k(x + 2y) = 7 \rightarrow x + 2y = \frac{7}{k}$ so $y = -\frac{1}{2}x + \frac{7}{2k}$

Set the slopes equal and solve for k:

$$-3(k-1) = -\frac{1}{2}$$

$$k - 1 = \frac{1}{6}$$

$$k = 1 + \frac{1}{6}$$

$$k = \frac{7}{6}$$

3. **(1/2)** If the system of equations

$$\frac{1}{3}r + 4s = 1$$

$$kr + 6s = -5$$

has no solution, then the slopes of the lines must be the same or, equivalently, when each equation is written in the form $ax + by = c$, the constants a and b are the same for each equation. Multiply the first equation by 6 and the second equation by 4 in order to get the coefficients of s to be the same:

$$6\left(\frac{1}{3}r + 4s\right) = 6(1) \;\rightarrow\; 2r + 24s = 6$$

$$4(kr + 6s) = 4(-5) \rightarrow 4kr + 24s = -20$$

Since the coefficients of r must be the same,

$4k = 2$ so $k = \frac{1}{2}$.

4. **(7)** If the graph of a line in the xy-plane passes through the points $(5, -5)$ and $(1, 3)$, then its slope is $\frac{3 - (-5)}{1 - 5} = \frac{8}{-4} = -2$ so its equation has the form $y = -2x + b$. Substituting the coordinates of $(1, 3)$ into the equation gives $3 = -2(1) + b$ so $b = 5$. If the graph of a second line has a slope of 6 and passes through the point $(-1, 15)$, then its equation has the form $y = 6x + b$. Substituting the coordinates $(-1, 15)$ into the equation gives $15 = 6(-1) + b$ so $b = 21$. The equations of the two lines are $y = -2x + 5$ and $y = 6x + 21$. Since the lines intersect, $6x + 21 = -2x + 5$ so $8x = -16$ and $x = -2$. To find y, substitute -2 for x into either equation: $y = 6(-2) + 21 = 9$. Hence, the coordinates of the point of intersection are $(p, q) = (-2, 9)$ so $p + q = -2 + 9 = 7$.

LESSON 3-12
Multiple-Choice

1. **(A)** If $f(x) = 3x + 2$ and $f(a) = 17$, then $f(a) = 3a + 2 = 17$, so $3a = 15$ and $a = \frac{15}{3} = 5$.

2. **(B)** Using the definition $f(1) = 2$, $f(2) = 5$, and $f(n) = f(n-1) - f(n-2)$, substitute 3 for n:

$$\begin{aligned} f(3) &= f(3-1) - f(3-2) \\ &= f(2) - f(1) \\ &= 5 - 2 \\ &= 3 \end{aligned}$$

Next, substitute 4 for n:

$$\begin{aligned} f(4) &= f(4-1) - f(4-2) \\ &= f(3) - f(3) \\ &= 3 - 5 \\ &= -2 \end{aligned}$$

3. **(D)** To figure out the number of x-values for which $y = f(x) = 2$, draw a horizontal line through $(0, 2)$ and count the number of points that the line intersects the graph:

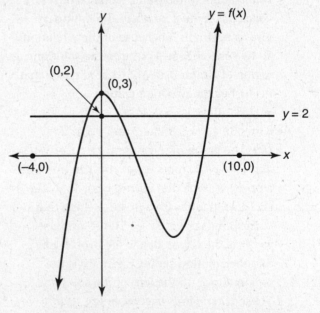

Since there are three points of intersection, there are three x-values for which $f(x) = 2$.

4. **(B)** If the function f is defined by $f(x) = 5x + 3$, then

$$2f(x) - 3 = 2(5x + 3) - 3$$
$$= 10x + 6 - 3$$
$$= 10x + 3$$

5. **(C)** If $k(h) = (h + 1)^2$, then

$$k(x - 2) = ((x - 2) + 1)^2$$
$$= (x - 1)^2$$
$$= (x - 1)(x - 1)$$
$$= x^2 - 2x + 1$$

6. **(C)** The accompanying tables define functions f and g. Since $f(3) = 5$, $g(f(3)) = g(5) = 8$.

x	1	2	3	4	5
$f(x)$	3	4	5	6	7

x	3	4	5	6	7
$g(x)$	4	6	8	10	12

7. **(D)** The postage function, $c(z)$, includes a fixed cost of $0.48 plus a rate of change in cost of $0.21 per ounce for each ounce over the first. For a total of z ounces, the function is $c(z) = 0.21(z - 1) + 0.48$. Verify that when $z = 1$ ounces, the cost is $0.48:

$$c(1) = 0.21(1 - 1) + 0.48$$
$$= 0.21(0) + 0.48$$
$$= 0.48$$

8. **(D)** The graphs of functions f and g, shown in the accompanying figure, are given. The region in which $f(x) \geq g(x)$ is the region where the graph of $y = f(x)$ is above the graph of $y = g(x)$, as indicated by the shaded region. Thus, $f(x) \geq g(x)$ for $-2 \leq x \leq 1$. Roman numeral choices I and II are included in this interval, which corresponds to choice (D).

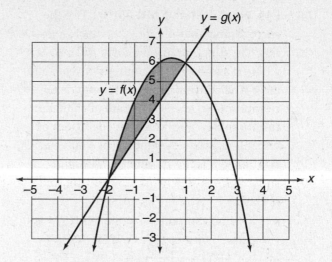

9. **(B)** Since $f(3) = 9$, $f(6) = 4f(3) = 4 \cdot 9 = 36$. Examine the answer choices for the equation that produces 9 for $n = 3$, and which equation produces 36 for $n = 6$. Since 9 is the square of 3 and 36 is the square of 6, the equation that defines function f could be $f(n) = n^2$.

10. **(B)** Reading from the graph, when p in (p, q) is between 0 and 5, the y-coordinates of all such points on the graph range from a minimum of -5 to a maximum of 10, which is represented by the inequality $-5 \leq q \leq 10$.

11. **(D)** If $x = -1$, $g(-1) = f(-1 + 4) = f(3)$. Reading from the given graph of function f, $f(3) = 5 = g(-1)$.

12. **(C)** According to the table, $f(5) = p = 1$, so $g(p) = g(1) = 3$.

x	$f(x)$	$g(x)$
1	2	3
2	4	5
3	5	1
4	3	2
5	1	4

13. **(B)** It is given that function h is defined by $h(x) = 2f(x) - 1$, where function f is defined in the accompanying table. If $h(k) = 5$, then $h(k) = 2f(k) - 1 = 5$, $2f(k) = 6$ so $f(k) = 3$. From the table, $f(4) = 3$, so $k = 4$. Again reading from the table, $g(k) = g(4) = 2$.

14. **(A)** From the first table, $f(5) = 2$. Use the second table to evaluate $g(2)$, which gives –1.

15. **(D)** The function has the form $f(n) = a + bn$, where n stands for the number of years after 2012. The starting constant amount is 68 restaurants so $a = 68$. The rate of change in restaurants is 9 per year, the coefficient of variable n, so $b = 9$. Hence, $f(n) = 68 + 9n$.

16. **(D)** Since n is the number of *thousands* of subscriptions sold, when 250,000 are sold, $n = 250$.

- First find the value of k. Since $n(15) = 250$,

$$n(15) = \frac{5{,}000}{4(15) - k} = 250$$

$$\frac{5{,}000}{60 - k} = 250$$

$$250(60 - k) = 5{,}000$$

$$\frac{250(60 - k)}{250} = \frac{5{,}000}{250}$$

$$60 - k = 20$$

$$k = 40$$

- Rewrite the original function with $k = 40$:

$$n(p) = \frac{5{,}000}{4p - 40}$$

- Find $n(20)$:

$$n(20) = \frac{5{,}000}{4(20) - 40}$$

$$= \frac{5{,}000}{40}$$

$$= 125$$

Because n is the number of *thousands* of subscriptions sold, the number of subscriptions that could be sold at \$20 for each subscription is 125,000.

Grid-In

1. **(0)** If $h(x) = x + 4^x$, then

$$h\!\left(-\frac{1}{2}\right) = -\frac{1}{2} + 4^{-\frac{1}{2}}$$

$$= -\frac{1}{2} + \frac{1}{\sqrt{4}}$$

$$= -\frac{1}{2} + \frac{1}{2}$$

$$= 0$$

2. **(4.3)** If $f(w + 1.7) = 0$, then $w + 1.7$ must be an x-intercept. Since w must be positive as the answer grid cannot accommodate negative numbers, use the positive x-intercept of 6 by setting $w + 1.7$ equal to 6, which makes $w = 4.3$.

3. **(2)** Since, $f(x) = x^2 + 12$, $f(n) = n^2 + 12$, and $f(3n) = (3n)^2 + 12 = 9n^2 + 12$. If $f(3n) = 3f(n)$, then

$$9n^2 + 12 = 3(n^2 + 12)$$

$$9n^2 + 12 = 3n^2 + 36$$

$$9n^2 - 3n^2 = 36 - 12$$

$$6n^2 = 24$$

$$n^2 = \frac{24}{6}$$

$$n = \pm\sqrt{4} = \pm 2$$

Since n must be a positive number, $n = 2$.

4. **(7)** Use the graph of each function to find the function values you need.

- Since $(3, 1)$ is a point on the graph of function g, $g(3) = 1$. Hence, $f(g(3)) = f(1)$.
- Because $(1, 7)$ is a point on the graph of function f, $f(1) = 7$.
- Therefore, $f(g(3)) = f(1) = 7$.

5. **(12)** From the graph, $f(-1) = 0$ and $f(1) = 4$. Hence, $2f(-1) + 3f(1) = 2 \cdot 0 + 3 \cdot 4 = 12$.

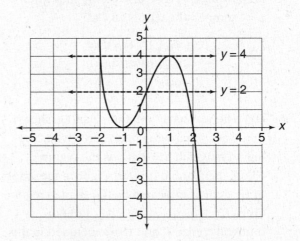

Graph for Exercises 5 and 6

6. **(6)** Use the accompanying graph.

- To find n, the number of values of x for which $f(x) = 2$, count the number of points at which the horizontal line $y = 2$ intersects the graph. As shown in the accompanying graph, $n = 3$.

- To find m, number of values of x for which $f(x) = 4$, count the number of points at which the horizontal line $y = 4$ intersects the graph. As shown in the accompanying graph, $m = 2$.
  Hence, $mn = 2 \times 3 = 6$.

7. **(17)** It is given that $g(x) = x - 1$ and $\frac{1}{2}g(c) = 4$. Hence,

$$\frac{1}{2}g(c) = \frac{1}{2}(c-1) = 4$$
$$c - 1 = 8$$
$$c = 9$$

Now find the value of $g(2c)$ where $c = 9$:

$$g(2c) = g(18) = 18 - 1 = 17$$

8. **(6)** Since $f(x) = h(2x) + 1$, $f(-1) = h(-2) + 1$. According to the given graph of function h, $h(-2) = 5$, so $f(-1) = 5 + 1 = 6$.

LESSON 4-1

Multiple-Choice

1. **(C)** Since Terry passed 80% of his science tests, he failed 20% or $\frac{1}{5}$ of the tests taken.

   Terry failed 4 science tests, so 4 represents $\frac{1}{5}$ of the total number of science tests. Hence, Terry took 5×4 or 20 science tests. Since he failed 4 tests, he passed $20 - 4$ or 16 science tests.

2. **(B)** Since the soccer team has won 60% of the 25 games it has played, it has won 0.6×25 or 15 games.

   Solution 1: If x represents the minimum number of additional games the team must win in order to finish the season winning 80% of the games it has played, then $15 + x =$ 80% of $(25 + x)$, so $15 + x = 0.80(25 + x)$. Remove the parentheses by multiplying each term inside the parentheses by 0.80:

   $$15 + x = 20 + 0.8x$$
   $$x - 0.8x = 20 - 15$$
   $$0.2x = 5$$
   $$x = \frac{5}{0.2} = 25$$

   Solution 2: For each answer choice, form and then evaluate the fraction

   $$\frac{\text{Wins}}{\text{Total games}} = \frac{\text{Answer Choice} + 15}{\text{Answer Choice} + 25}$$

   The correct answer choice is (B), which gives a fraction equal to 0.80:

   $$\frac{15}{25} = \frac{25 + 15}{25 + 25} = \frac{40}{50} = 0.80$$

3. **(C)** If 480 of 500 seats were occupied, then 20 seats were *not* occupied. Hence, the percent of the seats that were *not* occupied is

   $$\frac{20}{500} \times 100\% = \frac{1}{25} \times 100\% = 4\%$$

4. **(C)** Plug easy-to-use numbers into the problem. Assume there are 20 members of the class who do not belong to the math club. Then 50% of 20 = 10 members of the class belong to the math club. Hence, the total number of math club members in this class is $\frac{10}{10 + 20} \times 100\% = 33\frac{1}{3}\%$ of the entire class.

5. **(A)**

   $$\text{Percent decrease} = \frac{\text{Amount of decrease}}{\text{Original amount}} \times 100\%$$
   $$= \frac{168 - 147}{168} \times 100\%$$
   $$= \frac{21}{168} \times 100\%$$
   $$= 0.125 \times 100\%$$
   $$= 12.5\% \text{ or } 12\frac{1}{2}\%$$

6. **(C)** Since $\frac{3}{5} \times 3 = \frac{9}{5}$, the original 3-cup mixture contains $\frac{9}{5}$ cups of milk. After 1 cup of milk is added to the 3-cup mixture, the 4-cup mixture contains

   $$\frac{9}{5} + 1 = \frac{9}{5} + \frac{5}{5} = \frac{14}{5} \text{ cups of milk}$$

   Since

   $$\frac{\frac{14}{5}}{4} = \frac{4}{5} \times \frac{1}{4} = \frac{14}{20} = 0.70$$

   70% of the 4-cup mixture is milk.

7. **(B)** If the result of increasing a by 300% of a is b, then $a + 3a = 4a = b$. Dividing both sides of $4a = b$ by b gives $\frac{4a}{b} = 1$, so $\frac{a}{b} = \frac{1}{4}$. Since $\frac{1}{4} = 25\%$, a is 25% of b.

8. **(D)** After a 20% price increase, the new price of a radio is $78.00. Hence,

$$\text{original price} = \frac{\$78.00}{1 + 20\%}$$
$$= \frac{\$78.00}{1.2}$$
$$= \$78.00 \div 1.2$$
$$= \$65.00$$

9. **(C)** After a discount of 15%, the price of a shirt is $51. Hence,

$$\text{original price} = \frac{\$51}{1 - 15\%}$$
$$= \frac{\$51}{1 - 0.15}$$
$$= \$51 \div 0.85$$
$$= \$60.00$$

10. **(C)** Two students use the computer a total of 28% + 52% or 80% of the total time of 3 hours or 180 minutes. The third student uses the computer for 100% − 80% or 20% of 180 minutes, which equals $0.2 \times 180 = 36$ minutes.

11. **(C)** In the opinion poll, 70% of the 50 men or $0.70 \times 50 = 35$ men preferred fiction to nonfiction books. In the same poll, 25% of the 40 women or $0.25 \times 40 = 10$ women preferred fiction to nonfiction books. Thus, 45 (= 35 + 10) of the 90 (= 50 + 40) people polled preferred fiction to nonfiction books. Since $\frac{45}{90} = \frac{1}{2} = 50\%$, 50% of the people polled preferred to read fiction.

12. **(C)** 0.04 percent means 0.04 out of every 100 or, equivalently, 4 out of every 10,000. Since $\frac{4}{10,000} = \frac{1}{2,500}$, 1 out of every 2,500 light bulbs manufactured will, on the average, be defective. Maintaining the same ratio, on the average $1 \times 3 = 3$ out of every $2,500 \times 3 = 7,500$ light bulbs manufactured will be defective.

13. **(B)** Since 40% + 25% + 20% +10% = 95%, the percent of the vehicles that are jeeps is 100% − 95% = 5%. Hence, 5 out of every 100 cars are jeeps. If there are $5 \times 4 = 20$ jeeps, there must be a total of $100 \times 4 = 400$ vehicles.

14. **(C)** Suppose Jack's beginning weight was 100 pounds. After a 20% increase, his new weight is 100 + 20 = 120 pounds. Since 25% of 120 = 30, after a 25% weight *decrease*, his final weight is 120 − 30 = 90 pounds, which is $\frac{90}{100} \times 100\% = 90\%$ of his beginning weight.

15. **(C)** Suppose the original price of the stock was $100. If the price falls 25%, the new price of the stock is $0.75 \times \$100 = \$75$. To reach its original value, the price of the stock must rise $25, which is $\frac{1}{3}$ or $33\frac{1}{3}\%$ of its $75 price.

16. **(B)** To find the number of people who voted for candidate A in the actual election, add the number of people who chose candidate A in the poll to the number of people who voted for candidate A but were undecided in the poll.
    - Since 30% of 4,000 = $0.30 \times 4,000 = 1,200$, 1,200 people voted for candidate A in the poll as well as in the actual election.
    - Since 20% of 4,000 = $0.20 \times 4,000 = 800$, 800 people were undecided in the poll.
    - It is given that in the actual election, the undecided people voted for candidate A in the same proportion as the people who cast votes for candidates in the poll. Hence, of the 800 undecided people,
    $$\frac{30\%}{30\% + 50\%} \times 800 = \frac{3}{8} \times 800 = 300 \text{ voted}$$
    for candidate A in the actual election.
    - Hence, a total of 1,200 + 300 = 1,500 people voted for candidate A in the actual election.

17. **(A)** If a car traveled at an average speed of 65 miles per hour for 3 hours, then it travels a distance of $65 \times 3 = 195$ miles. If it consumes 1 gallon of gas for each 30 miles, then it consumes $\frac{195}{30} = 6.5$ gallons, which is $\frac{6.5}{20} \times 100\% = 32.5\%$ of the starting amount of 20 gallons of gas.

Grid-In

1. **(87.5)** The cash price of $84.00 reflects the new amount after a discount of 4%. Hence,

$$\frac{\text{Original amount}}{} = \frac{\text{New amount after decrease of } P\%}{1 - P\%}$$

$$= \frac{\$84}{1 - 4\%}$$

$$= \frac{\$84}{1 - 0.04}$$

$$= \frac{\$84}{0.96}$$

$$= \$87.5$$

The credit-card purchase price of the same item is $8.75.

2. **(50)** After three boys are dropped from the class, 25 students remain. Of the 25 students, 44% are boys, so 56% are girls. Since 56% of $25 = 0.56 \times 25 = 14$, 14 girls are enrolled in the class. Hence, 14 of the 28 students in the original class were girls. Thus, the number of girls in the original class comprised $\frac{1}{2}$ or 50% of that class.

3. **(12)** Since the team wins 60% of the first 15 matches, it wins 0.60×15 or 9 of its first 15 matches. Let x represent the number of additional matches it must win to finish the 28-match season winning 75% of its scheduled matches. Since $75\% = \frac{3}{4}$,

$$\frac{\text{Total wins}}{\text{Total games}} = \frac{9 + x}{28} = \frac{3}{4}$$

Solve the equation by cross-multiplying:

$$4(9 + x) = 3(28)$$
$$36 + 4x = 84$$
$$4x = 48$$
$$x = \frac{48}{4}$$
$$= 12$$

4. **(55)** or **(5/9)** Since

$$80\% \text{ of } 35 = 0.80 \times 35 = 28$$

and

$$25\% \text{ of } 28 = 0.25 \times 28 = 7$$

then 35 (= 28 + 7) of the 63 (35 + 28) boys and girls have been club members for more than 2 years. Since $\frac{35}{63} = 0.5555...$, 55% or 5/9% of the club have been members for more than 2 years.

LESSON 4-2
Multiple-Choice

1. **(C)** When the recipe is adjusted from 4 to 8 servings, the amounts of salt and pepper are *each* doubled, so the ratio of 2 : 3 remains the same.

2. **(D)** Solution 1: If x represents the unknown number of miles, then

$$\frac{\text{Inches}}{\text{Miles}} = \frac{\frac{3}{8}}{120} = \frac{1\frac{3}{4}}{x}$$

Cross-multiplying gives

$$\frac{3}{8}x = 1\frac{3}{4}(120) = \frac{7}{4}(120) = 210$$

Then

$$x = \frac{8}{3}(120) = 560 \text{ miles}$$

Solution 2: Since $\frac{3}{8}$ of an inch represents 120 miles, $\frac{1}{8}$ of an inch represents $\frac{1}{3} \times 120 = 40$ miles. Since $1\frac{3}{4} = \frac{14}{8}$, $1\frac{3}{4}$ inches represent 14×40 or 560 miles.

3. **(C)** Let p represent the initial population of bacteria. After 12 minutes the population is $2p$, after 24 minutes it is $4p$, after 36 minutes it is $8p$, after 48 minutes it is $16p$, and after 60 minutes or 1 hour the population is $32p$. Since

$$\frac{32p}{p} = \frac{32}{1}$$

the ratio of the number of bacteria at the end of 1 hour to the number of bacteria at the beginning of that hour is $32 : 1$.

4. **(D)** If x represents the number of losses, then

$$\frac{\text{Wins}}{\text{Losses}} = \frac{4}{3} = \frac{12}{x}$$
$$4x = 36$$
$$x = 9$$

The total number of games played was $12 + 9$ or 21.

5. **(C)** If $\frac{s}{t} = \frac{4}{7}$, then $s = \frac{4t}{7}$. Since s is an integer, t must be divisible by 7. In the interval $8 < t < 40$, there are four integers that are divisible by 7: 14, 21, 28, and 35.

6. **(D)** If x represents the number of sophomores in the school club, then $3x$ represents the number of juniors, and $2x$ represents the number of seniors. Since the club has 42 members,

$$x + 3x + 2x = 42$$
$$6x = 42$$
$$x = \frac{42}{6} = 7$$

The number of seniors in the club is $2x = 2(7) = 14$.

7. **(A)** If $\frac{c - 3d}{4} = \frac{d}{2}$, then

$$2(c - 3d) = 4d$$
$$2c - 6d = 4d$$
$$2c = 10d$$

Since $\frac{c}{d} = 5$, the ratio of c to d is $5 : 1$.

8. **(D)** If x represents the number of pairs of socks that can be purchased for \$22.50, then

$$\frac{\text{Pairs of socks}}{\text{Cost}} = \frac{4}{10} = \frac{x}{22.50}$$

So

$$10x = 4(22.50) = 90$$
$$x = \frac{90}{10} = 9$$

9. **(C)** The number of boys working and the time needed to complete the job are inversely related since, as one of these quantities increases, the other decreases. Let x represent the time three boys take to paint a fence; then

$$3x = 2(5) = 10$$
$$x = \frac{10}{3} = 3\frac{1}{3}$$

10. **(B)** If x represents the number of miles the car travels in 5 hours, then

$$\frac{\text{Distance}}{\text{Time}} = \frac{96}{2} = \frac{x}{5}$$

so

$$2x = 5(96) = 480$$
$$x = \frac{480}{2} = 240$$

11. **(C)** If x represents the number of kilograms of corn needed to feed 2,800 chickens, then $2x - 30$ is the number of kilograms needed to feed 5,000 chickens. If the amount of feed needed for each chicken is assumed to be constant,

$$\frac{x}{2,800} = \frac{2x - 30}{5,000}$$

so

$$5,000x = 2,800(2x - 30)$$
$$50x = 28(2x - 30)$$
$$= 56x - 840$$
$$6x = 840$$
$$x = \frac{840}{6} = 140$$

12. **(C)** If the number of calories burned while jogging varies directly with the number of minutes jogging, then the ratio of these quantities remains constant. If x represents the number of calories George burns while jogging for 40 minutes, then

$$\frac{180}{25} = \frac{x}{40}$$
$$25x = 7,200$$
$$x = \frac{7,200}{25} = 288$$

13. **(D)** If y varies directly as x and $y = 12$ when $x = c$, then when $x = 8$

$$\frac{12}{c} = \frac{y}{8}$$
$$cy = 96$$
$$y = \frac{96}{c}$$

14. **(D)** Since $\frac{x}{z} = \frac{1}{3}$, then

$$\frac{x^2}{z} = x\left(\frac{x}{z}\right) = x\left(\frac{1}{3}\right)$$

Determine whether each Roman numeral value is a possible value of $\frac{x^2}{z}$ when x and z are integers.

- I. If $\frac{x^2}{z} = x\left(\frac{1}{3}\right) = \frac{1}{9}$, then $x = \frac{1}{3}$. Since x must be an integer, $\frac{1}{9}$ is not a possible value of $\frac{x^2}{z}$.

- II. If $\frac{x^2}{z} = x\left(\frac{1}{3}\right) = \frac{1}{3}$, then $x = 1$. Hence, $\frac{1}{3}$ is a possible value of $\frac{x^2}{z}$.

- III. If $\frac{x^2}{z} = x\left(\frac{1}{3}\right) = 3$, then $x = 9$. Hence, 3 is a possible value of $\frac{x^2}{z}$.

Only Roman numeral values II and III are possible values of $\frac{x^2}{z}$.

15. **(A)** If $a - 3b = 9b - 7a$, then

$$a + 7a = 9b + 3b$$
$$8a = 12b$$
$$\frac{a}{b} = \frac{12}{8} = \frac{3}{2}$$

Hence, the ratio of a to b is $3 : 2$.

16. **(C)** Since the lowest common multiple of 8 and 12 is 24, multiply each term of the ratio $a : 8$ by 3 and multiply each term of the ratio $12 : c$ by 2. The ratio of A to B is then $3a : 24$, and the ratio of B to C is $24 : 2c$, so the ratio of A to C is $3a : 2c$. Since you are given that the ratio of A to C is $2 : 1$, then

$$\frac{2}{1} = \frac{3a}{2c}, \text{ so}$$
$$\frac{a}{c} = \frac{2(2)}{3(1)} = \frac{4}{3}$$

Hence, the ratio of a to c is $4 : 3$.

17. **(A)** If $8^r = 4^t$, then $2^{3r} = 2^{2t}$ so $3r = 2t$. Dividing both sides of this equation by $3t$ gives

$$\frac{3r}{3t} = \frac{2t}{3t} \quad \text{or} \quad \frac{r}{t} = \frac{2}{3}$$

The ratio of r to t is $2 : 3$.

18. **(B)** Rewrite each fraction by dividing each term of the numerator by the denominator:

$$\frac{a+b}{b} = \frac{a}{b} + \frac{b}{b} = \frac{a}{b} + 1 = 4, \frac{a}{b} = 3$$
$$\frac{a+c}{c} = \frac{a}{c} + \frac{c}{c} = \frac{a}{c} + 1 = 3, \frac{a}{c} = 2$$

or $\frac{c}{a} = \frac{1}{2}$.

Multiply corresponding sides of the two proportions:

$$\frac{a}{b} \times \frac{c}{a} = 3 \times \frac{1}{2}$$
$$\frac{c}{b} = \frac{3}{2}$$

The ratio of c to b is $3 : 2$.

19. **(C)** Since the ratio of mathematics majors to English majors is given as 3 : 8, suppose there are 30 mathematics majors and 80 English majors.

- If the number of mathematics majors increases 20%, then the new number of mathematics majors is $30 + (20\% \times 30) = 30 + 6 = 36$.
- If the number of English majors decreases 15%, then the new number of English majors is $80 - (15\% \times 80) = 80 - 12 = 68$.
- The new ratio of mathematics majors to English majors is $\dfrac{36}{68} = \dfrac{9}{17}$ or 9 : 17.

20. **(A)** The ratio of freshmen to juniors who attended the game is 3 : 4, and the ratio of juniors to seniors is 7 : 6. Since the least common multiple of 4 and 7 is 28, multiply the terms of the first ratio by 7 and multiply the terms of the second ratio by 4. Since 21 : 28 is then the ratio of freshmen to juniors and 28 : 24 is the ratio of juniors to seniors, 21 : 24 is the ratio of freshmen to seniors.
Since

$$21 : 24 = \frac{21}{24} = \frac{7}{8}$$

the correct choice is (A).

21. **(B)** The number of men working and the hours needed for the men to complete the job are inversely related. If x represents the number of men needed to complete the job in $5 - 1 = 4$ hours, then $4x = 5 \cdot 12 = 60$.
Hence, $x = \dfrac{60}{4} = 15$. This means that $15 - 12 = 3$ additional men are needed.

22. **(C)** If $4 - 3x < 6$, then $-3x < 2$, so $\dfrac{-3x}{-3} > \dfrac{2}{-3}$ and $x > -\dfrac{2}{3}$. Three of the six numbers in $\{-3, -2, -1, 0, 1, 2\}$ are greater than $-\dfrac{2}{3}$: 0, 1, and 2. Hence, the required probability is $\dfrac{3}{6}$ or $\dfrac{1}{2}$.

23. **(C)** Between points A and B, the run is $7 - (-2) = 9$ and the rise is $3 - 9 = -6$.
Since $AP : PB$ is 1 : 2, the distance from A to P is one-third of the distance from A to B.
Multiply the run and the rise by $\dfrac{1}{3}$ and then add the products to the corresponding coordinates of point A:

$$P\left(-2 + \frac{1}{3} \times 9, \ 9 + \frac{1}{3} \times (-6)\right) = P(-2 + 3, \ 9 - 2) = P(1, 7)$$

Grid-In

1. **(12)** Since the lengths of the two pieces of string are in the ratio 2 : 9, let $2x$ and $9x$ represent their lengths. Hence:

$$9x - 2x = 42$$
$$7x = 42$$
$$x = \frac{42}{7} = 6$$

Since $2x = 2(6) = 12$, the length of the shorter piece of string is 12 inches.

2. **(15)** Since $2^3 = 2 \times 2 \times 2 = 8$, then $a = 3$ and $b = 2$, so $\dfrac{a}{b} = \dfrac{3}{2}$. You are told that $\dfrac{a}{b} = \dfrac{c}{d}$.
Substituting $\dfrac{3}{2}$ for $\dfrac{a}{b}$ and 10 for d gives $\dfrac{3}{2} = \dfrac{c}{10}$. Since 10 is 5 times 2, c must be 5 times 3 or 15.

3. **(4)** Since jars A, B, and C each contain 8 marbles, there are 24 marbles in the three jars. Let x, $2x$, and $3x$ represent the new numbers of marbles in jars A, B, and C, respectively. Hence, $x + 2x + 3x = 24$ or $6x = 24$, so

$$x = \frac{24}{6} = 4$$

To achieve a ratio of 1 : 2 : 3, jars A, B, and C must contain 4, 8, and 12 marbles, respectively. Since jar B already contains 8 marbles, 4 of the 8 marbles originally in jar A must be transferred to jar C.

4. **(16)** If x and y are inversely related, their product is constant. The point on the graph has coordinates (50, 40), which means the constant term is $50 \times 40 = 2{,}000$. If y is the number of hours it would take to do the mailing using 125 workers, then $125y = 2{,}000$ so $y = \dfrac{2{,}000}{125} = 16$.

5. **(2/9)** To find the probability that a dart will land in the triangular region formed by the lines $y = 2$, $x = 6$, and $y = x$, find the ratio of the area of the triangular region to the area of the square.

   - The area of the square dartboard is $6 \times 6 = 36$.

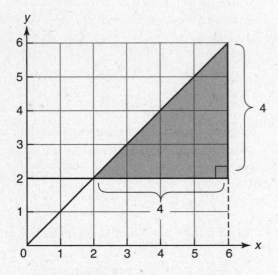

   - The area of the triangular region is $\dfrac{1}{2} \times 4 \times 4 = 8$.

   - Hence, the probability that the dart lands in the triangular region is $\dfrac{8}{36} = 2/9$.

LESSON 4-3
Multiple-Choice

1. **(C)** If 4 pens cost \$1.96, the cost of 1 pen is $\dfrac{\$1.96}{4} = \$0.49$. The greatest number of pens that can be purchased for \$7.68 is the greatest integer that is equal to or less than $\dfrac{\$7.68}{\$0.49}$. Since $\dfrac{\$7.68}{\$0.49} = 15.67$, is the greatest number of pens that can be purchased is 15.

2. **(A)** If the larger pipe can fill a swimming pool 1.25 times faster than the slower pipe, then the rate of the slower pipe is $\dfrac{1}{1.25x}$ and the rate of the faster pipe is $\dfrac{1}{x}$ where x represents the number of hours that it would take to fill the pool working alone. Since the larger pipe is working for 6 hours and the smaller pipe for 5 hours,

$$\left(\frac{1}{1.25x}\right)5 + \left(\frac{1}{x}\right)6 = 1$$

3. **(C)** If x represents the distance apart, in miles, then

$$\frac{120}{1.5} = \frac{x}{12} \leftarrow \text{change 1 foot to 12 inches}$$
$$1.5x = (12)(120)$$
$$x = \frac{1{,}440}{1.5} = 960$$

4. **(B)** Since the freight train leaves 1 hour before the passenger train, it travels 1 hour longer. Therefore, when the passenger train has traveled x hours, the freight train has traveled $x + 1$ hours. Make a table.

Rate	×	Time	=	Distance
50 mph		$x + 1$ hours		$50(x + 1)$
60 mph		x hours		$60x$

Since the two trains are traveling in opposite directions, the sum of their distances must equal 380 miles:

$$50(x + 1) + 60x = 380$$
$$50x + 50 + 60x = 380$$
$$110x = 330$$
$$x = \frac{330}{110} = 3$$

Since the passenger train leaves at 1:00 P.M., the two trains are 380 miles apart 3 hours later at 4:00 P.M.

5. **(D)** Let x represent the distance driven at each speed. Since time = distance ÷ rate of speed, the time driving at 60 miles per hour can be represented as $\frac{x}{60}$, and the time driving at 40 miles per hour can be represented as $\frac{x}{40}$.

- Since the total driving time was 1 hour,

$$\frac{x}{60} + \frac{x}{40} = 1$$

- Solve for x by multiplying each term on both sides of the equation by 2,400, the lowest common multiple of 60 and 40:

$$\overset{40}{\cancel{2400}}\left(\frac{x}{\cancel{60}}\right) + \overset{60}{\cancel{2400}}\left(\frac{x}{\cancel{40}}\right) = 2{,}400(1)$$

$$40x \quad + \quad 60x \quad = 2{,}400$$

$$\frac{100x}{100} = \frac{2{,}400}{100}$$

$$x = 24$$

- Because the one-way trip distance is 24 miles, the round trip distance is $24 + 24 = 48$ miles.

6. **(A)** If x people working together at the same rate can complete a job in h hours, then each person working alone can complete the same job in xh hours, so the rate of work is the reciprocal of xh or $\frac{1}{xh}$.

Hence, the part of the job one person working alone can complete in k hours is $\frac{k}{xh}$.

7. **(D)** Since the rate of work is constant, form a proportion where x stands for the time it takes to install 8 light fixtures:

$$\frac{\text{number of light fixtures}}{\text{time to install}} = \frac{5}{3} = \frac{8}{x}$$

$$5x = 24$$

$$\frac{5x}{5} = \frac{24}{5}$$

$$x = 4\frac{4}{5}\text{ hours}$$

8. **(D)** Let x be the rate of speed of the freight train. Make a table.

Rate	×	Time	=	Distance
x mph		3 hours		$3x$
$x + 20$ mph		3 hours		$3(x + 20)$

The total distance the two trains travel in 3 hours is $500 - 80$ or 420 miles.
Hence,

$$3x + 3(x + 20) = 420$$

$$3x + 3x + 60 = 420$$

$$6x = 360$$

$$x = \frac{360}{6} = 60$$

9. **(C)** If a machine can seal 360 packages per hour, then it seals $\frac{360}{60}$ or 6 packages per minute. If another machine can seal 140 packages per hour, then it seals $\frac{140}{60}$ or $\frac{7}{3}$ packages per minute. If the two machines working together can seal a total of 700 packages in x minutes, then

$$6x + \frac{7}{3}x = 700$$

$$18x + 7x = 2100$$

$$25x = 2100$$

$$x = \frac{2100}{25} = 84$$

10. **(B)** Make a table.

Rate	×	Time	=	Distance
18 mph		$x - \frac{1}{4}$ hours		$18\left(x - \frac{1}{4}\right)$
12 mph		x hours		$12x$

Since the length of the lake remains constant,

$$18\left(x - \frac{1}{4}\right) = 12(x)$$

$$18x - \frac{18}{4} = 12x$$

$$6x = \frac{9}{2}$$

$$x = \frac{9}{6(2)} = \frac{9}{12} = \frac{3}{4}$$

Hence,

$$\text{Rate} \times \text{Time} = \text{Distance}$$

$$12 \times \frac{3}{4} = 9 \text{ miles}$$

11. **(D)** Let x be the time for the outgoing trip. Make a table.

Rate	×	Time	=	Distance
x mph		$\frac{120}{x}$ hours		120
$x + 5$ mph		$\frac{120}{x+5}$ hours		120

Since the time for the return trip was one-third of an hour less than the time for the outgoing trip,

$$\frac{120}{x} = \frac{120}{x+5} + \frac{1}{3}$$

12. **(B)** The distance Jonathan drove to the airport is 45 mph × 3 hrs = 135 miles. If x represents the time, in hours, it took him to drive home, then $55x = 135$

so $x = \dfrac{135}{55} = 2.45 \approx 2.5$ hours.

13. **(A)** It is given that a plumber works twice as fast as his apprentice. If his apprentice takes $2x$ hours working alone to complete the job, then the plumber would take x hours working alone to complete the same job. The rates of work of the plumber and his apprentice are $\dfrac{1}{x}$ and $\dfrac{1}{2x}$, respectively.

Multiply these rates by the number of hours each worked to get 1 complete job:

$$\left(\frac{1}{x}\right)(3 + 4) + \left(\frac{1}{2x}\right)(4) = 1$$

$$\frac{7}{x} + \frac{2}{x} = 1$$

$$x\left(\frac{7}{2} + \frac{2}{x}\right) = x \cdot 1$$

$$9 = x$$

Grid-In

1. **(8.0)** or **(8)** If 5 pounds of fruit serve 18 people, then $\dfrac{5}{18}$ pound serves one person, so

$$24 \times \frac{5}{18} = 4 \times \frac{5}{3} = \frac{20}{3} \text{ pounds}$$

serve 24 people. Since the fruit costs $1.20 a pound, the cost of the fruit needed to serve 24 people is

$$\frac{20}{3} \times \$1.20 = 20 \times \$0.40 \text{ or } \$8.00$$

2. **(12)** Since the rate of work is constant, form a proportion where x stands for the time, in *minutes*, it takes to produce 920 flyers:

$$\frac{\text{Number of flyers}}{\text{Time to produce}} = \frac{4,600}{60 \text{ min } (=1 \text{ hour})} = \frac{920}{x}$$

$$\frac{460x}{460} = \frac{6 \times \overset{2}{\cancel{920}}}{\cancel{460}}$$

$$x = 12 \text{ minutes}$$

3. **(52)** According to the conversion rates given, U.S. dollar to Euro is 1.56 to 1.38. Hence, if x represents the cost, in U.S. dollars, of a shirt that costs 46 Euros, then

$$\frac{1.56}{1.38} = \frac{x}{46}$$

$$1.38x = (46)(1.56)$$

$$x = \frac{71.76}{1.38} = 52$$

4. **(12)** Since $\dfrac{1,200}{240} = 5$, Joseph can type five

240 word pages in 25 minutes. If x represents the number of 240-word pages he can type in 1 hour or 60 minutes, then

$$\frac{5}{25} = \frac{x}{60}$$

$$\frac{1}{5} = \frac{x}{60}$$

$$5x = 60$$

$$x = \frac{60}{5} = 12$$

LESSON 4-4
Multiple-Choice

1. **(C)** There are $\dfrac{1,000 \text{ meters}}{\text{km}}$ and $\dfrac{1 \text{ hour}}{60 \text{ min}}$.

   Verify that the correct pairs of units cancel out:

   $$8\,\frac{\cancel{\text{km}}}{\cancel{\text{hr}}} \times \frac{1000 \text{ meters}}{\cancel{\text{km}}} \times \frac{1 \cancel{\text{hr}}}{60 \text{ min}} = \dots \frac{\text{meters}}{\text{min}} \checkmark$$

2. **(D)**

   $$72\,\frac{\cancel{\text{km}}}{\text{hr}} \times 1,000\,\frac{\text{m}}{\cancel{\text{km}}} = 72 \times 10^3 = 7.2 \times 10^4\,\frac{\text{m}}{\text{hr}}$$

3. **(D)** Multiply

   $$\left(1.67 \times 10^{-24}\right) \times 1,000 = \left(1.67 \times 10^{-24}\right) \times 10^3$$
   $$= 1.67 \times 10^{-24+3}$$
   $$= 1.67 \times 10^{-21}$$

4. **(C)** Choose the product in which the units cancel out so that the answer is expressed in the correct units of rolls of tape as in choice (C):

   $$12 \cancel{\text{ players}} \cdot \frac{2 \cancel{\text{ ankles}}}{1 \cancel{\text{ player}}} \cdot \frac{1 \text{ roll}}{3 \cancel{\text{ ankles}}} = 8 \text{ rolls}$$

5. **(B)** Since the answer must be in $\dfrac{\text{miles}}{\text{hr}}$, the conversion factor involving miles must have miles in the numerator. Thus, the conversion factor of $\dfrac{5,280 \text{ ft}}{1 \text{ mile}}$ should read $\dfrac{1 \text{ mile}}{5,280 \text{ ft}}$.

6. **(A)** Multiply

   $$3.1 \times 10^4 \cancel{\text{ light yrs}} \times 5.9 \times 10^{12}\,\frac{\text{miles}}{1 \cancel{\text{ light yr}}} = 18.29 \times 10^{4+12}$$
   $$= 18.29 \times 10^{16}$$
   $$\approx 1.83 \times 10^{17}$$

7. **(B)** Convert 10 decaliters into milliliters:

   $$10 \cancel{\text{ decaliters}} \times 10\,\frac{\cancel{\text{liters}}}{\cancel{\text{decaliter}}} \times 1,000\,\frac{\text{milliliters}}{\cancel{\text{liter}}}$$
   $$= 10 \times 10^4 \text{ milliters}$$

   To find the number of 2.5-milliliter bottles that can be filled, divide the number of milliliters equivalent to 10 decaliters by 2.5:

   $$\frac{10 \times 10^4}{2.5} = 4 \times 10^4$$

Grid-In

1. **(27)** Find the number of ounces in six 1-liter bottles:

   $$6 \cancel{\text{ liters}} \times 1.1\,\frac{\cancel{\text{quarts}}}{1 \cancel{\text{ liter}}} \times 32\,\frac{\text{ounces}}{1 \cancel{\text{ quart}}} = 211.2 \text{ ounces}$$

   Then divide the total number of ounces by 8:

   $$\frac{211.2}{8} \approx 26.4$$

   You must round up to obtain the least number of cups that are needed, which gives 27 so the last cup will not be entirely filled.

2. **(8.64)** If 1 inch represents 2 kilometers, then a region on the map that is 1.5 inches by 4.0 inches is actually 3 kilometers by 8 kilometers. Since 1 kilometer is equivalent to 0.6 miles, the region is 1.8 miles by 4.8 miles, which has an area of $1.8 \times 4.8 = 8.64$ square miles.

3. **(332)** Since $\text{time} = \dfrac{\text{distance}}{\text{speed}}$,

   $$\frac{136,000,000}{28,500 \times 0.6} \approx 7,953.2 \text{ hours}$$

To find the number of days, divide the number of hours by 24, and round up:

$$\frac{7,953.2}{24} \approx 331.4 \approx 332$$

4. **(12)** Calculate the volume of box in cubic centimeters and then convert to liters:

$$(25\,\text{cm} \times 20\,\text{cm} \times 16\,\text{cm}) \times \frac{1\,\text{liter}}{1,000\,\text{cubic cm}}$$

$$= 8 \text{ liters}$$

Multiplying 8 liters by $1.5\,\dfrac{\text{hours}}{1\,\text{liter}}$ gives 12 hours.

5. **(8,512)** Convert nautical miles to feet per hour:

$$3.5\,\text{knots} = 3.5\,\frac{\text{nautical mile}}{\text{hour}} \times 6,080\,\frac{\text{feet}}{\text{nautical mile}}$$

$$= 21,280\,\frac{\text{feet}}{\text{hour}}$$

Since 24 minutes is $\dfrac{24}{60} = 0.4$ hours, multiply the speed in feet per hour by 0.4 hours, which gives the number of feet traveled in 24 minutes:

$$21,280\,\frac{\text{feet}}{\text{hour}} \times 0.4\,\text{hours} = 8,512\,\text{feet}$$

LESSON 4-5
Multiple-Choice

1. **(C)** Graph C shows that after the first month the price of the stock is $180 - \dfrac{1}{3}(180) = 120$ and continues to decay exponentially using $\dfrac{1}{3}$ as the decay factor.

2. **(C)** The given process is modeled by the exponential function $y = 16(2)^{\frac{d}{6}}$. If $d = 60$, then

$$y = 16(2)^{\frac{d}{6}} = 16(2)^{\frac{60}{6}}$$
$$= 16(2)^{10}$$
$$= 2^4 \cdot (2)^{10}$$
$$= 2^{14}$$

3. **(B)** The interval of pH that corresponds to the lowest rate of change is the interval that has the segment with the smallest positive slope, which is the interval from 6 to 7.

4. **(C)** Each time x, the number of folds, increases by 1, the number of layers of paper doubles, which corresponds to the equation $f(x) = 2^x$.

5. **(B)** In a linear model of the form $f(x) = a + bx$, the constant a is an initial amount or a beginning value of what is being modeled and b, the coefficient of x, is a rate of change. Hence, in the model $f(n) = 5 + 8n$, 5 is the initial number of restaurants and 8 is the rate or number of restaurants being added per year.

6. **(C)** In the model $C = 60 + 0.05d$, 0.05 is the coefficient of the variable d, expressed in megabytes, so it represents a rate of change—the change in cost per megabyte of data after the first gigabyte of data is used.

7. **(D)** During the 9 to 12 minute interval, the graph shows a horizontal segment, which indicates that there is no change in speed. Hence, she was jogging at a constant rate.

8. **(D)** In the sequence 54, 18, 6,..., each term after the first is being multiplied by $\dfrac{1}{3}$. If a_n represents the nth term of the sequence, then $a_n = 54\left(\dfrac{1}{3}\right)^{n-1}$. The exponent is $n-1$ rather than n since for $n = 1$, a_1 must be equal to 54.

9. **(B)** For the profit function $P(x) = 8,600 - 22x$, 22 is the hourly rate of pay so x must be the number of hours the employee works, which when multiplied together gives the amount of money that must be subtracted from the revenue of 8,600 to obtain the profit.

10. **(B)** For the exponential model $p(n) = 300(0.5)^n$, examine each statement in turn:
   - I. Since 300 represents the initial amount of the substance, Statement I is false. ✗
   - II. Since the decay factor is 0.5, at the end of each year, the amount present at the beginning of the year gets reduced by 0.5

times of that amount. For example, $p(1) = 300(0.5)^1 = 150$, $p(2) = 300(0.5)^2 = 75$, and so forth. Statement II is true. ✓

- III. Since Statement III states that the substance gets reduced each year by the same amount, one-half of 300, it is false. ✗

Hence, only Statement II is true.

11. **(C)** In the exponential model $y = 5,000(0.98)^x$, the coefficient of the exponential term, 5,000, always represents the starting amount, which in this case is the amount of money in the account initially.

12. **(C)** Since the car is losing its value at the rate of 14% per year, the decay factor is $1 - 14\% = 1 - 0.14 = 0.86$, which appears in the equation $v = 27,000(0.86)^4$.

13. **(B)** Profit is the difference between the current balance and the initial amount invested, which in this problem is

$$P = 10,000(1.06)^5 - 10,000$$
$$= 10,000\left[(1.06)^5 - 1\right]$$

14. **(C)** Miriam uses the growth function $f(t) = n^{2t}$. If she starts with 16 bacteria, then $f(t) = 16^{2t}$. Jessica uses the growth function $g(t) = \left(\dfrac{n}{2}\right)^{4t}$. To find the number of bacteria Jessica must start with to achieve the same growth over time, set the two functions equal to each other and solve for n:

$$16^{2t} = \left(\frac{n}{2}\right)^{4t}$$

$$\left(4^2\right)^{2t} = \left(\frac{n}{2}\right)^{4t}$$

$$\left(4\right)^{4t} = \left(\frac{n}{2}\right)^{4t} \leftarrow \text{since powers are equal,}$$

set bases equal

$$4 = \frac{n}{2}$$
$$8 = n$$

15. **(A)** If the area, A, covered by a bacteria culture is increasing at a rate of 20% every 7 days, then the growth factor is $1 + 20\% = 1.20$, and its exponent is $\dfrac{d}{7}$ where d is the

number of days. If the initial area of the bacteria culture is 10 square centimeters, then $A = 10(1.20)^{\frac{d}{7}}$.

16. **(A)**

Health Club Fees

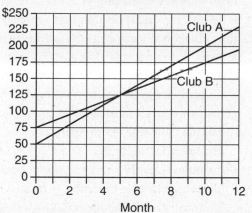

According to the graph:

- At month 0, the cost of Club A is $50, and the cost at month 5 is $125. This means that the monthly membership fee is $50 and the cost of 5 months without the membership fee included is $125 – $50 = $75. Therefore, the cost per month for Club A is $\dfrac{\$75}{5} = \$15$.

- At month 0, the cost of Club B is $75, and the cost at month 5 is $125. This means that the monthly membership fee is $75 and the cost of 5 months without the membership fee included is $125 – $75 = $50. Therefore, the cost per month for Club B is $\dfrac{\$50}{5} = \$10$.

- Thus, the monthly charge for Club A exceeds the monthly charge for Club B by $15 – $10 = $5.

17. **(C)** Pick any two ordered pairs from the table and calculate the average rate of change. Using (0.5, 9.0) and (1, 8.75):

$$\frac{\Delta y}{\Delta x} = \frac{8.75 - 9}{1 - 0.5} = -\frac{0.25}{0.5} = -\frac{1}{2}$$

This is the same for any two ordered pairs in the table.

18. **(D)** Since the decline is exponential and at a rate of 12% every 5 years, the decay factor is $1 - 12\% = 1 - 0.12 = 0.88$, and its exponent is $\frac{n}{5}$ where n stands for the number of years from now. If the starting amount of vehicles is 40,000, the equation that best represents the vehicle traffic projections for this bridge n years from now is $40,000(0.88)^{\frac{n}{5}}$.

19. **(B)** Examine each of the four tables in turn.
    - In Table I, the population decreases by a constant amount of 3,000 every 6 months so this table does not illustrate exponential decay. ✗
    - In Table II, every 6 months the population decreases by one-half of its previous amount. This table illustrates exponential decay. ✓
    - In Table III, the population *increases* by a factor of 2 every 6 months. This table illustrates exponential growth, not decay. ✗
    - In Table IV, the population increases by a constant amount every 6 months so this is linear growth, not exponential decay. ✗

20. **(D)** If a radioactive substance with an initial mass of 100 grams is reduced by 40% every 5 years, then this process is an example of exponential decay. The decay factor is $1 - 40\% = 1 - 0.40 = 0.60$. Hence, an equation that could be used to find the number of grams in the mass, y, that remains after x years is $y = 100(0.60)^{\frac{x}{5}}$.

21. **(B)** The graph that represents the fact that a car travels with a full tank of gas. If the car travels 75 miles on 4 gallons of gas, then the graph that represents the rate of change in the amount of gas in the tank will have a negative slope, which eliminates Graphs C and D. Since the gas tank starts with 16 gallons of gas, the y-intercept of the graph must be 16. The rate of change is $\frac{75 \text{ miles}}{4 \text{ gallons}} = 18.75 \frac{\text{miles}}{\text{gallon}}$. Using a full tank of gas, the car can travel a distance of $18.75 \times 16 = 300$ miles so the x-intercept of the graph must be 300, which is Graph B.

Grid-In

1. **(14.5)** The rate at which paint is being used is

$$\frac{4 - 13}{3} = \frac{-9}{3} = -3$$

Since paint is being used at a constant rate, 1.5 gallons were used during the first half hour of the job, which means that Jacob started the job at 12:00 noon with $13 + 1.5 = 14.5$ gallons of paint.

2. **(140)** The order pairs (100, 270) and (160, 410) must satisfy the equation $H = kX + q$. Since the rate of change is constant:

$$\frac{\Delta H}{\Delta X} = \frac{410 - 270}{160 - 100} = \frac{7}{3}$$

Each additional monitor takes $\frac{7}{3}$ hours or $\frac{7}{3} \times 60 = 140$ minutes to manufacture.

3. **(135)** Solve the equation $b(t) = 51,200 = 100(2)^t$ for t:

$$\frac{51,200}{100} = \frac{100(2)^t}{100}$$
$$512 = 2^t$$
$$2^9 = 2^t$$
$$9 = t$$

Since there are nine 15-minute periods, the total number of minutes it will take for the population of bacteria to reach 51,200 is $9 \times 15 = 135$ minutes.

4. **(82)** First, find $x(5)$:

$$x(5) = 512(0.7)^5 \approx 86$$

Then find $h(86)$:

$$h(86) = 65 + 0.2(86) = 82.2$$

5. **(2021)** Use trial and error:
    - If $t = 5$, then $p(5) = 300\left(\frac{1}{2}\right)^5 = 9.375$
    - If $t = 6$, then $p(6) = 300\left(\frac{1}{2}\right)^6 = 4.6875$

Since $t = 0$ corresponds to the year 2015, $t = 6$ corresponds to the year 2021.

6. **(20)** Sasha's account balance, y, after n years is given by the exponential function $A(n) = 1,200(1.016)^n$. You are asked to find $A(3) - A(2)$:

- $A(3) = 1,200(1.016)^3 = 1,258.53$
- $A(2) = 1,200(1.016)^2 = 1,238.71$
- $A(3) - A(2) = 19.82$ or 20 rounded to the *nearest dollar*.

LESSON 4-6

Multiple-Choice

1. **(B)** Since the slices that comprise a pie chart must add up to 100%, $10 + 15 + 30 + 25 + x = 100$, so $x = 20$.

   Solution 1: If N represents the total number of SAT math scores, then 30% of N is 72 so $N = 72 \div 0.30 = 240$. Since 20% of 240 $= 0.20 \times 240 = 48$, 48 SAT math scores are above 700.

   Solution 2: Since 30% of the total number of SAT scores represented by the graph is 72, 10% of the total number of SAT scores represented by the graph is $\frac{1}{3} \times 72 = 24$.

   Since 10% of the circle represents 24 scores, 20% of the circle represents 2×24 or 48 SAT math scores.

2. **(A)** Since 25% of the 240 SAT math scores are from 610 to 700, $25\% \times 240 = \frac{1}{4} \times 240 = 60$ students had scores from 610 to 700. If 20% of these students received scholarships, then $20\% \times 60 = 0.20 \times 60 = 12$ students with math scores from 610 to 700 received college scholarships.

3. **(C)** According to the given table, a person can obtain a driver's license at 14 years of age in 7 states and at 15 years of age in 12 states. Hence, there are $7 + 12$ or 19 states in which a person under the age of 16 can obtain a driver's license. Since the total number of states is 50, the percent of the states in which a person under the age of 16 can obtain a driver's license is $\frac{19}{50} \times 100\%$ or 38%.

4. **(D)** $P(\text{minimum age} \geq 16) = \frac{27 + 2 + 2}{50} = \frac{31}{50}$

5. **(D)** According to the height of the bars in the given graph, in 2010 the number of households with computers was 40 million and in 2014 the number was 100 million. Hence,

$$\text{Percent increase} = \frac{\text{Amount of increase}}{\text{Original amount}} \times 100\%$$
$$= \frac{100 - 40}{40} \times 100\%$$
$$= \frac{60}{40} \times 100\%$$
$$= 1.5 \times 100\%$$
$$= 150\%$$

6. **(A)** Calculate the percent of increase for each 2 consecutive years:

- From 2010 to 2011:

$$\text{Percent increase} = \frac{\text{Amount of increase}}{\text{Original amount}} \times 100\%$$
$$= \frac{55 - 40}{40} \times 100\%$$
$$= \frac{15}{40} \times 100\%$$
$$= 37.5\%$$

- From 2011 to 2012:

$$\text{Percent increase} = \frac{\text{Amount of increase}}{\text{Original amount}} \times 100\%$$
$$= \frac{65 - 55}{55} \times 100\%$$
$$= \frac{10}{55} \times 100\%$$
$$= 18.18\%$$

- From 2012 to 2013:

$$\text{Percent increase} = \frac{\text{Amount of increase}}{\text{Original amount}} \times 100\%$$
$$= \frac{80 - 65}{65} \times 100\%$$
$$= \frac{15}{65} \times 100\%$$
$$= 23.08\%$$

- From 2013 to 2014:

$$\text{Percent increase} = \frac{\text{Amount of increase}}{\text{Original amount}} \times 100\%$$

$$= \frac{100 - 80}{80} \times 100\%$$

$$= \frac{20}{80} \times 100\%$$

$$= 25\%$$

Hence, the greatest percent of increase occurred from 2010 to 2011.

7. **(A)** The sum of all of the sectors of a circle graph is 100%. Hence,

$$11\% + 35\% + 23\% + x\% + 7\% + 3x\% = 100\%$$
$$76\% + x\% + 3x\% = 100\%$$
$$4x\% = 100\% - 76\%$$
$$= 24\%$$
$$x = \frac{24\%}{4\%} = 6$$

Since $3x\% = 3(6)\% = 18\%$, municipal bonds make up 18% of the investment portfolio. To find the amount of money invested in municipal bonds, multiply the total value of the portfolio by 18%:

18% of $250,000 = 0.18 × $250,000 = $45,000. Hence, $45,000 is invested in municipal bonds.

8. **(D)** The original amount invested in health stocks is 11% of the total investment. You are told that 20% of the 23% that is invested in technology stocks is reinvested in health stocks. Since 20% of 23% = 0.2 × 15.6% is now invested in health stocks,

15.6% of $250,000 = 0.156 × $250,000
$$= \$39,000$$

The amount of money now invested in health stocks is $39,000.

9. **(D)** From data in the graph:
   - Percent increase from 2010 to 2011

$$= \frac{700 - 400}{400} \times 100\%$$

$$= \frac{300}{400} \times 100\%$$

$$= 75\%$$

- Percent increase from 2012 to 2013

$$= \frac{500 - 300}{300} \times 100\%$$

$$= \frac{200}{300} \times 100\%$$

$$\cong 66.7\%$$

- Percent increase from 2010 to 2011 exceeds the percent increase from 2012 to 2013 by about 75% − 66.7% or, approximately, 8%.

10. **(D)** From 2013 to 2014, the percent decrease was $\frac{500 - 400}{500} \times 100\%$ or 20%. If the percent increase from 2014 to 2015 was 20%, then the number of students enrolled in advanced mathematics courses in 2015 was 400 + 20% of 400 = 400 + 80 = 480.

11. **(D)** In 2012, about 450 cars were purchased and 500 cars leased. Since $\frac{450}{500} = 90\%$, the best approximation for x is 90.

12. **(B)** For 2011–2012, 2012–2013, and 2013–2014, the number of cars purchased decreased a total of 660 − 240 or 420 cars. Hence, the decrease in the number of cars per year was approximately $\frac{420}{3}$ or 140.

13. **(D)** The interval on which the heart rate is changing at a constant rate is the interval on which the graph is linear and not horizontal, which is the interval from 3 to 4 minutes.

14. **(A)** The greatest amount of change in heart rate over equal intervals of time occurred in the interval from 0 to 1 minutes.

15. **(A)** From years 1 to 5, Jacob's head circumference increases 4 + 2 = 6 cm so that in the first year his head circumference was 81 − 6 = 75 cm. Use the formula provided in the table to solve for height, h, when circumference equals 75:

$$75 = \frac{h + 12}{2}$$
$$150 = h + 12$$
$$138 = h$$

16. **(B)** For a cumulative histogram, the number of data items in an interval is the difference between the cumulative frequency of that interval and the interval that comes before it. For example, in the interval 71–80 there are $16 - 8 = 8$ test scores. Hence, if a student is selected at random, the probability that the student will have a score between 71 and 80 is $\frac{8}{24} = \frac{1}{3}$.

17. **(C)** P(female junior or male senior) =
$$\frac{5 + 11}{40} = \frac{16}{40} = \frac{2}{5}$$

Grid-In

1. **(44)** $\frac{60 + 50}{250} = \frac{11}{25} = 44\%$

2. **(204)** Of the 250 surveyed, $\frac{30}{250} = 12\%$ saw a romance movie. Assuming that the same proportion holds for the entire population of movie tickets sold, about $1{,}700 \times 0.12 = 204$ tickets for a romance movie were sold.

3. **(5)** Find the weighted average:

$$\frac{\begin{array}{l}(2 \times 5) + (3 \times 0) + (4 \times 10) + (5 \times 7) + \\ (6 \times 0) + (7 \times 3) + (8 \times 2)\end{array}}{27} = \frac{122}{27} \approx 4.52$$

Round up to 5.

4. **(7/9)** P(does not belong ≥ 3 clubs) =
$$\frac{212 + 138}{450} = \frac{350}{450} = \frac{7}{9}$$

5. **(400)** The number of students who belong to at least 2 clubs is $138 + 100 = 238$, which represents 28% of the school enrollment. Hence, if x represents the school enrollment, then $0.28x = 238$ so $x = \frac{238}{0.28} = 850$. Since a total of 450 students participated in the survey, $850 - 450 = 400$ students did not participate in the survey.

6. **(.4)** P(9th grade or ≥ 3 clubs)
$$= \frac{98 + (100 - 18)}{450} = \frac{180}{450} = \frac{2}{5} \text{ or } .4$$

7. **(56)** If x represents the number of males with blue eyes, then $4x$ is the number of males with brown eyes. If y represents the number of females with blue eyes, the $7y$ is the number of females with brown eyes. Hence, $4x + 7y = 133$.

   Since a total of 133 people have brown eyes and another 45 have hazel eyes, the number of people with blue eyes is $200 - (133 + 45) = 22$. Therefore, $x + y = 22$ so $x = 22 - y$. Substituting in the first equation gives

$$4(22 - y) + 7y = 133$$
$$88 - 4y + 7y = 133$$
$$88 + 3y = 133$$
$$3y = 45$$
$$y = \frac{45}{3} = 15$$
$$x = 22 - y = 22 - 15 = 7$$

You are asked to find the percent of the subjects who were either male with blue eyes or female with brown eyes:

$$\frac{\begin{array}{l}\text{Males with blue eyes} + \\ \text{Females with brown eyes}\end{array}}{200} = \frac{7 + 7(15)}{200}$$
$$= \frac{112}{200}$$
$$= \frac{56}{100} \text{ or } 56\%$$

8. **(15)** The total number of swimmers is 20 and the number of swimmers in the interval 200–249 is $15 - 12 = 3$. Hence, the percent of swimmers who swam between 200 and 249 yards is $\frac{3}{20}$ or 15%.

9. **(3/4)** The number of students who made 10 or fewer mistakes is 7, which means that $28 - 7 = 21$ students made more than 10 mistakes. If a student is selected at random, the probability that the student made more than 10 mistakes is $\frac{21}{28} = \frac{3}{4}$.

LESSON 4-7
Multiple-Choice

1. **(A)** A bias may arise in statistical surveys when the group surveyed is either insufficient in number or not truly representative of the entire target population. Surveying a sample of people leaving a movie theater to determine ice cream flavor preferences is most likely to have the least bias as the two activities— attending movies and eating ice cream—are unrelated. In each of the other choices, there is a relationship between the activity in which the surveyed subjects were engaged and the topic of the survey.

2. **(D)** Surveying people leaving a high school football game about a proposed increase in the sports budget of the school district would most likely contain the most bias as the subjects surveyed would mostly likely be sports enthusiasts and interested in supporting the sports program.

3. **(B)** Based on the number of drivers in the different age groups that were surveyed, the results would mostly likely reflect some bias since individuals 36 years old and older were underrepresented.

4. **(C)** Since the graph contains the points (13, 30,000) and (18, 42,500), the profit in the 18th month exceeded the profit in the 13th month by 42,500 − 30,000 = 12,500 dollars.

5. **(C)** The predicted number of people attending the class is found from the line of best fit by locating the y-coordinate of the point on the line that has 4 as its x-coordinate. The line of best fit contains the point (4, 780). The data point that has 4 as its x-coordinate is (4, 600). Since the predicted number is 780 and the actual observed number is 600,

$$\frac{780 - 600}{600} = \frac{180}{600} = 30\%$$

6. **(C)** The line passes through the points (0, 300) and (5, 900). Hence, over the five-month period, the average increase in people attending the class per month is

$$\frac{900 - 300}{5} = \frac{600}{5} = 120$$

7. **(A)** The month for which the vertical distance between a point on the line and a data point is greatest is month 2.

8. **(A)** If the scores for each student are recorded as ordered pairs of the form

   (math SAT score, verbal SAT score)

then there is only one ordered pair, corresponding to student A, in which the y-value would be greater than the corresponding x-value. If these ordered pairs are graphed as a scatterplot, then that data point would be the only one of the five data points that would lie above the line $y = x$.

Grid-In

1. **(38)** Since the line of best fit passes through the point (17, 38), a predicted neck circumference of 38 cm corresponds to a wrist circumference of 17 cm.

2. **(5)** To find how many of the 12 people have an actual neck circumference that differs by more than 1 centimeter from the predicted neck circumference, simply count the number of data points for which the vertical distance from the corresponding point on the best fit line is more than 1 centimeter. There are 5 such points.

3. **(2.2)** The slope of the best fit line represents the average increase in neck circumference per centimeter increase in wrist circumference. Pick any two convenient points on the best fit line, say (14, 31.5) and (17, 38), and calculate the slope of the line:

$$\frac{38.0 - 31.5}{17 - 14} = \frac{6.5}{3} \approx 2.17$$

Rounded to the *nearest tenth* of a centimeter, the answer is 2.2.

LESSON 4-8
Multiple-Choice

1. **(B)** If the average (arithmetic mean) of a set of seven numbers is 81, then the sum of these seven numbers is 7×81 or 567. Since, if one of the numbers is discarded, the average of the six remaining numbers is 78, the sum of these six numbers is 6×78 or 468. Since $567 - 468 = 99$, the value of the number that was discarded is 99.

2. **(A)** If the mean of a set of 20 scores is x, then the sum of those 20 scores is $20x$. If each score is increased by y points, then the new mean is

$$\frac{\text{Sum of scores}}{20} = \frac{20x + 20y}{20} = \frac{20(x + y)}{20} = x + y$$

3. **(C)** The radii of circles with areas of 16π and 100π are 4 and 10, respectively. The average of 4 and 10 is

$$\frac{4 + 10}{2} = \frac{14}{2} = 7$$

The area of a circle of radius 7 is $\pi(7)^2$ or 49π.

4. **(B)** Since there are a total of 30 scores, the median is contained in the interval in which the 15th and 16th scores are located, which is the interval 71–80.

5. **(B)** The average of 3^k and 3^{k+2} is their sum divided by 2:

$$\frac{3^k + 3^{k+2}}{2} = \frac{3^k + 3^2 \cdot 3^k}{2}$$
$$= \frac{1 \cdot 3^k + 9 \cdot 3^k}{2}$$
$$= \frac{10 \cdot 3^k}{2}$$
$$= 5 \cdot 3^k$$

6. **(C)** When x is subtracted from $2y$ the difference is equal to the average of x and y, so $2y - x = \frac{x + y}{2}$ or $2(2y - x) = x + y$. Hence,

$$4y - 2x = x + y$$
$$4y - y = x + 2x$$
$$3y = 3x$$

Dividing both sides of $3y = 3x$ by $3y$ gives

$$\frac{x}{y} = 1.$$

7. **(D)** If the average of x, y, and z is 32, then $\frac{x + y + z}{3} = 32$, so

$$x + y + z = 3 \times 32 = 96$$

Since the average of y and z is 27, then $\frac{y + x}{2} = 27$, so

$$y + z = 2 \times 27 = 54$$

Substituting the value of $y + z$ in the equation $x + y + z = 96$ gives $x + 54 = 96$, so

$$x = 96 - 54 = 42$$

and

$$2x = 84$$

Hence, the average of x and $2x$ equals

$$\frac{42 + 84}{2} = \frac{126}{2} = 63$$

8. **(B)** Consider each statement in turn.
   - I. The mean salary for company 1 is 33,750 so Statement I is false. ✗
   - II. The sum of the ages of workers in both companies is 133 so the mean age of workers in both companies is the same. Statement II is false. ✗
   - III. The salary range in company 1 is $38,000 - 30,000 = 8,000$; the salary range in company 2 is $65,000 - 29,000 = 36,000$. Statement III is true. ✓

   Hence, only Statement III is true.

9. **(D)** Let x represent the average rate of speed in miles per hour for the second part of the trip, which lasts $7 - 3$ or 4 hours. Since rate multiplied by time equals distance, the man drives 45×3 or 135 miles during the first part of the trip and $4x$ miles during the second part. Total distance divided by total

time gives the average rate of speed for the entire trip, so

$$\frac{135 + 4x}{7} = 53$$

$$135 + 4x = 7 \times 53 = 371$$

$$4x = 371 - 135 = 236$$

$$x = \frac{236}{4} = 59$$

10. **(B)** If each value is reduced by 3, then the ordering of the data values remains the same so the median of the revised set is $m - 3$.

11. **(A)** If x represents the lowest score Susan can receive on her next math exam and have an average of at least 85 on the five exams, then

$$\frac{78 + 93 + 82 + 76 + x}{5} = 85$$

$$329 + x = 5 \times 85$$

$$= 425$$

$$x = 425 - 329$$

$$= 96$$

12. **(D)** Since

$$(x + y)^2 = x^2 + 2xy + y^2$$

and

$$(x - y)^2 = x^2 - 2xy + y^2$$

the average of $(x + y)^2$ and $(x - y)^2$ is

$$\frac{\left(x^2 + 2xy + y^2\right) + \left(x^2 - 2xy + y^2\right)}{2}$$

or

$$\frac{2\left(x^2 + y^2\right)}{2} = x^2 + y^2$$

13. **(B)** The sum of the scores of the group of x students is $76x$ and the sum of the scores of the group of y students is $90y$. Since there are $x + y$ students in the combined group, the average of the combined group is $\frac{76x + 90y}{x + y}$. Hence:

$$\frac{76x + 90y}{x + y} = 85$$

$$76x + 90y = 85(x + y)$$

$$76x + 90y = 85x + 85y$$

$$90y - 85y = 85x - 76x$$

$$\frac{5\cancel{y}}{\cancel{y}} = \frac{9x}{y}$$

$$\frac{5}{9} = \frac{\cancel{9}x}{\cancel{9}y}$$

$$\frac{5}{9} = \frac{x}{y}$$

14. **(C)** From the graph, the set of temperatures are 20, 25, 25, 30, 25, 20, and 35. Arrange the values in ascending order: 20, 20, 25, 25, 25, 30, 35. Hence, the median is 25, the mode is 25, and the mean is about 25.7. Hence, the only correct choice is the median = mode.

15. **(D)** If the average of $a, b, c, d,$ and e is 28, then

$$\frac{a + b + c + d + e}{5} = 28$$

or

$$a + b + c + d + e = 5 \times 28 = 140$$

If the average of $a, c,$ and e is 24, then

$$\frac{a + c + e}{3} = 24$$

or

$$a + c + e = 3 \times 24 = 72$$

Substituting 72 for $a + c + e$ in $a + b + c + d + e = 140$ gives

$$b + d + 72 = 140$$

$$b + d = 140 - 72 = 68$$

$$\frac{b + d}{2} = \frac{68}{2} = 34$$

16. **(A)** Adding corresponding sides of the given equations, $2a + b = 7$ and $b + 2c = 23$, gives $2a + 2b + 2c = 30$ or $a + b + c = 15$. Hence:

$$\frac{a + b + c}{3} = \frac{15}{3} = 5$$

17. **(D)** Since a weighted average is required, eliminate choices (A) and (B) in which the minutes were not multiplied by their frequencies. Eliminate choice (C) since the 16-minute entry was not multiplied by its frequency of 16 as in choice (D).

18. **(B)** The average of a, b, c, and d is p, so

$$\frac{a+b+c+d}{4} = p$$

or $a + b + c + d = 4p$. Similarly, since the average of a and c is q, then $a + c = 2q$, so $b + d + 2q = 4p$. Since $b + d = 4p - 2q$, then

$$\frac{b+d}{2} = \frac{4p}{2} - \frac{2q}{2} = 2p - q$$

19. **(A)** Consider each statement in turn.
   - I. If each score is increased by k, then the mean must also be increased by k since the revised sum is increased by the number of scores in the set times k. Statement I is true. ✓
   - II. The range remains unchanged: $(y + k) - (x + k) = y - x$. Statement II is false. ✗
   - III. If each score is increased by k, then the size order of the scores is not changed, but the median is increased by k. Statement III is false. ✗

20. **(B)** The standard deviation is a statistic that measures how spread out the data are from the mean. Based on the frequencies, the mean grade will be close to a B so most of the scores are clustered about the mean. For class Y the grades are more spread out relative to its mean grade so the standard deviation of grades is greater for class Y than it is for class X. Hence, Statement I is false and Statement II is true. The median letter grade is B for class X and C for class Y so statement III is false. Only Statement II is true.

21. **(B)** Since not every data value is being increased by the same dollar amount, the mean does not increase by 10%. The median and mode are each increased by 10%.

22. **(A)** If the team signs an additional player for a salary of 7.5 million dollars per year, the sum of the salaries for the 25 players is now 69 million. The new mean is $\frac{69}{25} = 2.76$ million while the old mean is

$\frac{69 - 7.5}{24} = 2.5625$ million so the difference is 0.1975 million or 197,500 dollars.

Grid-In

1. **(18)** It is given that $\frac{r+s}{2} = 7.5$, so $r + s = 2(7.5) = 15$. It is also given that $\frac{r+s+t}{3} = 11$, so $r + s + t = 3(11) = 33$. Hence, $15 + t = 33$, so $t = 33 - 15 = 18$.

2. **(36)** Solution 1: Since the average of x, y, and z is 12, then $x + y + z = 3 \times 12 = 36$. Hence,

$$3x + 3y + 3z = 3(36) = 108$$

The average of $3x$, $3y$, and $3z$ is their sum, 108, divided by 3 since three values are being added: $\frac{108}{3}$ or 36.

Solution 2: The average of x, y, and z is 12, so the average of any constant multiple of x, y, and z is 12 times that constant multiple, or $12(3) = 36$.

3. **(64)** If x represents the old mean, then $\frac{1}{2}x + 50 = 82$ so $\frac{1}{2}x = 32$ and $x = 64$.

4. **(76)** If x represents the mean and y the standard deviation, then $x - 2y = 58$ and $x + y = 85$. Subtracting corresponding sides of the two equations gives $3y = 27$ so $y = 9$. Hence, the mean is $85 - 9 = 76$.

LESSON 5-1
Multiple-Choice

1. **(B)** $b^{-\frac{1}{2}} = \dfrac{1}{b^{\frac{1}{2}}} = \dfrac{1}{\sqrt{b}} \cdot \dfrac{\sqrt{b}}{\sqrt{b}} = \dfrac{\sqrt{b}}{b}$

2. **(A)** $\left(9x^2 y^6\right)^{-\frac{1}{2}} = \dfrac{1}{\sqrt{9x^2 y^6}} = \dfrac{1}{3xy^3}$

3. **(A)** It is given that $4^y + 4^y + 4^y + 4^y = 16^x$. On the left side of the equation, four identical terms are being added together. Thus, $4 \cdot 4^y = (4^2)^x$, so $4^{y+1} = (4^{2x})$. Hence, $y + 1 = 2x$ and $y = 2x - 1$.

4. **(D)** If $\sqrt{m} = 2p$, then $m^{\frac{1}{2}} = 2p$, so

$$\left(m^{\frac{1}{2}}\right)^3 = (2p)^3$$

$$m^{\frac{3}{2}} = 8p^3$$

5. **(C)** Solve each exponential equation.
   - Since $3^x = 81 = 3^4$, $x = 4$.
   - Since $2^{x+y} = 64 = 2^6$, $x + y = 6$. Because $x = 4$, $4 + y = 6$, so $y = 2$.
   - Hence, $\dfrac{x}{y} = \dfrac{4}{2} = 2$.

6. **(B)** $y^{\frac{3}{2}} = \left(y^3\right)^{\frac{1}{2}} = \sqrt{y^3}$

7. **(D)** $\dfrac{(2xy)^{-2}}{4y^{-5}} = \dfrac{y^5}{4(2xy)^2} = \dfrac{y^5}{4(4x^2 y^2)} = \dfrac{y^3}{16x^2}$

8. **(C)** If $10^k = 64$, then $\left(10^k\right)^{\frac{1}{2}} = 64^{\frac{1}{2}} = \sqrt{64} = 8$.

   Hence, $10^{\frac{k}{2}} = 8$. Multiply both sides of $10^{\frac{k}{2}} = 8$ by 10:

$$10^1 \times \left(10^{\frac{k}{2}}\right) = 10^1 \times 8$$

$$10^{\frac{k}{2}+1} = 80$$

9. **(C)** To find how much greater $x^{\frac{5}{2}}$ is than x^2, subtract x^2 from $x^{\frac{5}{2}}$: $x^{\frac{5}{2}} - x^2$. Then factor out the GCF of x^2 from the difference:

$$x^{\frac{5}{2}} - x^2 = x^2\left(x^{\frac{1}{2}} - 1\right)$$

10. **(C)** $\dfrac{x^2}{\sqrt{x^3}} = \dfrac{x^2}{x^{\frac{3}{2}}} = x^{\frac{1}{2}} = \sqrt{x}$

11. **(C)** If $8(2^p) = 4^n$, then $(2^3)(2^p) = (2^2)^n$, and $2^{p+3} = 2^{2n}$. Thus, $p + 3 = 2n$, and $n = \dfrac{p+3}{2}$.

12. **(D)** If $k = 3$, then $2\sqrt{x-3} = x - 6$. Squaring both sides gives $4(x - 3) = x^2 - 12x + 36$. Hence,

$$x^2 - 16x + 48 = 0$$

$$(x - 4)(x - 12) = 0$$

$$x = 4 \ or \ x = 12$$

   Check for extraneous roots. The solution $x = 4$ doesn't work since the right side of the equation must be negative. Hence, the only solution is 12.

13. **(B)** If $x^{-1} - 1$ is divided by $x - 1$, then

$$\dfrac{x^{-1} - 1}{x - 1} = \left(\dfrac{\frac{1}{x} - 1}{x - 1}\right) \cdot \left(\dfrac{x}{x}\right)^{-1} = \dfrac{1 - x}{x(x - 1)} = -\dfrac{1}{x}$$

14. **(C)** If n is a negative integer, then it must be the case that $6n^{-1} < 4n^{-1}$ since

$$\dfrac{6n^{-1}}{n^{-1}} \boxed{<} \dfrac{4n^{-1}}{n^{-1}}$$

   Because n^{-1} is negative, reverse the direction of the inequality:

$$\dfrac{6n^{-1}}{n^{-1}} > \dfrac{4n^{-1}}{n^{-1}}$$

$$6 > 4 \ \checkmark$$

15. **(A)** If $g(x) = a\sqrt{a(1 - x)}$ and $g(-8) = 5$, then

$$a\sqrt{a(1 - (-8))} = 375$$

$$a\sqrt{9a} = 375$$

$$3a\sqrt{a} = 375$$

$$a^{\frac{3}{2}} = 125$$

$$a = (125)^{\frac{2}{3}} = \left(\sqrt[3]{125}\right)^2 = (5)^2 = 25$$

16. **(A)** If $27^x = 9^{y-1}$, then $3^{3x} = 3^{2(y-1)}$. Hence, $3x = 2(y-1)$, so $3x = 2y - 2$ and $3x + 2 = 2y$. Solving for y:

$$y = \frac{3x+2}{2}$$
$$= \frac{3}{2}x + \frac{2}{2}$$
$$= \frac{3}{2}x + 1$$

Grid-In

1. **(5)** If $\sqrt{3p^2 - 11} - x = 0$ and $x = 8$, then $\sqrt{3p^2 - 11} = 8$. Raising both sides to the second power makes $3p^2 - 11 = 64$ so

   $p^2 = \frac{75}{3} = 25$. Since $p > 0$, $p = 5$.

2. **(16)** If $x^{-\frac{1}{2}} = \frac{1}{8}$, then $x^{\frac{1}{2}} = 8$. Since $\sqrt{x} = 8$, $x = 64$.

   Hence,

   $$x^{\frac{2}{3}} = 64^{\frac{2}{3}} = \left(\sqrt[3]{64}\right)^2 = 4^2 = 16$$

3. **(27/2)** $\dfrac{6(2y)^{-2}}{(3y)^{-2}} = \dfrac{6(3y)^2}{(2y)^2} = \dfrac{54y^2}{4y^2} = \dfrac{27}{2}$

4. **(1/6)** If $(2rs)^{-1} = 3s^{-2}$, then

   $$\frac{1}{2rs} = \frac{3}{s^2}$$
   $$\frac{1}{2 \cdot 3} = \frac{rs}{s^2}$$
   $$\frac{1}{6} = \frac{r}{s}$$

5. **(3/10)** If $\left(2\sqrt{2}\right)^m = 32^p$, then

   $$\left(2 \cdot 2^{\frac{1}{2}}\right)^m = (2)^{5p}$$
   $$2^{\frac{3m}{2}} = 2^{5p}$$
   $$\frac{3m}{2} = 5p$$
   $$\frac{3m}{10m} = \frac{10p}{10m}$$
   $$\frac{3}{10} = \frac{p}{m}$$

6. **(1/4)** If a, b, and c are positive numbers such that $\sqrt{\dfrac{a}{b}} = 8c$, then

   $$\left(\sqrt{\frac{a}{b}}\right)^2 = (8c)^2$$
   $$\frac{a}{b} = 64c^2$$

   Since it is also given that $ac = b$,

   $$\frac{a}{ac} = 64c^2$$
   $$\frac{1}{c} = 64c^2$$
   $$64c^3 = 1$$
   $$c^3 = \frac{1}{64}$$
   $$\left(c^3\right)^{\frac{1}{3}} = \left(\frac{1}{64}\right)^{\frac{1}{3}}$$
   $$c = \frac{1}{4}$$

7. **(32/3)** If $k = 8\sqrt{2}$ then $\frac{1}{2}k = 4\sqrt{2} = \sqrt{3h}$.

   Hence, $\left(4\sqrt{2}\right)^2 = \left(\sqrt{3h}\right)^2$ so $(16 \cdot 2) = 3h$ and

   $h = \dfrac{32}{3}$.

8. **(5/2)** If $64^{2n+1} = 16^{4n-1}$, then $(4^3)^{2n+1} = (4^2)^{4n-1}$ so $3(2n + 1) = 2(4n - 1)$. Simplifying gives

   $6n + 3 = 8n - 2$ so $2n = 5$ and $n = \dfrac{5}{2}$.

9. **(7/6)** If $\dfrac{\sqrt[3]{a^8}}{\left(\sqrt{a}\right)^3} = a^x$, then

   $$\frac{a^{\frac{8}{3}}}{a^{\frac{3}{2}}} = a^x$$
   $$\frac{a^{\frac{16}{6}}}{a^{\frac{9}{6}}} = a^x$$
   $$a^{\frac{7}{6}} = a^x$$
   $$\frac{7}{6} = x$$

10. **(9/4)** or **(2.25)** Using the given formula, $t(d) = 0.08d^{\frac{3}{2}}$, solve for d when the storm lasts 16.2 minutes. Since the formula works when t is expressed in hours, change 16.2 minutes to hours: $\frac{16.2}{60} = 0.27$ hours.

$$0.27 = 0.08d^{\frac{3}{2}}$$
$$\frac{0.27}{0.08} = d^{\frac{3}{2}}$$
$$\left(\frac{27}{8}\right)^{\frac{3}{2}} = \left(d^{\frac{3}{2}}\right)^{\frac{3}{2}}$$
$$\left(\frac{3}{2}\right)^2 = d$$
$$\frac{9}{4} = d$$

LESSON 5-2

Multiple-Choice

1. **(B)** Factoring by group pairs of terms:

$$\begin{aligned} x^3 - 2x^2 - 9x + 18 &= (x^3 - 2x^2) + (-9x + 18) \\ &= x^2\underline{\underline{(x-2)}} + (-9)\underline{\underline{(x-2)}} \\ &= (x-2)(x^2 - 9) \\ &= (x-2)(x-3)(x+3) \end{aligned}$$

2. **(D)** Simplify the complex fraction by multiplying the numerator and the denominator by R_1R_2:

$$\frac{R_1R_2}{R_1R_2}\left(\frac{1}{\frac{1}{R_1} + \frac{1}{R_2}}\right) = \frac{R_1R_2}{R_1 + R_2}$$

3. **(B)** To find the number of different points at which the graph of $f(x) = x^2 - 2x^2 + x - 2$ intersects the x-axis, find the x-intercepts:

$$x^2 - 2x^2 + x - 2 = 0$$
$$x^2(x-2) + (x-2) = 0$$
$$(x-2)(x^2+1) = 0$$

- If $x - 2 = 0$, then $x = 2$ is an x-intercept.
- If $x^2 + 1 = 0$, then there is no real value of x that satisfies the equation.

Hence, the graph intersects the x-axis at one point.

4. **(D)** Change to a multiplication problem and factor where possible:

$$\begin{aligned} \frac{x^2 + 9x - 22}{x^2 - 121} \div (2 - x) &= \frac{\cancel{(x+11)}\,\overset{-1}{\cancel{(x-2)}}}{\cancel{(x+11)}(x-11)} \cdot \frac{1}{\cancel{2-x}} \\ &= -\frac{1}{x-11} \\ &= \frac{1}{11-x} \end{aligned}$$

5. **(B)** In general, if $p(x)$ is a polynomial function and $p(c) = 0$, then $x - c$ is a factor of $p(x)$ so $p(x)$ is divisible by $x - c$. If $p(4) = 0$, then $x - 4$ is a factor of $p(x)$.

6. **(A)** Begin by factoring the first term of the given expression:

$$\begin{aligned} \left(\frac{9}{4}x^2 - 1\right) - \left(\frac{3}{2}x - 1\right)^2 &= \underline{\underline{\left(\frac{3}{2}x - 1\right)}}\left(\frac{3}{2}x + 1\right) - \underline{\underline{\left(\frac{3}{2}x - 1\right)}}^2 \\ &= \left(\frac{3}{2}x - 1\right)\left[\left(\frac{3}{2}x + 1\right) - \left(\frac{3}{2}x - 1\right)\right] \\ &= \left(\frac{3}{2}x - 1\right)(2) \\ &= 3x - 2 \end{aligned}$$

7. **(A)** Simplify the complex fraction by first changing to positive exponents:

$$\begin{aligned} \frac{\frac{x-y}{y}}{y^{-1} - x^{-1}} &= \frac{\frac{x-y}{y}}{\frac{1}{y} - \frac{1}{x}} \quad \leftarrow \text{LCD} = xy \\ &= \left(\frac{xy}{xy}\right) \cdot \left(\frac{\frac{x-y}{y}}{\frac{1}{y} - \frac{1}{x}}\right) \\ &= \frac{(xy)\left(\frac{x-y}{y}\right)}{(xy)\left(\frac{1}{y} - \frac{1}{x}\right)} \\ &= \frac{x\cancel{(x-y)}}{\cancel{x-y}} \\ &= x \end{aligned}$$

8. **(B)** To find the zeros of $f(x) = 3x^3 - 5x^2 - 48x + 80$, express $f(x)$ in factored form and then find the values of x that make $f(x) = 0$.

$$f(x) = x^2(3x-5) + (-16)(3x-5)$$
$$= (3x-5)(x^2-16)$$
$$= (3x-5)(x-4)(x+4)$$

Since the zeros of the function are –4, 4, and $\frac{5}{3}$, the sum of the zeros is $\frac{5}{3}$.

9. **(A)** Factor the numerator and the denominator of the given fraction:

$$\frac{y^3 + 3y^2 - y - 3}{y^2 + 4y + 3} = \frac{y^2(y+3) - (y+3)}{(y+3)(y+1)}$$
$$= \frac{(y+3)(y^2-1)}{(y+3)(y+1)}$$
$$= \frac{(y-1)(y+1)}{(y+1)}$$
$$= y - 1$$

10. **(D)** The given table is:

x	$f(x)$	$g(x)$
–3	3	0
–1	0	3
0	–4	4
2	0	–2

Consider each statement in turn.
- $f(0) = -4$ and $g(0) = 4$ so $f(0) + g(0) = 0$. Statement I is true. ✓
- $f(x)$ is divisible by $x + 2$ if $f(-2) = 0$. Since the function value for $x = -2$, this statement is not necessarily true. ✗
- $g(x)$ is divisible by $x + 3$ if $g(-3) = 0$ which it is. Statement III is true. ✓

11. **(B)** Since the x-intercepts of the graph are –2, 1, and 3, the corresponding factors of the function are $x + 2$, $x - 1$, and $x - 3$. Hence, $y = (x+2)(x-1)(x-3)$. Consider each statement in turn.
- Since $y = (x+2)(x-1)(x-3) = (x+2)(x^2-4x+3)$. Statement I is false.
- Because $y = (x+2)(x-1)(x-3) = (x-3)[(x+2)(x-1)] = (x-3)(x^2+x-2)$. Statement II is true.
- Since $y = (x+2)(x-1)(x-3) = (x-1)[(x+2)(x-3)] = (x-1)(x^2-x-6)$. Statement III is false.

12. **(C)** The function $f(x) = (x+1)(x-1)(x-4)$ has zeros –1, 1, and 4. Multiplying two of the factors together gives

$$f(x) = (x-1)(x+1)(x-4)$$
$$= (x-1)(x^2 - 3x - 4)$$

13. **(A)** Change to a multiplication example:

$$\left(\frac{10x^2y}{x^2 + xy}\right) \cdot \left(\frac{(x+y)^2}{2xy}\right) \div \left(\frac{x^2 - y^2}{y^2}\right)$$
$$= \left(\frac{10x^2y}{x^2 + xy}\right) \cdot \left(\frac{(x+y)^2}{2xy}\right) \cdot \left(\frac{y^2}{x^2 - y^2}\right)$$
$$= \left(\frac{10x^2y}{x(x+y)}\right) \cdot \left(\frac{(x+y)^2}{2xy}\right) \cdot \left(\frac{y^2}{(x-y)(x+y)}\right)$$
$$= \frac{10x^2y^2}{2x^2(x-y)}$$
$$= \frac{5y^2}{x-y}$$

14. **(C)** If $f(x) = (2-3x)(x+3) + 4(x^2-6)$, then

$$f(x) = (-3x^2 + 2x - 9x + 6) + 4x^2 - 24$$
$$= x^2 - 7x - 18$$
$$= (x-9)(x+2)$$

The zeros of the function are –2 and 9 so their sum is 7.

15. **(A)** If the factors of a function are x, $x - 2$, and $x + 5$, then the graph of the functions intercepts the x-axis at –5, 0, and 2, which corresponds to only Graph I.

16. **(C)** The line $y = 6$ intersects the graph in three different points.

17. **(C)** The remainder when $f(x)$ is divided by $x + 3$ is $f(-3)$, which is approximately 6.5.

18. **(D)** A circle with center at the origin can intersect the graph of $f(x)$ in a maximum of 6 points. For example, consider such a circle with radius 6:

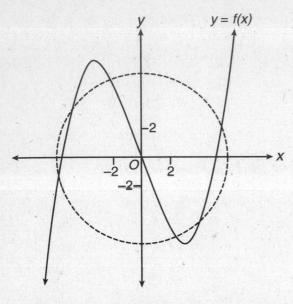

19. **(B)** Multiply the left side of the equation $(y^2 + ky - 3)(y - 4) = y^3 + by^2 + 5y - 12$, and then compare terms:

$$(y^2 + ky - 3)(y - 4) = y^3 + by^2 + 5y + 12$$
$$y^3 + (k-4)y^2 + y(-3-4k) + 12 = y^3 + by^2 + 5y + 12$$

Since the coefficients of the y-term must agree, $5 = -3 - 4k$ so $4k = -8$ and $k = -2$.

20. **(D)** Factor the numerator and the denominator of the given fraction:

$$\frac{16a^4 - 81b^4}{8a^3 + 12a^2b + 18ab^2 + 27b^3}$$
$$= \frac{(4a^2 - 9b^2)(4a^2 + 9b^2)}{4a^2(2a+3b) + 9b^2(2a+3b)}$$
$$= \frac{\cancel{(2a+3b)}(2a-3b)\cancel{(4a^2+9b^2)}}{\cancel{(2a+3b)}\cancel{(4a^2+9b^2)}}$$
$$= 2a - 3b$$

Grid-In

1. **(3)** Clear the given equation of its fractions by multiplying each term by 12, the LCD of its denominators:

$$12\left(\frac{k}{6}\right) + 12\left(\frac{3(1-k)}{4}\right) = 12\left(\frac{k-5}{2}\right)$$
$$2k + 9(1-k) = 6(k-5)$$
$$-7k + 9 = 6k - 30$$
$$-13k = -39$$
$$k = \frac{-39}{-13} = 3$$

2. **(1/2)** Clear the given equation of its fractions by multiplying each term by $12m$, the LCD of its denominators:

$$12m\left(\frac{3}{2}\right) = 12m\left[\frac{-(5m-3)}{3m}\right] + 12m\left(\frac{7}{12m}\right)$$
$$18m = -4(5m-3) + 7$$
$$18m = -20m + 19$$
$$38m = 19$$
$$m = \frac{19}{38} = \frac{1}{2}$$

3. **(2)** First find the zeros of the function $f(x) = x^3 + 5x^2 - 4x - 20$:

$$f(x) = x^2(x+5) - 4(x+5)$$
$$= (x+5)(x^2-4)$$
$$= (x+5)(x-2)(x+2)$$

The zeros of the function are -5, -2, and 2. Two of these zeros are located in the interval $-4 \le x \le 4$.

4. **(3/2)** Eliminate the fractions in the given equation:

$$\frac{t}{t-3} - \frac{t-2}{2} = \frac{5t-3}{4t-12}$$
$$\frac{t}{t-3} - \frac{t-2}{2} = \frac{5t-3}{4(t-3)}$$

Multiply each term by $4(t-3)$:

$$4(t-3)\left(\frac{t}{t-3}\right) - 4(t-3)\left(\frac{t-2}{2}\right) = 4(t-3)\left(\frac{5t-3}{4(t-3)}\right)$$
$$4t - 2(t-3)(t-2) = 5t - 3$$
$$4t - 2(t^2 - 5t + 6) = 5t - 3$$
$$4t - 2t^2 + 10t - 12 = 5t - 3$$
$$2t^2 - 9t + 9 = 0$$
$$(2t-3)(t-3) = 0$$
$$t = \frac{3}{2} \ or \ t = 3$$

The difference in the roots is $3 - \frac{3}{2} = \frac{3}{2}$.

5. **(6)** If $x^3 + 150 = 6x^2 + 25x$, then
$x^3 - 6x^2 - 25x + 150 = 0$. Solve by factoring:

$$(x^3 - 6x^2) + (-25x + 150) = 0$$
$$x^2(x - 6) - 25(x - 6) = 0$$
$$(x^2 - 25)(x - 6) = 0$$
$$(x - 5)(x + 5)(x - 6) = 0$$

Hence, the roots are –5, 5, and 6. The sum of these roots is 6.

6. **(3/7)** If $p(t) = t^5 - 3t^4 - kt + 7k^2$ is divisible by $t - 3$, then $p(3) = 0$:

$$p(3) = 3^5 - 3(3^4) - 3k + 7k^2 = 0$$
$$7k^2 - 3k = 0$$
$$k(7k - 3) = 0$$
$$k = 0 \text{ or } 7k - 3 = 0$$

Since it is given that the k is a nonzero constant, the only solution is $7k = 3$ so $k = \dfrac{3}{7}$.

LESSON 5-3
Multiple-Choice

1. **(D)** $i^{50} + i^0 = (i^2)^{25} + 1 = (-1)^{25} + 1 = -1 + 1 = 0$

2. **(A)** $2i^2 + 3i^3 = 2(-1) + 3(-i) = -2 - 3i$

3. **(C)** Change to i-form and then simplify the radicals:

$$2\sqrt{-50} - 3\sqrt{-8} = 2\sqrt{50}(\sqrt{-1}) - 3\sqrt{8}(\sqrt{-1})$$
$$= 2i\sqrt{25}\sqrt{2} - 3i\sqrt{4}\sqrt{2}$$
$$= 10i\sqrt{2} - 6i\sqrt{2}$$
$$= 4i\sqrt{2}$$

4. **(C)** Make the substitutions for x, y, and z. Then multiply the factors together and simplify:

$$xy^2z = (3i)(2i)^2(m + i)$$
$$= (3i)(4i^2)(m + i)$$
$$= -12i(m + i)$$
$$= -12mi - 12i^2$$
$$= -12mi - 12(-1)$$
$$= 12 - 12mi$$

5. **(B)** If $g(x) = \left(x\sqrt{1 - x}\right)^2$, then

$$g(10) = \left(10\sqrt{1 - 10}\right)^2$$
$$= \left(10\sqrt{-9}\right)^2$$
$$= (30i)^2$$
$$= 900i^2$$
$$= -900$$

6. **(D)** To simplify $(x + i)^2 - (x - i)^2$, square each binomial and combine like terms:

$$(x + i)^2 - (x - i)^2 = (x^2 + 2xi + i^2) - (x^2 - 2xi + i^2)$$
$$= 4xi$$

7. **(C)** If $i^{13} + i^{18} + i^{31} + n = 0$, then

$$i^{13} + i^{18} + i^{31} + n = 0$$
$$(i^{12} \cdot i) + (i^2)^9 + (i^{30} \cdot i) + n = 0$$
$$(i^2)^6 \cdot i + (-1)^9 + (i^2)^{15} \cdot i + n = 0$$
$$i - 1 - i + n = 0$$
$$-1 + n = 0$$
$$n = 1$$

8. **(B)** $2i(xi - 4i^2) = 2i^2x - 8i^3 = -2x - 8(-1)i = -2x + 8i$

9. **(A)** If $x = 2i$, $y = -4$, and $z = 3i$, then

$$\sqrt{x^3yz} = \sqrt{(2i)^3(-4)(3i)}$$
$$= \sqrt{(-8i)(-12i)}$$
$$= \sqrt{96i^2}$$
$$= \sqrt{-96}$$
$$= i\sqrt{16} \cdot \sqrt{6}$$
$$= 4\sqrt{6}i$$

10. **(C)** Multiply $(13 + 7i)(4 - 9i)$ using FOIL:

$$(13 + 7i)(4 - 9i) = 52 - 117i + 28i - 63i^2$$
$$= 52 - 89i + 63$$
$$= 115 - 89i$$

11. **(A)** If $(x + yi) + (a + bi) = 2x$, then $a = x$ and $b = -y$. Hence,

$$(x + yi)(a + bi) = (x + yi)(x - yi)$$
$$= x^2 + y^2$$

12. **(D)** To change $\dfrac{3+i}{4-7i}$ so that the denominator is real, multiply the numerator and the denominator by $4+7i$:

$$\frac{3+i}{4-7i}\cdot\frac{4+7i}{4+7i}=\frac{12+21i+4i+7i^2}{4^2+7^2}$$
$$=\frac{12-7+25i}{16+49}$$
$$=\frac{5+25i}{65}$$
$$=\frac{\cancel{5}(1+5i)}{\cancel{5}(13)}$$
$$=\frac{1}{13}+\frac{5}{13}i$$

13. **(A)** If $E = IZ$ where $I = 3 + i$ and $E = -7 + i$, then

$$Z=\frac{E}{I}=\frac{-7+i}{3+i}\cdot\frac{3-i}{3-i}$$
$$=\frac{-21+7i+3i-i^2}{9+1}$$
$$=\frac{-21+10i+1}{10}$$
$$=\frac{-20+10i}{10}$$
$$=-2+i$$

14. **(C)** First, factor out $(4-3i)$:

$$(9+2i)(4-3i)-(5-i)(4-3i)=(4-3i)[(9+2i)-(5-i)]$$
$$=(4-3i)(4+3i)$$
$$=16+9$$
$$=25$$

Grid-In

1. **(21/4)** $\left(\dfrac{1}{2}+i\sqrt{5}\right)\left(\dfrac{1}{2}-i\sqrt{5}\right)=\dfrac{1}{4}-i^2(5)$

$$=\frac{1}{4}+5=\frac{21}{4}$$

2. **(39)**

$$\left(2-\sqrt{-25}\right)\left(-7+\sqrt{-4}\right)=(2-5i)(-7+2i)$$
$$=-14+4i+35i-10i^2$$
$$=-14+39i+10$$
$$=-4+39i$$

If $-4 + 39i = x + yi$, then $y = 39$.

3. **(8)**

$$(1-3i)(7+5i+i^2)=(1-3i)(7+5i-1)$$
$$=(1-3i)(6+5i)$$
$$=6+5i-18i-15i^2$$
$$=6-13i+15$$
$$=21-13i$$

If $21 - 13i = a + bi$, then $a = 21$, $b = -13$, and $a + b = 21 + (-13) = 8$.

4. **(8/5)** To change $\dfrac{6+4i}{1-3i}$ so that the denominator is real, multiply the numerator and the denominator by $1 + 3i$:

$$\frac{6+4i}{1-3i}\cdot\frac{1+3i}{1+3i}=\frac{6+18i+4i+12i^2}{1+9}$$
$$=\frac{-6+22i}{10}$$
$$=\frac{-6}{10}+\frac{22}{10}i$$
$$=-\frac{3}{5}+\frac{11}{5}i$$

If $-\dfrac{3}{5}+\dfrac{11}{5}i = a + bi$, then $a = -\dfrac{3}{5}$, $b = \dfrac{11}{5}$, and $a + b = \dfrac{8}{5}$.

5. **(1/3)** If $g(x) = a\sqrt{41-x^2}$ and $g(2i) = \sqrt{5}$, then

$$g(2i)=a\sqrt{41-(2i)^2}=\sqrt{5}$$
$$a\sqrt{41-4i^2}=\sqrt{5}$$
$$a\sqrt{41+4}=\sqrt{5}$$
$$a\sqrt{45}=\sqrt{5}$$
$$a=\frac{\sqrt{5}}{\sqrt{45}}=\frac{1}{\sqrt{9}}=\frac{1}{3}$$

LESSON 5-4

Multiple-Choice

1. **(A)** If $3x^2 - 33 = 18x$, then before using the quadratic formula, simplify the equation:

$$\frac{3x^2}{3} - \frac{18x}{3} - \frac{33}{3} = 0$$
$$x^2 - 6x - 11 = 0$$

$$x = \frac{-(-6) \pm \sqrt{6^2 - 4(1)(-11)}}{2(1)}$$

$$= \frac{6 \pm \sqrt{80}}{2}$$

$$= \frac{6 \pm \sqrt{16} \cdot \sqrt{5}}{2}$$

$$= \frac{6 \pm 4 \cdot \sqrt{5}}{2}$$

$$= 3 \pm 2\sqrt{5}$$

2. **(A)** If $2x^2 - 8x - 5 = 0$, then

$$x = \frac{-(-8) \pm \sqrt{(-8)^2 - 4(2)(-5)}}{2(2)}$$

$$= \frac{8 \pm \sqrt{64 + 40}}{4}$$

$$= \frac{8 \pm \sqrt{104}}{4}$$

$$= \frac{8 \pm \sqrt{26} \cdot \sqrt{4}}{4}$$

$$= \frac{8 \pm 2\sqrt{26}}{4}$$

$$= 2 \pm \frac{\sqrt{26}}{2}$$

If p and q are the solutions, then $p = 2 + \frac{\sqrt{26}}{2}$

and $q = 2 - \frac{\sqrt{26}}{2}$ so

$$p - q = \left(2 + \frac{\sqrt{26}}{2}\right) - \left(2 - \frac{\sqrt{26}}{2}\right)$$

$$= \frac{\sqrt{26}}{2} + \frac{\sqrt{26}}{2}$$

$$= \sqrt{26}$$

3. **(C)** If $\frac{x+5}{4} = \frac{1-x}{3x-4}$, then cross-multiplying gives

$$(3x - 4)(x + 5) = 4(1 - x)$$
$$3x^2 + 15x - 4x - 20 = 4 - 4x$$
$$3x^2 + 15x - 24 = 0$$
$$x^2 + 5x - 8 = 0$$

$$x = \frac{-5 \pm \sqrt{25 - 4(1)(-8)}}{2}$$

$$= \frac{-5 \pm \sqrt{57}}{2}$$

The two solutions are $r = \frac{-5 + \sqrt{57}}{2}$ and

$s = \frac{-5 - \sqrt{57}}{2}$. Hence,

$$r - s = \left(\frac{-5 + \sqrt{57}}{2}\right) - \left(\frac{-5 - \sqrt{57}}{2}\right) = \sqrt{57}$$

4. **(B)** Complete the square for $y = 3x^2 + 18x - 13$:

$$y = 3(x^2 + 6x + 9) - 13 - 27$$
$$= 3(x + 3)^2 - 40$$

Hence, $h = -3$ and $k = -40$.

5. **(C)** Complete the square for $x^2 + 6x + y^2 - 8y = 56$ for both variables:

$$(x^2 + 6x + 9) + (y^2 - 8y + 16) = 56 + 25$$
$$(x + 3)^2 + (y - 4)^2 = 81$$

Since $r^2 = 81$, $r = 9$.

6. **(A)** To solve $\frac{4}{x-3} + \frac{2}{x-2} = 2$, eliminate the fractions by multiplying each term by $(x-3)(x-2)$:

$$(x-3)(x-2)\left(\frac{4}{x-3}\right) + (x-3)(x-2)\left(\frac{2}{x-2}\right)$$
$$= 2(x-3)(x-2)$$
$$4(x-2) + 2(x-3) = 2(x-3)(x-2)$$
$$2(x-2) + (x-3) = (x-3)(x-2)$$
$$3x - 7 = x^2 - 5x + 6$$
$$0 = x^2 - 8x + 13$$

Use the quadratic formula:

$$x = \frac{-(-8) \pm \sqrt{64 - 4(1)(13)}}{2(1)}$$

$$= \frac{8 \pm \sqrt{12}}{2}$$

$$= \frac{8 \pm 2\sqrt{3}}{2}$$

$$= 4 \pm \sqrt{3}$$

The roots have form $a \pm \sqrt{b}$ with $a = 4$ and $b = 3$.

7. **(B)** If the roots are $2 + 3i$ and $2 - 3i$, then

$$(x - (2 + 3i))(x - (2 - 3i)) = 0$$
$$x^2 - x(2 - 3i) - x(2 + 3i) + 13 = 0$$
$$x^2 - 4x + 13 = 0$$

8. **(A)** If $ax^2 + 6x - 9 = 0$ has imaginary roots, then the discriminant is less than 0:

$$b^2 - 4ac < 0$$
$$36 - 4a(-9) < 0$$
$$36 + 36a < 0$$
$$36a < -36$$
$$a < -1$$

9. **(D)** If x represents the resistance in ohms of R_1, then $x + 2$ is the resistance of R_2. Since $R_T = 1.5 = \dfrac{3}{2}$ ohms:

$$\frac{1}{x} + \frac{1}{x + 2} = \frac{2}{3}$$
$$3x(x + 2)\left(\frac{1}{x}\right) + 3x(x + 2)\left(\frac{1}{x + 2}\right) = 3x(x + 2)\left(\frac{2}{3}\right)$$
$$3x + 6 + 3x = 2(x^2 + 2x)$$
$$6x + 6 = 2x^2 + 4x$$
$$2x^2 - 2x - 6 = 0$$
$$x^2 - x - 3 = 0$$

Use the quadratic formula:

$$x^2 - x - 3 = 0$$
$$x = \frac{1 \pm \sqrt{1 - 4(1)(-3)}}{2}$$
$$= \frac{1 \pm \sqrt{13}}{2}$$

Since x must be positive, $x = \dfrac{1 + \sqrt{13}}{2}$.

10. **(B)** At time $t = 0$, the bathtub has 128 liters of water. You are asked to find t when $L = 64$:

$$-4t^2 - 8t + 128 = 64$$
$$-4t^2 - 8t + 64 = 0$$
$$t^2 + 2t - 16 = 0$$
$$t = \frac{-2 \pm \sqrt{(2)^2 - 4(1)(-16)}}{2(1)}$$
$$= \frac{-2 \pm \sqrt{68}}{2}$$
$$= \frac{-2 \pm 2\sqrt{17}}{2}$$
$$= -1 \pm \sqrt{17}$$

Since the answer must be positive, $t = -1 + \sqrt{17}$.

LESSON 5-5
Multiple-Choice

1. **(A)** The x-coordinate of the vertex of a parabola $y = ax^2 + bx + c$ is $-\dfrac{b}{2a}$. For the parabola $h(t) = -16t^2 + kt + 3$,

$$t = \frac{-k}{-32} = 4 \text{ so } k = 128$$

2. **(B)** To find the maximum height of the rocket described by the function $h(t) = -16t^2 + 80t + 10$, find the vertex of the parabola by completing the square:

$$h(t) = -16\left(t^2 - 5t + \frac{25}{4}\right) + 10 + (16)\left(\frac{25}{4}\right)$$
$$= -16\left(t - \frac{5}{2}\right)^2 + 110$$

The maximum value of the function is 110, which is the y-coordinate of the vertex.

3. **(C)** The ball will hit the ground when $h(t) = 0$:

$$h(t) = 54t - 12t^2 = 0$$
$$6t(9 - 2t) = 0$$
$$6t = 0 \text{ or } 9 - 2t = 0$$
$$t = \frac{9}{2} = 4.5$$

4. **(A)** If $y + 3 = (x - 4)^2 - 6$, then set y equal to 0 to find the x-intercepts:

$$0 + 3 = (x - 4)^2 - 6$$
$$\pm\sqrt{9} = \sqrt{(x - 4)^2}$$
$$\pm 3 = x - 4$$
$$-3 = x - 4 \quad or \quad 3 = x - 4$$
$$1 = x \qquad\qquad 7 = x$$

The x-intercepts are 1 and 7.

5. **(B)** To determine which statements are true about the graph of $y = (2x - 4)(x - 8)$, consider each statement in turn.

 ■ If $y = (2x - 4)(x - 8)$, then $y = 2x^2 - 20x + 32$ so its line of symmetry is

$$x = -\frac{b}{2a}$$
$$= -\frac{-20}{2(2)}$$
$$= 5$$

so Statement I is true. ✓

 ■ Since the x-coordinate of the vertex is 5, find the y-coordinate of the vertex by replacing x with 5 in its equation. Using the equation $y = (2x - 4)(x - 8)$:

$$y = (2(5) - 4)(5 - 8)$$
$$= 6(-3)$$
$$= -18$$

Statement II is false since it states that the minimum value of the function is –7. ✗

 ■ To find the y-intercept of the parabola, set $x = 0$:

$$y = (2(0) - 4)(0 - 8)$$
$$= (-4)(-8)$$
$$= 32$$

Statement III is true. ✓

Hence, only Statements I and III are true.

6. **(C)** If a parabola turns down, the vertex is the highest point on the parabola. Since it is given that $f(x) \leq b$ for all x, $P(a,b)$ must be the vertex of the given parabola. The vertical line of symmetry of a parabola passes through its vertex and bisects every

horizontal line segment that connects matching points on the parabola such as $(-1, 0)$ and $(6, 0)$.

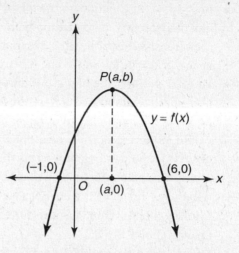

Since the midpoint of the segment is $\left(\dfrac{-1+6}{2}, \dfrac{0+0}{2}\right) = (2.5, 0) = (a, 0)$, the x-coordinate of the vertex must be 2.5. The only answer choice that has 2.5 as the x-coordinate of the vertex is (**2.5**, 4).

7. **(B)** Because of symmetry, $f(5) = f(-3) = f(c)$, so $c = -3$.

8. **(B)** It is given that the graph of a quadratic function f intersects the x-axis at $x = -2$ and $x = 6$. Because the parabola's line of symmetry bisects every horizontal segment that joins two points on the parabola, the equation of the line of symmetry is

$$x = \frac{-2+6}{2} = 2.$$

Since $x = 8$ is 6 units to the right of $x = 2$, the corresponding point on the parabola is 6 units to the left of $x = 2$, which makes its x-coordinate $2 - 6 = -4$. Thus, $f(8) = f(-4)$.

9. **(C)** If in the quadratic function $f(x) = ax^2 + bx + c$, a and c are both negative, the graph is a parabola that opens down ($a < 0$) and intersects the negative y-axis at a negative value ($c < 0$), as shown in choice (C).

10. **(D)** The given points on the parabola, $(0, 0)$ and $(6, 0)$, are its x-intercepts. Since the midpoint of the horizontal segment connecting these points is $(3, 0)$, the line of

symmetry is $x = 3$. As the line of symmetry contains the vertex, $T(h, 4) = T(3, 4)$.

- Since $h = 3$, Choice I is false.
- Because $(1, 2)$ is 2 units to the left of the line of symmetry, its matching point on the opposite side of the line of symmetry is $(3 + 2, 2) = (5, 2)$. So Choice II is true.
- Vertex T is either the highest or lowest point of the parabola. If T were the lowest point on the parabola, then the parabola would *not* intersect the x-axis. Since it is given that the parabola intersects the x-axis, T must be the highest point on the parabola. Choice III is true. Because Choices II and III are true, the correct answer choice is (D).

11. **(B)** Since the three graphs have two points in common, the system has two solutions.

12. **(B)** The x-intercepts of the parabola are -3 and $\frac{1}{2}$ so the corresponding factors of the parabola equation are $(x + 3)$ and $(2x - 1)$. Thus, a possible equation of the parabola is $y = (x + 3)(2x - 1)$.

13. **(B)** Find the vertex of the parabola $y = 2x^2 - 12x + 11$ by completing the square:

$$y = 2(x^2 - 6x + 9) + 11 - 18$$
$$= 2(x - 3)^2 - 7$$

The vertex of the parabola is $(3, -7)$. The vertical distance from the vertex to the point $(3, 1)$ is $1 - (-7) = 1 + 7 = 8$.

14. **(D)** It is given that the vertex of a parabola $y = f(x)$ is $(3, 2)$ and that $f(-1) = p$ so $(-1, p)$ is a point on the parabola. Since this point is 4 units horizontally to the left of the vertex, the matching point on the parabola must be 4 units to the right of 3 so its x-coordinate is $3 + 4 = 7$. These points have the same y-coordinate so $f(-1) = f(7) = p$.

15. **(A)** Since the parabola passes through the points $(-3, -40)$, $(0, 29)$, and $(10, -1)$, the coordinates of these points must satisfy the parabola equation $y = ax^2 + bx + c$. Thus,
- Using the point $(0, 29)$, $29 = a(0^2) + (0) + c$ so $c = 29$.

- Using the point $(-3, -40)$, $-40 = 9a - 3b + 29$ so $9a - 3b = -69$ and $3a - b = -23$.
- Using the point $(10, -1)$, $10 = a(-1)^2 + b(-1) + 29$, which simplifies to $-a + b = 19$.

Solve the system of equations $3a - b = -23$ and $-a + b = 19$ by adding their corresponding sides. Thus, $2a = -4$ so $a = -2$. Substituting into the last equation, $-(-2) + b = 19$ so $b = 17$.

Since the parabola equation is $y = -2x^2 + 17x + 29$, the line of symmetry is $x = -\frac{17}{2(-2)} = \frac{17}{4}$.

16. **(D)** If $x^2 + y^2 = 416$ and $y + 5x = 0$, then substitute $-5x$ for y in the second-degree equation:

$$x^2 + (-5x)^2 = 416$$
$$x^2 + 25x^2 = 416$$
$$26x^2 = 416$$
$$x^2 = \frac{416}{26} = 16$$
$$x = \sqrt{16} = 4$$

If $x = 4$, then $y + 5(4) = 0$ so $y = -20$. Hence, $x - y = 4 - (-20) = 24$.

17. **(D)** To find the maximum height of the rocket described by the function $h(t) = -4.9t^2 + 68.6t$, write the equation in vertex form by completing the square:

$$h(t) = -4.9(t^2 + 14t)$$
$$= -4.9(t^2 + 14t + 49) + 240.1$$

Hence, the maximum height of the toy rocket, to the *nearest meter*, is 240.

18. **(C)** Rewrite the parabola equation in vertex form by completing the square:

$$y = k(x - 1)(x + 9)$$
$$y = k(x^2 + 8x - 9)$$
$$= k(x^2 + 8x + ?) - 9k$$
$$= k(x^2 + 8x + 16) - 9k - 16k$$
$$= k(x + 4)^2 - 25k$$

The vertex of the parabola is $(-4, -25k)$. Since $k > 0$, the y-coordinate of the vertex, $-25k$, is the minimum value of y.

Grid-In

1. **(9/2)** If $x^2 - y^2 = 18$ and $y = x - 4$, substitute $x - 4$ for y in the second-degree equation, which gives

$$x^2 - (x-4)^2 = 18$$
$$x^2 - (x^2 - 8x + 16) = 18$$
$$8x - 16 = 18$$
$$8x = 34$$
$$x = \frac{34}{8} = \frac{17}{4}$$
$$y = x - 4 = \frac{17}{4} - 4 = \frac{1}{4}$$

Hence, $x + y = \frac{17}{4} + \frac{1}{4} = \frac{18}{4} = \frac{9}{2}$.

2. **(49)** To find the maximum height of the diver above the water, find the maximum value of the function $d(t) = -16t^2 + 40t + 24$ by writing it in vertex form:

$$d(t) = -16\left(t^2 - \frac{5}{4}t\right) + 24$$
$$= -16\left(t^2 - \frac{5}{2}t + \frac{25}{16}\right) + 24 + (16)\left(\frac{25}{16}\right)$$
$$= -16\left(t - \frac{5}{4}\right)^2 + 49$$

The maximum height is 49 feet.

3. **(75)** If x represents the number of \$5 reductions in the price of a pair of running shoes, then $50x$ represents the number of *additional* pairs of running shoes that will be sold during the month. Hence, $600 + 50x$ is the total number of pairs of running shoes that will be sold at the reduced price of $90 - 5x$ for each pair of running shoes.

- The store's monthly revenue R from the sale of running shoes is the price of each pair times the number sold:

$$R(x) = (90 - 5x)(600 + 50x)$$
$$= 54{,}000 + 1{,}500x - 250x^2$$

- The maximum value of $R(x)$ occurs at

$$x = -\frac{b}{2a} = -\frac{1{,}500}{2(-250)} = 3$$

- When $x = 3$, the reduced price of a pair of running shoes is

$$90 - 5x = 90 - (5 \times 3) = 75$$

4. **(150)** To determine the number of feet the football travels horizontally before it hits the ground, set $h(x) = -\frac{1}{225}x^2 + \frac{2}{3}x$ equal to 0 and solve for x:

$$h(x) = -\frac{1}{225}x^2 + \frac{2}{3}x = 0$$
$$x\left(-\frac{1}{225}x + \frac{2}{3}\right) = 0$$
$$\frac{1}{225}x = \frac{2}{3}$$
$$x = \left(\frac{2}{3}\right)(225)$$
$$= 150$$

5. **(25)** To find the maximum height of the football, write the parabola function in vertex form by completing the square:

$$h(x) = -\frac{1}{225}x^2 + \frac{2}{3}x$$
$$= -\frac{1}{225}(x^2 + 150x)$$
$$= -\frac{1}{225}(x^2 + 150x + 5{,}625) + \frac{5{,}625}{225}$$
$$= -\frac{1}{225}(x + 75)^2 + 25$$

The maximum height of the football is the y-coordinate of the vertex of the parabola equation, which is 25 feet.

6. **(1)** Change 45 yards to $45 \times 3 = 135$ feet. Then evaluate $h(135)$ in order to find the height of the football when it has traveled 135 feet horizontally:

$$h(135) = -\frac{1}{225}(135)^2 + \frac{2}{3}(135)$$
$$= -\frac{182{,}525}{225} + 90$$
$$= -81 + 90$$
$$= 9$$

Since the goal post is 10 feet high, the football will fail to pass over the goal by $10 - 9 = 1$ foot.

LESSON 5-6
Multiple-Choice

1. **(A)** If c is a positive number, the graph of $y = f(x)$ can be shifted left c units by graphing $y = f(x + c)$ and shifted right c units by graphing $y = f(x - c)$. If $y = f(x) = 2^x$, then graphing $y = f(x - 3) = 2^{x-3}$ shifts the graph 3 units to the right.

2. **(A)** In general, $y = f(-x)$ is the graph of $y = f(x)$ reflected over the y-axis. Since it is given that $y = f(x) = 2x - 3$, $f(-x) = 2(-x) - 3 = -2x - 3$.

3. **(D)** If $f(x - 5) = f(x)$ for all values of x, then $f(19) = f(14) = f(9) = f(4) = 3$.

4. **(D)** Given that the endpoints of $\overline{AB}$ are $A(0, 0)$ and $B(9, -6)$, the slope of $\overline{AB}$ is $\dfrac{-6}{9} = -\dfrac{2}{3}$, its equation is $y = -\dfrac{2}{3}x$. In general, $y = f(-x)$ is the graph of $y = f(x)$ reflected over the y-axis. Hence, the equation of the line that contains the reflection of $\overline{AB}$ in the y-axis is $y = f(-x) = -\dfrac{2}{3}(-x) = \dfrac{2}{3}x$.

5. **(B)** If $g(x) = -f(x)$, then $y = g(x)$ is the reflection of the given graph in the y-axis, which is represented by the graph in choice (B).

6. **(C)** The graph of $y = f(x + 2)$ can be obtained from the graph of $y = f(x)$ by shifting it to the left 2 units, since $y = f(x + 2) = f(x - (-2))$. Hence, the point $(2, -1)$ on the graph $y = f(x)$ is shifted to the point $(2 - 2, -1) = (0, -1)$ on the graph of $y = f(x + 2)$.

7. **(D)** Compare the coordinates of the vertex of the two function graphs. The vertex of function f is at $(0, 0)$, and the vertex of function g is at $(1, 2)$. This means that function g is the graph of function f shifted 1 unit to the right and 2 units up. Hence, an equation of function g is $g(x) = f(x - 1) + 2$. By comparing this equation to $g(x) = f(x + h) + k$, you know that $h = -1$ and $k = 2$. Thus, $h + k = -1 + 2 = 1$.

8. **(D)** The graph of $g(x) = -2$ is a horizontal line 2 units below the x-axis. If the graph of $g(x) = -2$ intersects the graph of $y = f(x) + k$ at one point, then the point of intersection must be one of the turning points. Consider the turning point located between 1 and 2. Since the y-coordinate of that turning point is approximately -3.5, the graph of $y = f(x)$ would need to be shifted up approximately 1.5 units. Hence, a possible value of k is 1.5.

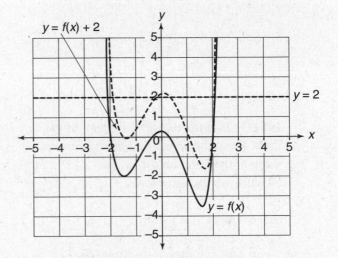

9. **(D)** The graph of $y = f(x + 1)$ is the graph of $y = f(x)$ shifted horizontally to the left 1 unit, shown by the graph in choice (D). You should confirm that this is the correct graph by checking a few specific points. The original graph includes the points $(-3, 1)$, $(0, 4)$ and $(3, -2)$. Since the original graph is shifted to the left 1 unit, the matching points on the graph of $y = f(x + 1)$ must have their x-coordinates reduced by 1: $(-4, 1)$, $(-1, 4)$, and $(2, -2)$. Only the graph in choice (D) contains all three of these points.

10. **(D)** METHOD 1: Since the line contains the points $(0, 3)$ and $(2, 0)$, its slope is $\dfrac{3 - 0}{0 - 2} = -\dfrac{3}{2}$. The y-intercept of the line is 3, so its equation is $y = f(x) = -\dfrac{3}{2}x + 3$. In general, $y = -f(x)$ is the graph of $y = f(x)$ reflected over the x-axis. Since $y = f(-x) = -\dfrac{3}{2}(-x) + 3 = \dfrac{3}{2}x + 3$, the slope of the reflected line is $\dfrac{3}{2}$.

METHOD 2: Find that the slope of the given line is $-\dfrac{3}{2}$. By sketching the reflected line, it is easy to calculate its slope directly or to see that the slopes of the two lines must be opposite in sign:

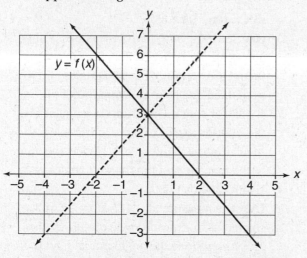

11. **(C)** The graph of $y = f(x - 2) + 1$ is the graph of $y = f(x)$ shifted two units to the right and 1 unit up. Since the original graph contains $(0, 0)$ and $(2, 2)$, the translated graph must contain $(0 + 2, 0 + 1) = (2, 1)$ and $(2 + 2, 2 + 1) = (4, 3)$. Only the graph in choice (C) contains both of these points.

12. **(C)** According to the definition of absolute value, when $f(x) \geq 0$, $y = f(x)$, and when $f(x) < 0$, $y = -f(x)$. Hence, the graph of $y = |f(x)|$ includes the portion of the graph of function f that is in quadrant I, where $f(x) > 0$, and the reflection in the x-axis of the portion of the original graph of function f that is in quadrant III, where $f(x) < 0$. This is the graph in choice (C).

Grid-In

1. **(2)** When $k > 0$, the graph of $y = f(x - k)$ can be obtained from the graph of $y = f(x)$ by shifting it to the right k units. Hence, we need to find k so that $(-1, 0)$, the x-intercept of $y = f(x)$, is mapped onto $(1, 0)$. This means that $-1 + k = 1$, so $k = 2$.

2. **(5)** The graph of $g(x) = 3$ is a horizontal line 3 units above the x-axis.
   - Since $g(x) = 3$ and $y = f(x)$ intersect at 3 points, $m = 3$.
   - The graph of $y = f(x) - 1$ is the graph of $y = f(x)$ shifted down 1 unit, which shifts the turning point at $(1, 4)$ to $(1, 3)$. Hence, $g(x) = 3$ intersects $y = f(x) - 1$ at 2 points, so $n = 2$.
   - Thus, $m + n = 3 + 2 = 5$.

LESSON 6-1
Multiple-Choice

1. **(B)** Since vertical angles are equal, the angle opposite the 40° angle also measures 40°. The measure of an exterior angle of a triangle is equal to the sum of the measures of the two nonadjacent interior angles of the triangle. Hence:

$$x° = 80° + 40° = 120°$$

and

$$y° = 40° + 70° = 110°$$

so

$$x + y = 120 + 110 = 230$$

2. **(B)** Since vertical angles are equal, the angle opposite the angle marked $3x°$ also measures $3x°$. Further, the angle opposite the angle marked $2x°$ also measures $2x°$. The sum of the measures of the angles about a point is 360°. Hence:

$$3x + 2x + x + 3x + 2x + y = 360$$

or $11x + y = 360$. Since vertical angles are equal, $y = x$. Thus:

$$11x + x = 360$$
$$12x = 360$$
$$x = \frac{360}{12} = 30$$

Hence, $y = x = 30$.

3. **(B)** First find the measures of the two angles that lie above ℓ_2 and that have the same vertex as angle x. The vertical angle opposite the 58° angle also measures 58°. The acute angle above line ℓ_2 that is adjacent to the 58° angle measures 37° since alternate interior angles formed by parallel lines have equal measures. Since the sum of

the measures of the angles that form a straight line is 180°,

$$37° + 58° + x° = 180°$$

so $x = 180 - 95 = 85$.

4. **(D)** Since the sum of the measures of the angles about a point is 360, the sum of the measures of the marked angles and the unmarked angles at the four vertices is 4×360 or 1,440. The unmarked angles are the interior angles of the two triangles, so their sum is 2×180 or 360. Hence, the sum of the measures of the marked angles is $1,440 - 360$ or 1,080.

5. **(A)** The measure of $\angle ACE$ is $x + 2x$ or $3x$ since it is an exterior angle of $\triangle ABC$. Also, $3y = 3x + y$ since the angle marked $3y°$ is an exterior angle of the triangle in which $3x°$ and $y°$ are the measures of the two nonadjacent interior angles. Hence, $2y = 3x$, so $y = \frac{3}{2}x$.

6. **(D)** Since acute angle ABD is supplementary to obtuse angle CDB:

$$(y + 2y + y) + 5y = 180$$
$$9y = 180$$
$$y = \frac{180}{9} = 20$$

7. **(C)** In the smaller triangle, $a + b + 140 = 180$, so $a + b = 40$. In $\triangle RST$, $x + 2a + 2b = 180$. Dividing each member of this equation by 2 gives $\frac{x}{2} + a + b = 90$. Since $a + b = 40$, then

$$\frac{x}{2} + 40 = 90$$
$$\frac{x}{2} = 50$$
$$x = 100$$

8. **(B)** If two angles of a triangle each measure 60°, then the third angle also measures 60° because $180 - 60 - 60 = 60$. Since an equiangular triangle is also equilateral, the

length of each side of the acute triangle in the figure is 10. Since the hypotenuse of the right triangle in the figure is 10, the lengths of the sides of the right triangle form a (6, 8, 10) Pythagorean triple, where $x = 6$.

9. **(D)** In $\triangle JKL$:

$$\angle L = 180 - 50 - 65 = 65$$

Since $\angle K = 65$ and $\angle L = 65$, then $\angle K = \angle L$, so the sides opposites these angles, JK and JL, must be equal in length. Hence:

$$JK = JL$$
$$3x - 2 = x + 10$$
$$2x = 12$$
$$x = \frac{12}{2} = 6$$

10. **(C)** Let the length of WS be any convenient number. If $WS = 1$, then $TS = 1 \times \sqrt{3} = \sqrt{3}$. In right triangle RST, RS is the side opposite the $60°$ angle, so

$$RS = \sqrt{3} \times TS = \sqrt{3} \times \sqrt{3} = 3$$

Hence, $RW = RS - WS = 3 - 1 = 2$. Since $RW = 2$ and $WS = 1$, the ratio of RW to WS is 2 to 1.

11. **(C)** Draw a diagram in which the four key points on Katie's trip are labeled A through D.

- To determine how far, in a straight line, Katie is from her starting point at A, find the length of $\overline{AD}$.

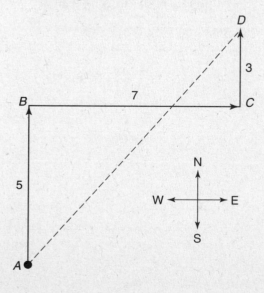

- Form a right triangle in which $\overline{AD}$ is the hypotenuse by completing rectangle $BCDE$, as shown in the accompanying diagram.

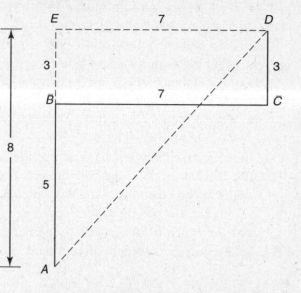

- Because opposite sides of a rectangle have the same length, $ED = BC = 7$, and $BE = CD = 3$. Thus, $AE = 5 + 3 = 8$.
- Since AD is the hypotenuse in right triangle AED,

$$(AD)^2 = (AF)^2 + (FD)^2$$
$$= 8^2 + 7^2$$
$$= 64 + 49$$
$$= 113$$
$$AD = \sqrt{113}$$

12. **(B)** Consider each statement in turn.
- II. Since $\angle BDC$ is an exterior angle of $\triangle DBA$, it is greater in measure than either of the nonadjacent interior angles ($\angle DBA$ and $\angle A$) of that triangle. Statement II is true. ✓
- I. There is no information provided that allows you to make a conclusion about how x and z compare. Statement I is not necessarily true. ✗
- III. $\angle BDA$ is an exterior angle of $\triangle CDB$, so $\angle BDA > \angle C$. Since the angles opposite equal sides are equal, and since m$\angle A = x°$, $\angle A > \angle C$, which means that $BC > AB$. Hence, Statement III is false. ✗

Only Statement II is true.

13. **(A)** Since n is an integer and $3 < n < 8$, n may be equal to 4, 5, 6, or 7. Test whether a triangle can be formed for each possible value of n.
- If $n = 4$, then $4 < 3 + 8$ and $3 < 4 + 8$, but 8 is not less than $3 + 4$, so $n \neq 4$.
- If $n = 5$, then $5 < 3 + 8$ and $3 < 5 + 8$, but 8 is not less than $3 + 5$, so $n \neq 5$.
- If $n = 6$, then $6 < 3 + 8$, $3 < 6 + 8$, *and* $8 < 3 + 6$, so n may be equal to 6.
- If $n = 7$, then $7 < 3 + 8$, $3 < 7 + 8$, *and* $8 < 3 + 7$, so n may be equal to 7.

Hence, two different triangles are possible.

14. **(B)** Since the right angles and the vertical angles at E are equal, triangles AEC and BED are similar. The lengths of corresponding sides of similar triangles are in proportion. Let x represent the length of AE. Then

$$\frac{AC}{DB} = \frac{AE}{BE}$$

$$\frac{3}{4} = \frac{x}{14 - x}$$

$$4x = 3(14 - x)$$

$$= 42 - 3x$$

$$7x = 42$$

$$x = \frac{42}{7} = 6$$

Hence, $AE = 6$.

15. **(A)** In any polygon the sum of the degree measures of the exterior angles is 360. If the sum of the degree measures of the interior angles of a polygon is 4 times the sum of the degree measures of the exterior angles, then

$$S = (n - 2)180 = 4 \times 360$$

or $(n - 2)180 = 1{,}440$. Dividing both sides of the equation by 180 gives

$$(n - 2) = \frac{1{,}440}{180} = 8$$

Since $n - 2 = 8$, then $n = 8 + 2$ or 10.

16. **(C)** Determine whether each Roman numeral statement is true, given that $ABCD$ is a parallelogram and $AB > BD$.
- I. Since opposite sides of a parallelogram have the same length, $CD = AB$. Substituting CD for AB in $AB > BD$ gives $CD > BD$. Hence, Statement I is false.
- II. Since $AB > BD$, the measures of the angles opposite these sides have the same size relationship. Thus, $\angle ADB > \angle A$. Since opposite angles of a parallelogram are equal, $\angle A = \angle C$, so $\angle ADB > \angle C$. Hence, Statement II is true.
- III. Alternate interior angles formed by parallel lines have equal measures, so $\angle CBD = \angle ADB$. Since $\angle ADB > \angle A$, then $\angle CBD > \angle A$, so Statement III is true.

Only Statements II and III must be true.

17. **(D)** Right triangles CDE and BAE are similar, so the lengths of corresponding sides are in proportion. Since $CD = 1$ and $AB = 2$, each side of $\triangle BAE$ is 2 times the length of the corresponding side of $\triangle CDE$. Since $AD = 6$, it must be the case that $AE = 4$ and $DE = 2$. Since $BC = CE + BE$, use the Pythagorean theorem to find the lengths of CE and BE.
- In right triangle CDE,

$$(CE)^2 = 1^2 + 2^2 = 1 + 4 = 5$$

so $CE = \sqrt{5}$.
- In right triangle BAE,

$$(BE)^2 = 4^2 + 2^2 = 16 + 4 = 20$$

so $BE = \sqrt{20} = \sqrt{4} \cdot \sqrt{5} = 2\sqrt{5}$.
Hence,

$$BC = CE + BE = \sqrt{5} + 2\sqrt{5} = 3\sqrt{5}$$

18. **(C)** The sum of the degree measures of the three adjacent angles at each of the marked vertices is equivalent to the sum of the measures of four straight angles, which equals 4×180 or $720°$. Since the sum of $720°$ includes the sum of the measures of the four angles of the inscribed quadrilateral, which is 360, the sum of the degree measures of only the marked angles is $720 - 360$ or 360.

19. **(C)** If each interior angle of a regular polygon measures $140°$, then each exterior angle measures $180° - 140° = 40°$. For any regular polygon with n sides, each exterior angle measures $\dfrac{360°}{n}$. Hence, if an exterior angle measures $40°$, then $40° = \dfrac{360°}{n}$ so $40n = 360$ and $n = \dfrac{360}{40} = 9$.

Grid-In

1. **(115)** The measure of interior angle A is $(180 - x)°$. The sum of the measures of the interior angles of a 5-sided polygon is $(5-2) \times 180° = 540°$. Hence,

$$(180 - x) + 132 + x + y + 113 = 540$$
$$425 + y = 540$$
$$y = 540 - 425 = 115$$

2. **(35)** In $\triangle ABC$,

$$\angle ACB = 180 - 25 - 45 = 110$$

Since angles ACB and DCE form a straight line,

$$\angle DCE = 180 - 110 = 70$$

Angle BED is an exterior angle of $\triangle ECD$. Hence:

$$3x = 70 + x$$
$$2x = 70$$
$$x = \frac{70}{2} = 35$$

3. **(25/2)** Since the two triangles are similar, the lengths of their corresponding sides are in proportion. If x represents the length of $\overline{DE}$, then

$$\frac{DE}{AE} = \frac{CD}{AB}$$
$$\frac{x}{30 - x} = \frac{15}{21}$$
$$\frac{x}{30 - x} = \frac{5}{7}$$
$$7x = 5(30 - x)$$
$$7x = 150 - 5x$$
$$12x = 150$$
$$x = \frac{150}{12} = \frac{25}{2}$$

4. **(45)** Since $AC = BC$, m$\angle A$ = m$\angle B$ = n. It is given that m$\angle C = 30$, so $n + n + 30 = 180$ and $n = 75$.

■ Because $\triangle ADE \sim \triangle ABC$, corresponding angles E and C have the same measure, so m$\angle E$ = m$\angle C$ = 30.

■ Angle ABC is an exterior angle of $\triangle EBF$. Hence,

$$m\angle ABC = 75 = m\angle E + \angle EFB$$
$$75 = 30 + m\angle EFB$$
$$45 = m\angle EFB$$

■ Since vertical angles are equal in measure, x = m$\angle EFB = 45$.

5. **(5)** Draw a diagram in which the starting point is labeled point X.

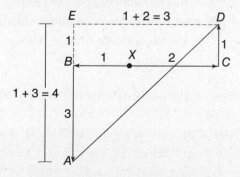

■ Label the diagram with the given information. Complete rectangle $BCDE$.

■ Because opposite sides of a rectangle have the same length, $ED = BC = 3$, and $BE = CD = 1$. Thus, $AE = 1 + 3 = 4$.

- Since AD is the hypotenuse of a right triangle in which the legs measure 3 and 4, $AD = 5$.

LESSON 6-2
Multiple-Choice

1. **(D)** The height of parallelogram $ABCD$ is the length of perpendicular segment BE. The length of base AD is $1 + 5$ or 6. In right triangle AEB:

$$(BE)^2 + 1^2 = 2^2$$

or $(BE)^2 + 1 = 4$, so $BE = \sqrt{3}$. Hence: Area of parallelogram $ABCD = BE \times AD$

$$= \sqrt{3} \times 6 \text{ or } 6\sqrt{3}$$

2. **(B)** The area of a square is one-half the product of the lengths of its equal diagonals. If the length of a diagonal of a square is $\sqrt{2}$, the area of the square is $\frac{1}{2}(\sqrt{2})(\sqrt{2})$ or $\frac{1}{2}(2)$, which equals 1.

3. **(C)** To figure out the area of quadrilateral $ABCD$, add the areas of right triangles DAB and DBC.

- The area of right triangle DAB is $\frac{1}{2} \times 3 \times 4$ or 6.
- The lengths of the sides of right triangle DAB form a $(3, 4, 5)$ Pythagorean triple in which hypotenuse $BD = 5$. The lengths of the sides of right triangle DBC form a $(5, 12, 13)$ Pythagorean triple in which $BC = 12$. Hence, the area of right triangle DBC is $\frac{1}{2} \times 5 \times 12$ or 30.

The area of quadrilatertal $ABCD$ is $6 + 30$ or 36.

4. **(D)** If the area of square $ABCD$ is 64, then the length of each side is $\sqrt{64}$ or 8. Since $BC = 8$, the length of each side of equilateral triangle BEC is 8. Hence:

$$\text{Area equilateral } \triangle BEC = \frac{(\text{side})^2}{4} \times \sqrt{3}$$

$$= \frac{(8)^2}{4} \times \sqrt{3}$$

$$= \frac{64}{4} \times \sqrt{3}$$

$$= 16\sqrt{3}$$

5. **(A)** Since the ratio of AD to DC is 3 to 2, DC is $\frac{2}{3+2}$ or $\frac{2}{5}$ of AC. The base of $\triangle BDC$ is $\frac{2}{5}$ of the base of $\triangle ABC$, and the heights of triangles ABC and DBC are the same. Hence, the area of $\triangle BDC$ is $\frac{2}{5}$ of the area of $\triangle ABC$, so

$$\text{Area } \triangle BDC = \frac{2}{5} \times 40 = 16$$

6. **(B)** Draw $\overline{DE} \perp \overline{AB}$.

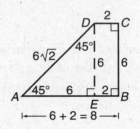

Since quadrilateral $DEBC$ is a rectangle, $EB = DC = 2$ and $DE = CD = 6$. In isosceles right triangle $\triangle AED$, $AE = DE = 6$, so

$$AB = AE + EB = 6 + 2 = 8$$

Also, hypotenuse $AD = 6\sqrt{2}$. Hence:

Perimeter $ABCD = AD + DC + BC + AB$

$$= 6\sqrt{2} + 2 + 6 + 8$$

$$= 16 + 6\sqrt{2}$$

7. **(C)** Area quad $ABCD$ = Area rect $DEBC$ + Area $\triangle AED$

$$= (2 \times 6) + \left(\frac{1}{2} \times 6 \times 6\right)$$

$$= 12 \quad + 18$$

$$= 30$$

8. **(B)** Since the perimeter of the triangle is 18, the length of the third side is 18 − 7 − 4 or 7. Draw a segment perpendicular to the shortest side from the opposite vertex. In an isosceles triangle, the perpendicular drawn to the base bisects the base.

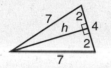

Hence, if h represents the height, then

$$h^2 + 2^2 = 7^2$$
$$h^2 = 49 - 4$$
$$= 45$$
$$h = \sqrt{45} = \sqrt{9} \cdot \sqrt{5} = 3\sqrt{5}$$

Since the base is 4, the area of the triangle is $\frac{1}{2} \times 4 \times 3\sqrt{5} = 6\sqrt{5}$.

9. **(C)**

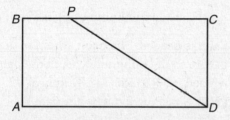

Quadrilateral $ABPD$ is a trapezoid with bases BP and AD, and height AB. Hence,

$$\text{Area trapezoid } ABPD = AB \times \frac{(BP + AD)}{2} = \frac{3}{4}$$

Try to transform the left side of the equation into $AB \times BC$, which represents the area of rectangle $ABCD$.

- Substitute $\frac{1}{4}BC$ for BP since BP is 25% of BC:

$$\frac{AB}{2} \times \left(\frac{1}{4}BC + AD \right) = \frac{3}{4}$$

- Since $AD = BC$, replace AD with BC:

$$\frac{AB}{2} \times \left(\frac{1}{4}BC + BC \right) = \frac{3}{4}$$

- Simplify:

$$\frac{AB}{2} \times \left(\frac{5}{4}BC \right) = \frac{3}{4}$$
$$\frac{5}{8} \times (AB \times BC) = \frac{3}{4}$$
$$\underbrace{AB \times BC}_{\text{Area rect. } ABCD} = \frac{8}{5} \times \frac{3}{4}$$
$$= \frac{6}{5}$$

10. **(D)** Since $ABCD$ is a square, C has the same x-coordinate as D and the same y-coordinate as B. Hence, $p = 7$ and $q = 6$. Point A has the same x-coordinate as B and the same y-coordinate as point D. Hence, $r = -2$ and $s = -3$.

- To find an equation of the line that contains diagonal $\overline{AC}$, first find the slope of $\overline{AC}$:

$$m = \frac{\Delta y}{\Delta x} = \frac{6 - (-3)}{7 - (-2)} = \frac{9}{9} = 1$$

Hence, the equation of $\overline{AC}$ has the form $y = x + b$.

NOTE: Instead of using the slope formula, you could reason that since the diagonal of a square forms two right triangles in which the vertical and horizontal sides always have the same length, their ratio is always 1. Since $\overline{AC}$ rises from left to right, its slope is positive. Therefore, the slope of $\overline{AC}$ is 1 (and the slope of diagonal $\overline{BD}$ is −1).

- To find b, substitute the coordinates of $C(7, 6)$ into $y = x + b$, which makes $6 = 7 + b$, so $b = -1$.
- An equation of $\overline{AC}$ is $y = x - 1$.

11. **(A)** Since $\triangle AEB$ is a 45-45 right triangle, $AB = EA = 3$. Quadrilateral $EABC$ is a trapezoid whose area, A, is given by the formula

$A = \frac{1}{2}h$(sum of bases). The bases of trapezoid

$EABC$ are AB and EC, the sum of whose lengths is $3 + 7 = 10$. The height of the trapezoid is EA, which is 3 units in length.

Hence, $A = \frac{1}{2}(3)(10) = 15$.

12. **(A)** Pick an easy number for the starting length of a side of the square, say 10. The original area is $10 \times 10 = 100$. If opposites are increased by 20%, then their new lengths are 12. If the other pair of sides are increased 50% in length, then their new lengths are 15. The area of the new rectangle thus formed is $12 \times 15 = 180$, which represents an 80% increase in area.

13. **(B)** You are given that $AB = BE = 8$ and $\angle A = 60°$. Since $\triangle ABE$ is isosceles, $m\angle E = 60°$, so

$$m\angle B = 180° - 60° - 60° = 60°$$

Since $\triangle ABE$ is equiangular, it is also equilateral, $AE = 8$ and $ED = 10 - 8 = 2$.

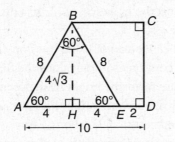

From B draw a perpendicular to AD, intersecting it at point H. Since BH is the side opposite the 60° angle in right triangle BHA,

$$BH = \frac{1}{2} \times 8 \times \sqrt{3} = 4\sqrt{3}$$

Also, since HE is the side opposite the 30° angle in right triangle BHE,

$$HE = \frac{1}{2} \times 8 = 4$$

so $HD = 4 + 2 = 6$.

Area quad $BCDE$ = Area rect $BHDC$ – Area right $\triangle BHE$

$$= BH \times HD - \frac{1}{2} \times BH \times HE$$

$$= 4\sqrt{3} \times 6 \quad - \frac{1}{2} \times 4\sqrt{3} \times 4$$

$$= 24\sqrt{3} \quad - 8\sqrt{3}$$

$$= 16\sqrt{3}$$

Grid-In

1. **(42)** First find the sum of the areas of the four walls:

$$12 \times 8 = 96$$
$$12 \times 8 = 96$$
$$16 \times 8 = 128$$
$$\underline{16 \times 8 = 128}$$
$$\text{Sum of areas} = 448$$

Since 1 gallon of paint provides coverage of an area of at most 150 square feet and

$\frac{448}{150} = 2.9...$, a minimum of 3 gallons of

paint is needed. The paint costs \$14 per gallon, so the minimum cost of the paint needed is $3 \times \$14$ or \$42. Grid in as 42.

2. **(1/9)** Since the figure is a square,

$$x = 4x - 1$$
$$3x = 1$$
$$x = \frac{1}{3}$$

The area of the square is

$$x^2 = \left(\frac{1}{3}\right)^2 = \frac{1}{9}$$

Grid in as 1/9.

3. **(27/8)** The area of quadrilateral $QBEF$ equals the area of right triangle PEF minus the area of right triangle PBQ.

- The area of square $ABCD$ is 9, so the length of each side of the square is $\sqrt{9}$ or 3. Since P and Q are midpoints,

$$PB = QB = \frac{3}{2}, \text{ so}$$

$$\text{Area right } \triangle PBQ = \frac{1}{2} \times \frac{3}{2} \times \frac{3}{2} \text{ or } \frac{9}{8}$$

- Since $PB = QB$, right triangle PBQ is isosceles, so $\angle EPF = 45°$. Hence, right triangle PEF is also isosceles, so $EF = PE = 3$, and

$$\text{Area right } \triangle PEF = \frac{1}{2} \times 3 \times 3 = \frac{9}{2}$$

Area $QBEF$ = Area right $\triangle PEF$ − Area right $\triangle PBQ$

$$= \frac{9}{2} - \frac{9}{8}$$

$$= \frac{36}{8} - \frac{9}{8}$$

$$= \frac{27}{8}$$

Grid in as 27/8.

4. **(13.5)** The area of the shaded region is equal to the difference in the areas of the two overlapping right triangles formed by the coordinate axes and the slanted lines:

$$\text{Area of shaded region} = \frac{1}{2}(6)(5) - \frac{1}{2}(1)(3)$$

$$= 15 - 1.5$$

$$= 13.5$$

5. **(49)** The area of a square can be determined by multiplying the square of the length of a diagonal by one-half. If the coordinates of the endpoints of a square are (−2, −3) and (5, 4), the length of the diagonal is

$$\sqrt{(5-(-2))^2 + (4-(-3))^2} = \sqrt{(5+2)^2 + (4+3)^2}$$

$$= \sqrt{7^2 + 7^2}$$

$$= \sqrt{98}$$

Hence, the area of the square is

$$\frac{1}{2}\left(\sqrt{98}\right)\left(\sqrt{98}\right) = \frac{1}{2}(98) = 49.$$

6. **(45/2)** The region bounded in the xy-plane by $y = |x| + 2$, $x = 5$, and the positive axes is a trapezoid:

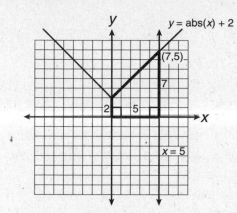

The lengths of the bases of the trapezoid are 2 and 7, and the altitude is 2. Hence, the area, A, is

$$A = \frac{1}{2}h(\text{sum of bases})$$

$$= \frac{1}{2}(5)(2+7)$$

$$= \frac{45}{2}$$

7. **(42)** Since $AB = BC$ and $m\angle B = 60$, $\angle CAB = \angle ACB = 60°$ so $\triangle ABC$ is equilateral. If s represents the length of a side of $\triangle ABC$, then

$$\triangle ABC = \frac{s^2}{4}\sqrt{3} = 9\sqrt{3}$$

Thus, $s^2 = 36$ and $s = 6$. Since the polygon has 7 sides, each having a length of 6 cm, the perimeter of ABCDEFG is $6 \times 12 = 72$ centimeters.

8. **(1.5)** The area of $\triangle MBA$ is $\frac{1}{2}(4.8)(15) = 36$.

Since the area of the rectangle is $6 \times 15 = 90$, the area of the shaded region is $90 - 36 = 54$. Compare the area of the shaded region to the area of $\triangle MBA$ by finding their ratio:

$$\frac{\text{Area shaded region}}{\text{Area } \triangle MBA} = \frac{54}{36} = \frac{3}{2} = 1.5 \text{ to } 1$$

9. **(72)** The area of an equilateral triangle with side length s is $\frac{s^2}{4}\sqrt{3}$. Find the length of a

side of the equilateral triangle by setting $16\sqrt{3}$ equal to $\dfrac{s^2}{4}\sqrt{3}$ and solving for s:

$$\frac{s^2}{4}\sqrt{3} = 16\sqrt{3}$$
$$s^2 = 16 \times 4$$
$$= 64$$
$$s = \sqrt{64} = 8$$

Since the polygon has 9 sides each having a length of 8, the perimeter of the polygon is $9 \times 8 = 72$.

LESSON 6-3

Multiple-Choice

1. **(B)** Since the radius of circle O is 2 and the radius of circle P is 6, $OP = 2 + 6$ or 8. Hence, the circumference of any circle that has OP as a diameter is 8π.

2. **(C)** If the circumference of a circle is 10π, its diameter is 10 and its radius is 5. Hence, its area is $\pi(5^2) = 25\pi$.

3. **(B)** If the tires are 24 inches in diameter, then the circumference of the wheel, in feet, is 2π. For each complete rotation of the pedals, the wheel travels $3 \times 2\pi = 6\pi$ feet. The minimum number of complete rotations of the pedals needed for the bicycle to travel at least 1 mile ($= 5{,}280$) feet is $\dfrac{5{,}280}{6\pi} \approx 280.11$, which, when rounded up, is 281.

4. **(B)** The distance between cars 4 and 8, $\dfrac{84}{\pi}$, represents the diameter of the ferris wheel so the circumference of the ferris wheel is

   $\pi \times \dfrac{84}{\pi} = 84$ meters. When car 4 travels

   clockwise to the base at position 7, it travels $\dfrac{5}{8}$ of the circumference of the circle or a

   distance of $\dfrac{5}{8} \times 84 = 52.5$ meters.

5. **(C)** The circle has a radius, r, of between 3 and 4 units so r^2 must be between 9 and 16.

Therefore, you eliminate choices (A) and (B). The center radius form of the equation of a circle with center at (h, k) is $(x - h)^2 + (y - k)^2 = r^2$. Since $(h, k) = (4, -2)$, the equation of the circle must have the form $(x - 4)^2 + (y + 2)^2 = r^2$ where $9 < r^2 < 16$, which corresponds to choice (C).

6. **(B)** Write the equation $x^2 - 10x + y^2 + 6y = -9$ in center radius form:

   $$\left(x^2 - 10x + 25\right) + \left(y^2 + 6y + 9\right) = -9 + 34$$
   $$(x - 5)^2 + (y + 3)^2 = 25$$

   A diameter of the circle contains the center $(5, -3)$ so the coordinates of this point must satisfy the equation of any line that contains a diameter of the circle. Substitute 5 for x and -3 for y in each equation until you find the one that works, as in choice (B):

   $$y = -2x + 7$$
   $$-3 = -2(5) + 7$$
   $$-3 = -10 + 7$$
   $$-3 = -3 \checkmark$$

7. **(C)** To find the length of segment XY, first find the length of segment OX. Draw radii OA and OB.

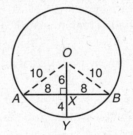

Since all radii of a circle have the same length, $\triangle AOB$ is isosceles, so perpendicular segment OX bisects chord AB. In right triangle AXO, radius $OA = 10$ and

$$AX = \frac{1}{2}(16) = 8$$

The lengths of the sides of right triangle AXO form a (6, 8, 10) Pythagorean triple where $OX = 6$. Since radius $OY = 10$,

$$XY = OY - OX = 10 - 6 = 4$$

8. **(A)** Let d represent the length in feet of the diameter of the bicycle wheel; then πd is the circumference of the wheel. After completing n revolutions, the wheel has traveled $n \times \pi d$ feet. Since it is given that the bicycle wheel has traveled $\dfrac{f}{\pi}$ feet after n complete revolutions:

$$n\pi d = \frac{f}{\pi}$$

$$\frac{\overset{1}{\cancel{n\pi d}}}{\cancel{n\pi}} = \frac{f}{n\pi \cdot \pi}$$

$$d = \frac{f}{n\pi^2}$$

9. **(B)** To find the area of the circle whose equation is $x^2 + y^2 - 6x + 8y = 56$, find the radius of the circle by rewriting the equation in center radius form:

$$x^2 + y^2 - 6x + 8y = 56$$
$$\left(x^2 - 6x + 9\right) + \left(y^2 + 8y + 16\right) = 56 + 9 + 16$$
$$(x - 3)^2 + (y + 4)^2 = 81$$

Since $r^2 = 81$, the area of the circle is πr^2 or 81π.

10. **(D)** Since the area of each circle is 7, $7 = \pi r^2$, so the radius of each circle is $\sqrt{\dfrac{7}{\pi}}$.

- AD = Vertical diameter of circle $X = 2\sqrt{\dfrac{7}{\pi}}$

- AB = Horizontal diameter of circle X + Horizontal radius of circle

$$Y = 2\sqrt{\frac{7}{\pi}} + \sqrt{\frac{7}{\pi}} = 3\sqrt{\frac{7}{\pi}}$$

- Area $ABCD = AD \times AB$

$$= 2\sqrt{\frac{7}{\pi}} \times 3\sqrt{\frac{7}{\pi}}$$

$$= 6 \times \frac{7}{\pi}$$

$$= \frac{42}{\pi}$$

11. **(D)** The area of the shaded region is the area of the rectangle minus the sum of the areas of the two quarter circles, BP and CP.
- Since the area of quarter circle BP is π,

$$\frac{1}{4} \times \pi (AB)^2 = \pi$$

Hence, $(AB)^2 = 4$ so $AB = AP = 2$.
- The area of quarter circle CP is also π, so $DP = AP = 2$.
- Since $AD = AP + PD = 2 + 2 = 4$, the area of rectangle $ABCD = AB \times AD = 2 \times 4 = 8$.
- The sum of the areas of the two quarter circles is $\pi + \pi$ or 2π.
- The area of the shaded region is $8 - 2\pi$.

12. **(B)** If R stands for the radius of the smaller circle, then $R + w$ represents the radius of the larger circle. Since the circumference of the larger circle exceeds the circumference of the smaller circle by 12π,

$$2\pi(R + w) - 2\pi R = 12\pi$$
$$(2\pi R + 2\pi w) - 2\pi R = 12\pi$$
$$2\pi w = 12\pi$$
$$w = \frac{12\pi}{2\pi} = 6$$

13. **(C)** Since $\overline{OP}$ and $\overline{OB}$ are radii and P is the midpoint of hypotenuse $\overline{OA}$, $OB : OA = 1 : 2$. Thus, the angle opposite $\overline{OB}$, $\angle A$, measures $30°$ so that central $\angle AOB = 90 - 3 = 60$.
- The central angle of the shaded region measures $360 - 60 = 300°$.
- Since the radius of circle O is 12, the area of circle O is $\pi \times 12^2 = 144\pi$.
- Hence:

$$\text{Area of shaded region} = \frac{300}{360} \times 144\pi$$

$$= \frac{5}{\cancel{6}} \times \overset{24}{\cancel{144\pi}}$$

$$= 120\pi$$

14. **(A)** The perimeter of the unbroken figure is the sum of the lengths of AC, BC, and major arc AB.
- The area of square $OACB$ is $4x^2$, so

$$AC = BC = \sqrt{4x^2} = 2x$$

- Since *OACB* is a square, it is equilateral, so radius $OA = AC = 2x$. The circumference of circle *O* is $2 \times \pi \times 2x$ or $4\pi x$. Angle *AOB* measures 90°, so the central angle that intercepts major arc *AB* is $360 - 90$ or 270°. Hence:

$$\text{Length major arc } AB = \frac{270°}{360°} \times 4\pi x$$
$$= \frac{3}{4} \times 4\pi x$$
$$= 3\pi x$$

- The perimeter of the unbroken figure is $2x + 2x + 3\pi x$ or $4x + 3\pi x$ or, factoring out *x*,
$$\text{Perimeter} = x(4 + 3\pi)$$

15. **(B)** The area of the shaded region is the area of square *OABC* minus the area of sector *AOC*.
- Since the area of circle *O* is given as 2π,

$$\text{Area sector } AOC = \frac{90°}{360°} \times 2\pi$$
$$= \frac{1}{4} \times 2\pi$$
$$= \frac{\pi}{2}$$

- The area of circle $O = 2\pi = \pi(OA)^2$, so $2 = (OA)^2$. Then

$$\text{Area square } OABC = (OA)^2 = 2$$

- The area of the shaded region $= 2 - \dfrac{\pi}{2}$.

16. **(C)** The area of the shaded region is the area of the square minus the sum of the areas of the four quarter circles.
- The area of the square is 2×2 or 4.
- Since the four quarter circles have equal areas, they have equal radii and the sum of their areas is equivalent to the area of one whole circle with the same radius length. The length of side *GE* is given as 2. Then the length of the radius of each quarter circle is 1 since *B*, *D*, *F*, and *H* are midpoints of the sides of the square. Hence, the sum of the areas of the four quarter circles is $\pi(1^2)$ or π.

- The area of the shaded region $= 4 - \pi$. Using $\pi = 3.14$, you find that the best approximation for the area of the shaded region is $4 - 3.14$ or 0.86.

17. **(C)** Determine whether each of the Roman numeral statements is always true.
- I. Since all four sides of a square have the same length, the area of the square is the product of the lengths of any two of its sides. Hence, the left side of the inequality $AB \times CD < \pi \times r \times r$ represents the area of square *ABCD*, and the right side of the inequality represents the area of inscribed circle *O*. Since the area of the square is greater, not less, than the area of the inscribed circle, Statement I is not true.
- II. Since a diameter of the circle can be drawn in which endpoints are points at which circle *O* intersects two sides of the square, the length of a side of the square is $2r$, so the area of the square can be represented as $2r \times 2r$ or $4r^2$. Hence, Statement II is true.
- III. Since the circumference of the circle is less than the perimeter of the square:

$$2\pi r < 4(CD)$$
$$r < \frac{4(CD)}{2\pi}$$
$$r < \frac{2(CD)}{\pi}$$

Hence, Statement III is true.
Only Roman numeral Statements II and III must be true.

18. **(D)** Draw radius *OB*. Since $\triangle OAB$ is an isosceles right triangle and side $AB = 6$, hypotenuse $OB = 6\sqrt{2} = $ radius of circle *O*. Thus:

$$\begin{array}{ccc} \text{Area of} & \text{Area of} & \text{Area of} \\ \text{shaded region} = \text{quarter circle} & - & \text{square } OABC \end{array}$$
$$= \frac{1}{4} \times \pi \times \left(6\sqrt{2}\right)^2 - (6 \times 6)$$
$$= \frac{1}{4} \times \pi \times 72 \quad -36$$
$$= 18\pi - 36$$
$$= 18(\pi - 2)$$

19. **(D)** The perimeter of the figure that encloses the shaded region is the length of arc PBQ plus the sum of the lengths of segments AP, CQ, AB, and BC.

- Draw radius OB. You are told that $OABC$ is a rectangle. Since the diagonals of a rectangle have the same length, $OB = AC = 8$. The circumference of the circle that has its center at O and that contains arc PBQ is $2 \times \pi \times 8$ or 16π. Since a 90° central angle intercepts arc PBQ:

$$\text{Length arc } PBQ = \frac{90°}{360°} \times 16\pi$$
$$= \frac{1}{4} \times 16\pi$$
$$= 4\pi$$

- In isosceles right triangle AOC,

$$OA = OC = \frac{1}{2} \times 8 \times \sqrt{2} = 4\sqrt{2}$$

Hence:

$$AP = OP - OA = 8 - 4\sqrt{2}$$

and

$$CQ = OQ - OC = 8 - 4\sqrt{2}$$

- In isosceles right triangle ABC,

$$AB = BC = \frac{1}{2} \times 8 \times \sqrt{2} = 4\sqrt{2}$$

- The perimeter of the figure that encloses the shaded region = arc $PBQ + AP + CQ + AB + BC = 4\pi + \left(8 - 4\sqrt{2}\right) + \left(8 - 4\sqrt{2}\right) + 4\sqrt{2} + 4\sqrt{2} = 4\pi + 16$ or $16 + 4\pi$.

20. **(C)** It is given that the center of circle Q is $(3, -2)$ and $R(7, 1)$ is the endpoint of a radius of the circle. Use the distance formula to find the radius length:

$$QR = \sqrt{(7-3)^2 + (1-(-2))^2}$$
$$= \sqrt{4^2 + 3^2}$$
$$= 5$$

Since $(h, k) = (3, -2)$ and $r^2 = 25$, the equation of the circle is

$$(x - 3)^2 + (y + 2)^2 = 25$$

21. **(A)** By drawing $\overline{AP}$, a 30-60 right triangle is formed in which the radius of $\frac{12}{\pi}$ is opposite the 30° angle so the side opposite the 60° angle, $\overline{NP}$, is $\frac{12\sqrt{3}}{\pi}$.

- Since $NP = PS = JP = PK$,

$$NP + PS + JP + PK = 4\left(\frac{12\sqrt{3}}{\pi}\right) = \frac{48\sqrt{3}}{\pi}.$$

- The circumference of each circle is $2\pi\left(\frac{12}{\pi}\right) = 24$ inches. Central angle NAJ and angle NPJ are supplementary so the degree measure of $\angle NAJ$ is 120, which means major arc NJ is $\frac{240}{360} = \frac{2}{3}$ of the circumference of the circle or $\frac{2}{3} \times 24 = 16$ inches. Hence, the sum of the lengths of major arcs NJ and KS is $16 + 16 = 32$ inches.

- The length of the belt is $32 + \frac{48\sqrt{3}}{\pi}$.

Grid-In

1. **(2)** The length of the radius of each semicircle above the line is 2, so

$$\text{Area each semicircle} = \frac{1}{2} \times \pi \times 2^2$$
$$= \frac{1}{2} \times 4\pi$$
$$= 2\pi$$

Hence,

$$X = 2\pi + 2\pi + 2\pi = 6\pi$$

The diameter of the larger semicircle below the line is 8, so its radius is 4. Then

$$\text{Area larger semicircle} = \frac{1}{2} \times \pi \times 4^2$$
$$= \frac{1}{2} \times 16\pi$$

so $Y = 8\pi$.
Hence,

$$Y - X = 8\pi - 6\pi = 2\pi = k\pi$$

so $k = 2$.

2. **(2/32)** or **(1/16)** Pick an easy number for the length of the diameter of the smaller semicircle. Then find the areas of semicircles PS and PR.

- If diameter $PS = 4$, the length of the radius of the smaller semicircle is 2, so:

$$\text{Area smaller semicircle} = \frac{1}{2} \times \pi \times 2^2$$
$$= \frac{1}{2} \times 4\pi$$
$$= 2\pi$$

- Since $PS = 4$ and S is the midpoint of PQ, the length of radius PQ of the larger semicircle is 8. Then:

$$\text{Area larger semicircle} = \frac{1}{2} \times \pi \times 8^2$$
$$= \frac{1}{2} \times 64\pi$$
$$= 32\pi$$

- The ratio of the area of semicircle PS to the area of semicircle PR is $\frac{2\pi}{32\pi}$ or $\frac{2}{32}$ $\left(\text{or } \frac{1}{16}\right)$. Grid in as 2/32 (or 1/16).

3. **(13 feet)** Rewrite the circle equation $x(x + 4) + y(y - 12) = 9$ in center-radius form:

$$x(x + 4) + y(y - 12) = 9$$
$$x^2 + 4x + y^2 - 12y = 9$$
$$(x^2 + 4x + 4) + (y^2 - 12y + 36) = 9 + 4 + 36$$
$$(x + 2)^2 + (y - 6)^2 = 49$$

The coordinates of the center of the circle are $(-2, 6)$. The distance from $(3, -6)$ to $(-2, 6)$ is $\sqrt{25 + 144} = 13$.

4. **(8)** First find the radius of the circle.

- Since a radius of a circle is perpendicular to a tangent at the point of contact, angles OAP and OBP measure 90°. The sum of the measures of the four angles of a quadrilateral is 360°. Thus, $120° + 90° + 90° + \angle P = 360°$, so $\angle P = 2a = 60°$ and $a = 30°$.

- In right triangle OAP, $a = 30°$, so $\angle AOP = 60°$. In a 30°-60° right triangle, the length of the side opposite the 30° angle is one-half the length of the hypotenuse. Hence, radius $OA = \frac{1}{2}OP = \frac{1}{2} \times \frac{24}{\pi} = \frac{12}{\pi}$.

- The cirumference of the circle O is

$$2\pi r = 2\pi \times \frac{12}{\pi} = 24$$

- Since $\frac{120°}{360°} = \frac{1}{3}$, the length of minor arc AB is $\frac{1}{3}$ of the circumference of the circle. Hence, the length of minor arc

$$AB = \frac{1}{3} \times 24 = 8.$$

5. **(13.1)** Draw the radius from the center of the semicircle to point A thereby forming a right triangle in which the hypotenuse, the radius drawn, is 14, one leg is 5, and the height of the banner is the other leg. If x represents the height of the banner, use the Pythagorean theorem to find x:

$$x^2 + 5^2 = 14^2$$
$$x^2 + 25 = 196$$
$$x^2 = 171$$
$$x = \sqrt{171}$$
$$x \approx 13.07699$$

To the *nearest tenth* of a foot, the height of the banner is 13.1 feet.

LESSON 6-4
Multiple-Choice

1. **(A)** The volume of a pyramid is one-third the area of its base times it height. If x represents the length of a side of a square that is the base of a pyramid that has a volume of 256 cubic centimeters and a height of 12 centimeters, then

$$V = \frac{1}{3}Bh$$

$$256 = \frac{1}{3}x^2(12)$$

$$256 = 4x^2$$

$$\frac{256}{4} = \frac{4x^2}{4}$$

$$64 = x^2$$

$$8 = x$$

2. **(B)** The volume of a rectangular box is the product of its length, width, and height. If the volume of a rectangular box is 144 cubic inches and its height is 8 inches, then the area of its base must be $\frac{144}{8} = 18$. Check each answer choice in turn until you find the one in which the product of the dimensions of the base is 18 as in choice (B): $2.5 \times 7.2 = 18$.

3. **(C)** The lateral surface area of the cylindrical tank is $2\pi rh$ where $r = 12$ and $h = 22$. Divide the lateral surface area by 500 square feet to find the number of cans of paint that must be purchased:

$$\frac{2\pi(12)(22)}{500} = \frac{528\pi}{500} \approx 3.31$$

Round the answer up to **4** cans.

4. **(A)** Draw a line from the vertex of the cone to the center. Also draw a radius thereby forming a 30-60 right triangle in which the slant height of 6 cm is the hypotenuse and the height is the side opposite the 60° angle so it measures $3\sqrt{3}$ cm. The radius of the base is the side opposite the 30° angle so it measures one-half of the hypotenuse or 3 cm.

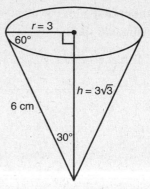

The volume of a circular cone is one-third the area of its circular base times its height:

$$V = \frac{1}{3}Bh$$

$$= \frac{1}{3}(\pi)(3)^2(3\sqrt{3})$$

$$\approx 49\,\text{cm}^3$$

5. **(C)** Since the circumference of the ball measures 13 inches, the diameter of the ball is $\pi D = 13$ so $D = \frac{13}{\pi}$. Hence, the side of the cube-shaped box must be at least $\frac{13}{\pi} \approx 4.13$ inches. If the box has integer dimensions, then the cube-shaped box with the smallest integer dimensions is $5 \times 5 \times 5 = 125$ cubic inches.

6. **(A)** To determine the amount of light produced by the cylindrical light bulb, multiply the lateral surface area of the bulb by 0.283 using $r = \frac{1}{2} \times 1 = \frac{1}{2}$ inch for the radius and $h = 32$ inches for the height of the bulb:

$$(2\pi rh) \times 0.283 = (2\pi)(0.5)(36) \times 0.283$$

$$= 36\pi \times 0.283$$

$$\approx 32$$

7. **(A)** When a solid cube with an edge length of 1 foot is placed in the fish tank, it displaces a volume of water that is equal to the volume of the cube, which is $1 \times 1 \times 1$ or 1 cubic foot. If h represents the change in

the number of feet in the height of the water in the fish tank, then

Volume of displaced water $= 2 \times 3 \times h = 1$

$$h = \frac{1}{6} \text{ foot}$$

Hence, the number of *inches* the level of the water in the tank will rise is $\frac{1}{6} \times 12$ or 2.

8. **(B)** Let h represent the equal heights of the cylinder and the rectangular box. Since you are given that the volume of the cylinder of radius r is $\frac{1}{4}$ of the volume of the rectangular box with a square base of side length x,

$$\overbrace{\pi r^2 h}^{\text{Volume of cylinder}} = \frac{1}{4} \overbrace{(x \cdot x \cdot h)}^{\text{Volume of box}}$$

$$\pi r^2 \cancel{h} = \frac{x^2}{4} \cancel{h}$$

$$r^2 = \frac{x^2}{4\pi}$$

$$r = \frac{\sqrt{x^2}}{\sqrt{4\pi}} = \frac{x}{2\sqrt{\pi}}$$

9. **(D)** You are given that the height of the sand in the cylinder-shaped can drops 3 inches or $\frac{1}{4}$ foot when 1 cubic foot of sand is poured out. If r represents the length in feet of the radius of the circular base, then the volume of the sand poured out must be equal to 1. Hence:

Volume of sand $= \pi r^2 \left(\frac{1}{4} \right) = 1$

$$\pi r^2 = 1 \times 4$$

$$r^2 = \frac{4}{\pi}$$

$$r = \frac{\sqrt{4}}{\sqrt{\pi}} = \frac{2}{\sqrt{\pi}} \text{feet}$$

$$= \frac{2}{\sqrt{\pi}} \text{feet} \times 12 \text{inches/feet}$$

$$= \frac{24}{\sqrt{\pi}} \text{inches}$$

Hence the radius of the cylinder is $\frac{2}{\sqrt{\pi}}$ feet $\times$ 12 or $\frac{24}{\sqrt{\pi}}$ inches, and the diameter in *inches* is $2 \times \frac{24}{\sqrt{\pi}}$ or $\frac{48}{\sqrt{\pi}}$.

10. **(A)** If the height h of a cylinder equals the circumference of the cylinder with radius r, then $h = 2\pi r$, so $r = \frac{h}{2\pi}$. Hence:

Volume of cylinder $= \pi r^2 h$

$$= \pi \left(\frac{h}{2\pi} \right)^2 h$$

$$= \pi \left(\frac{h^2}{4\pi^2} \right) h$$

$$= \frac{\pi h^3}{4\pi^2}$$

$$= \frac{h^3}{4\pi}$$

11. **(B)** In one complete rotation, the roller covers $2\pi rh = 2\pi(9)(42) \approx 2,375$.

12. **(A)** To find the volume of a lead ball that has a 5 inch diameter, use the formula for the volume of a sphere where $r = \frac{1}{2}(5) = 2.5$:

$$V = \frac{4}{3}\pi r^3$$

$$= \frac{4}{3}\pi(2.5)^3$$

$$\approx 65.45$$

To find the mass, m, of the lead ball, use the fact that density is mass divided by volume:

$$\frac{m}{65.45} = 0.41$$

$$m = 0.41 \times 65.45$$

$$m \approx 26.8$$

13. **(B)** *Solution 1.* The shortest distance from A to D is the length of AD. Draw AD, which is the hypotenuse of a right triangle whose legs are AE and ED. Since the edge length of the

cube is given as 4, $ED = AF = 4$. Segment AE is the diagonal of square $AFEH$, so $AE = 4\sqrt{2}$.

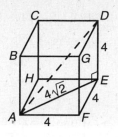

In right triangle AED,

$$(AD)^2 = (AE)^2 + (ED)^2$$
$$= \left(4\sqrt{2}\right)^2 + (4)^2$$
$$= (16 \cdot 2) + 16$$
$$= 32 + 16$$
$$AD = \sqrt{48} = \sqrt{16} \cdot \sqrt{3} = 4\sqrt{3}$$

Solution 2.

$$\text{Diagonal } d = \sqrt{\ell^2 + w^2 + h^2}$$
$$\text{Let } \ell = w = h = 4: \ = \sqrt{4^2 + 4^2 + 4^2}$$
$$= \sqrt{16 + 16 + 16}$$
$$= \sqrt{48}$$
$$= \sqrt{16} \cdot \sqrt{3}$$
$$= 4\sqrt{3}$$

14. **(D)** Let $\ell = 4$, $w = 3$, $h = 8$:
$$\text{Diagonal } d = \sqrt{\ell^2 + w^2 + h^2}$$
$$= \sqrt{4^2 + 3^2 + 8^2}$$
$$= \sqrt{16 + 9 + 64}$$
$$= \sqrt{89}$$
$$\approx 9.4$$

15. **(A)** Since G is the center of the square base, the four sides of the pyramid are congruent isosceles triangles so $EA = ED = EC = EB$. Hence, Statements I and II are true. By drawing $\overline{FG}$, a right triangle is formed in which $\overline{EF}$ is the hypotenuse so $EF > EG$. Hence, Statement III is false. Only Statements I and II are true.

16. **(D)** If a pyramid with a square base with side length s and a right cone with radius r have equal heights and volumes, then

$$\frac{1}{3}s^2h = \frac{1}{3}\pi r^2 h$$
$$s^2 = \pi r^2$$
$$s = \sqrt{\pi r^2}$$
$$= r\sqrt{\pi}$$

17. **(A)** If a cube with side s has a surface area of 150 square centimeters, then $6s^2 = 150$, $s^2 = \dfrac{150}{6} = 25$ so $s = 5$. The volume of the cube is $5^3 = 125$ cubic centimeters. If it is melting at the rate of 13 cubic centimeters per minute, it will melt completely in $\dfrac{125}{13} \approx 9.62$ minutes, which is closest in value to choice (A).

18. **(A)** The volume of the hot water tank in the shape of a right circular cylinder is 85 gallons or $\dfrac{85}{7.48} \approx 11.36$ cubic feet. Since the diameter of the cylinder is 1.8 feet, its radius, r, is 0.9 feet. To find the height, h, of the cylinder, set the volume of the cylinder equal to 11.36 and solve for h:

$$\pi r^2 h = 11.36$$
$$\pi (0.9)^2 h = 11.36$$
$$h = \frac{11.36}{0.81\pi}$$
$$h \approx 4.46$$

19. **(C)** The smaller of the two right triangles in the diagram is similar to the larger one so the lengths of their corresponding sides are in proportion. If x represents the length of

the hypotenuse of the smaller right triangle located at the top of the cone, then

$$\frac{x}{x+10} = \frac{12}{8+12}$$

$$\frac{x}{x+10} = \frac{12}{20}$$

$$\frac{x}{x+10} = \frac{3}{5}$$

$$5x = 3x + 30$$

$$2x = 30$$

$$x = \frac{30}{2} = 15$$

The lengths of the sides of the larger right triangle are R, 20, and 25. These lengths are the lengths of a 3-4-5 right triangle in which each side is multiplied by 5. Hence, $R = 3 \times 5 = 15$.

20. **(D)** To find the volume of the frustrum, subtract the volume of the small cone at the top from the volume of the larger cone.

- The lengths of the sides of the smaller right triangle are lengths of a 3-4-5 right triangle in which each side is multiplied by 3. Hence, $r = 3 \times 3 = 9$. The volume of the smaller cone with height 12 is

$$\frac{1}{3}(\pi 9^2)(12) = 324\pi.$$

- The area of the large cone with height 8 + 12 = 20 is $\frac{1}{3}\left(\pi(15)^2\right)(20) = 1{,}500\pi.$

- The area of the frustrum is $1{,}500\pi - 324\pi$ = $1{,}176\pi.$

Grid-In

1. **(60)** You are given that all the dimensions of a rectangular box are integers greater than 1. Since the area of one side of this box is 12, the dimensions of this side must be either 2 by 6 or 3 by 4. The area of another side of the box is given as 15, so the dimensions of this side must be 3 by 5. Since the two sides must have at least one dimension in common, the dimensions of the box are 3 by 4 by 5, so its volume is $3 \times 4 \times 5$ or 60.

2. **(8)** If h represents the height of the rectangular box, then

$$12 \times 8 \times h \geq 700$$

$$96h \geq 700$$

$$h \geq \frac{700}{96}$$

$$h \geq 7.3$$

Since it is given that h must be a whole number, $h = 8$.

3. **(110)** The original volume was $100 \times 75 \times 30$ cubic feet. If the width and height remain the same while the volume is increased by 10%, the remaining factor, the length, must be increased by 10%. Since 10% of 100 is 10, the new length is $100 + 10 = 110$ feet.

4. **(121)**

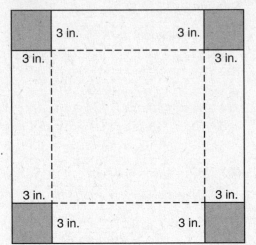

- Since 3-inch squares from the corners of the square sheet of cardboard are cut and folded up to form a box, the height of the box thus formed is 3 inches.

- If x represents the length of a side of the square sheet of cardboard, then the length and width of the box is $x - 6$.

- Since the volume of the box is 75 cubic inches:

length × width × height = Volume of box

$$(x - 6)(x - 6)3 = 75$$

$$(x - 6)^2 = \frac{75}{3} = 25$$

$$x - 6 = \sqrt{25} = 5$$

$$x = 5 + 6 = 11$$

Because the length of each side of the original square sheet of cardboard is 11 inches, the area of the square sheet of cardboard is $11 \times 11 = 121$ square inches.

5. **(24)** The greatest possible distance between a point on one sphere and a second point on another sphere is the sum of the lengths of their diameters.

    ■ If the volume of the smaller sphere is 36π, then $\frac{4}{3}\pi r^3 = 36\pi$ so $r^3 = \frac{3(36)}{4} = 27$ and $r = 3$. The diameter of the smaller sphere is 6 centimeters.

    ■ If the volume of the larger sphere is 972π, then $\frac{4}{3}\pi r^3 = 972\pi$ so $r^3 = \frac{3(972)}{4} = 729$ and $r = 9$. The diameter of the larger sphere is 18 centimeters.

    ■ The greatest distance between the two points is $6 + 18 = 24$ centimeters.

6. **(12.3)** The volume of unoccupied space is the difference between the sum of the volumes of the 3 balls and the tennis can.

    ■ If the diameter of a tennis ball is 2.5 inches, then the height of the can must be at least $3 \times 2.5 = 7.5$ inches. Hence, the smallest volume of the can is $\pi(1.25)^2(7.5) \approx 11.72\pi$.

    ■ The sum of the volumes of the 3 tennis balls is $3\left(\frac{4}{3}\pi(1.25)^3\right) \approx 7.81\pi$.

    ■ The difference in the volumes is $11.72\pi - 7.81\pi = 3.91\pi \approx 12.3$.

7. **(1,728)** Draw $\overline{FG}$ thereby forming a right triangle in which the hypotenuse is 15 and leg $FG = \frac{1}{2} \times 24 = 12$. The lengths of the sides of right $\triangle FGE$ are the lengths of a 3-4-5 right triangle in which each side is multiplied by 3. Hence, $h = 3 \times 3 = 9$. Calculate the volume, V, of the pyramid by multiplying one-third of the area of the square base times the height:

$$V = \frac{1}{3} \times (24 \times 24) \times 9 = 1,728$$

8. **(3.75)** Since each side of the pyramid's base and height are being multiplied by 2.5, the ratio of the volumes of the two pyramids is $\left(\frac{2.5}{1}\right)^3$. Hence, if x represents the weight of the larger pyramid, then

$$\frac{x}{0.24} = \left(\frac{2.5}{1}\right)^3$$
$$x = 0.24 \times (2.5)^3$$
$$= 3.75$$

LESSON 6-5
Multiple-Choice

1. **(C)** If $\frac{\sin A}{\cos B} = 1$, then $\sin A = \cos B$ so acute angles A and B must be complementary. If two angles are complementary, the sum of their measures is 90:

$$(x - 3) + (2x + 6) = 90$$
$$3x + 3 = 90$$
$$3x = 87$$
$$x = \frac{87}{3} = 29$$

2. **(D)** Use the tangent ratio:

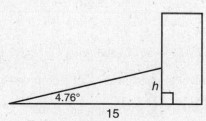

$$\tan 4.76° = \frac{\text{side opposite} \angle}{\text{side adjacent} \angle} = \frac{h}{15} \text{ so } h = 15\tan 4.76°$$

3. **(D)** Twenty-four minutes is $\frac{24}{60} = \frac{2}{5}$th of an hour. A complete rotation is 2π radians, so the angle through which the minute hand of a clock rotates in 24 minutes is $\frac{2}{5} \times 2\pi = 0.8\pi$.

4. **(A)** If $x = 1.75$ radians, then $\angle x$ is located in Quadrant II since $\frac{\pi}{2} < 1.75 < \pi$. The reference angle is $\pi - 1.75 \approx 1.39$. Since cosine is negative in Quadrant II, $\cos 1.75 = -\cos 1.39$.

5. **(B)** If $\sin\frac{2}{9}\pi = \cos x$, then angles must be complementary. Hence,

$$\frac{2}{9}\pi + x = \frac{\pi}{2}$$
$$x = \frac{\pi}{2} - \frac{2\pi}{9}$$
$$= \frac{9\pi}{18} - \frac{4\pi}{18}$$
$$= \frac{5}{18}\pi$$

6. **(B)** The length of the pendulum corresponds to the radius of the circle the pendulum sweeps out as it swings back and forth. In general, if s represents the arc length in a circle of radius r with central angle θ, $s = r\theta$. Substitute 3 for s, $\frac{1}{2}$ for θ, and solve for r:

$$s = r\theta$$
$$3 = r\left(\frac{1}{2}\right)$$
$$6 = r$$

7. **(C)** At 7 o'clock the hands of a clock form a central angle that is $\frac{5}{12}$ of a complete rotation. Thus, the angle formed is $\frac{5}{12} \times 2\pi = \frac{5\pi}{6}$.

8. **(B)** The circumference of a wheel with a radius of 18 inches is $2\pi \times 18 = 36\pi$ inches. If it rotates through an angle of $\frac{2\pi}{5} = 0.4\pi$, then it is rotating $\frac{0.4\pi}{2\pi} = 0.2$ of a complete rotation. Hence, the distance the wheel travels is $0.2 \times 36\pi = 7.2\pi \approx 22.6$, which is closest to 23.

9. **(B)** If the radius of a round pizza is 14 inches, then its circumference is 28π. Since 0.35 radians is $\frac{0.35}{2\pi}$ of the entire circle, the length of the rounded edge of the crust is

$$\frac{0.35}{2\pi} \times 28\pi = 0.35 \times 14 = 4.9 \text{ inches}$$

10. **(B)** Consider each statement in turn given that angles x and y are acute and $\sin x = \cos y$.
    - I. The measures of angles x and y add up to 90, but x does not necessarily equal y. Statement I is false. ✗
    - II. If $2(x + y) = \pi$, then $x + y = \frac{\pi}{2}$, which must be the case since $\sin x = \cos y$. Statement II is true. ✓
    - III. Since angles x and y are complementary, the cosine of either angle is equal to the sine of the other angle. Statement III is true. ✓

    Only Statements II and III are true.

11. **(D)** Cosine and tangent are both negative in Quadrant II while sine is positive in Quadrant II. If $\cos\theta = -\frac{3}{4} = \frac{x}{r}$, then

$$x^2 + y^2 = r^2$$
$$9 + y^2 = 16$$
$$y = \pm\sqrt{7}$$

Use $y = \sqrt{7}$ since y is positive in Quadrant II. Hence, $\sin\theta = \frac{y}{r} = \frac{\sqrt{7}}{4}$.

12. **(A)** In addition to being positive in Quadrant I, cosine is positive in Quadrant IV. If $\cos A = \frac{4}{5} = \frac{x}{r}$, then $x = 4$, $y = -3$ (since $y < 0$ in Quadrant IV), and $r = 5$. Hence,

$$\sin A = \frac{y}{r} = \frac{-3}{5} = -0.6.$$

13. **(D)** Simplify the product and then substitute b for $\sin A$:

$$\sin A \cdot \cos A \cdot \tan A = \sin A \cdot \cancel{\cos A} \cdot \frac{\sin A}{\cancel{\cos A}}$$
$$= (\sin A) \cdot (\sin A)$$
$$= (b)(b)$$
$$= b^2$$

14. **(C)** The equatorial radius of the Earth is 4,000 miles so the radius of the circular orbit is $4,000 + 1,600 = 5,600$ miles and the circumference of the circular path of the orbit is $2\pi \times 5,600 = 11,200\pi$. Since the satellite completes one orbit every 5 hours, in one hour it completes $\frac{1}{5}$ of an orbit and travels a distance of $\frac{1}{5} \times 11,200\pi = 2,240\pi$.

15. **(D)** In one revolution, a rod with a 6-inch radius rotates a distance of $2\pi \times 6 = 12\pi$, and in a one minute it rotates 165 revolutions for a distance of $12\pi \times 165$. In one second, the rod travels a distance of

$$\frac{\cancel{12}\pi \times 165}{\cancelto{5}{60}} = 33\pi$$

Grid-In

1. **(7/17)** Since the base angles of the triangle are equal in measure, the triangle is isosceles. An altitude drawn from B to $\overline{AC}$ bisects the base thereby forming an 8-15-17 right triangle with the side opposite $\angle A$ having a length of 15 and the side adjacent to $\angle A$ having a length of 8. Hence,

$$\sin A = \frac{15}{17} \text{ and } \cos A = \frac{8}{17} \text{ so } \sin A - \cos A = \frac{7}{17}$$

2. **(589)**

$$\text{Area of lawn} = \frac{\frac{5\pi}{3}}{2\pi} \times \pi(15)^2$$
$$= \frac{5\pi}{6} \times 225$$
$$\approx 589.05$$

To the nearest square foot, 589 square feet of the lawn get sprayed.

3. **(.923)** Draw a line segment from D and perpendicular to $\overline{AC}$, intersecting $\overline{AC}$ at F. Hence $AF = 15 - 10 = 5$ and $DF = DE = 12$.

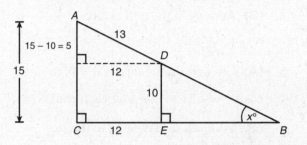

The lengths of the sides of right triangle AFD are lengths of a 5-12-**13** right triangle in hypotenuse $AD = 13$. Hence,

$\sin A \approx \frac{12}{13} = 0.923$. Angles A and B are complementary, so $\cos x = \sin A = 0.923$.

4. **(.8)** If $\tan x = 0.75$, then $\frac{12}{AB} \approx 0.75$ so

$AB = \frac{12}{0.75} = 16$. The lengths of the sides of right triangle ABC are lengths of a 3-4-**5** right triangle in which the lengths are multiplied by 4 so the length of hypotenuse $AC = 4 \times 5 = 20$. Then $\cos x = \frac{16}{20} = 0.8$.

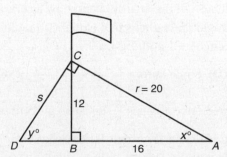

5. **(35)** Since angles x and y are complementary, $\cos x = \sin y$. In right triangle DBC, $\sin y = \dfrac{12}{s} = 0.8$ so $s = \dfrac{12}{0.8} = 15$. Hence, $r + s = 20 + 15 = 35$.

LESSON 6-6
Multiple-Choice

1. **(D)** The maximum value of y occurs when $\sin 13x$ has its maximum value which is 1. Hence, the maximum altitude is $y = 27(1) + 30 = 57$ meters.

2. **(B)** Working in the unit circle,
$$\sin w = \frac{CD}{OC} = \frac{CD}{1} = CD.$$

3. **(A)** Use a test value for x, say $60°$, so $\cos 60° = \dfrac{1}{2}$. Then consider each choice in turn until you find the one that is not equivalent to $\cos x$.

- $-\cos(-60°) = -(\cos 300°) = -\left(\dfrac{1}{2}\right) = -\dfrac{1}{2}.$ ✗

  You should also recall that, in general, $\cos(-x) = \cos x$.

- $\sin\left(\dfrac{\pi}{2} - 60\right) = \sin 30° = \cos 60° = \dfrac{1}{2}.$ ✓

- $-\cos(60 + 180)° = -\cos 240° = -\left(-\dfrac{1}{2}\right) = \dfrac{1}{2}.$ ✓

- $\cos(60 - 360)° = \cos(-300)° = \cos 60° = \dfrac{1}{2}.$ ✓

4. **(A)** Since $\tan\theta = \dfrac{y}{x} = \dfrac{\sqrt{3}}{2} \div \left(-\dfrac{1}{2}\right) = -\sqrt{3}$, the reference angle is $\dfrac{\pi}{3}$ so $\theta = \pi - \dfrac{\pi}{3} = \dfrac{2}{3}\pi$.

5. **(B)** Use the relationship $s = r\theta$ where s is the length of the arc intercepted by a central angle of θ radians in a circle with radius r. Since $r = 1$ and $\theta = 4$, $s = 4(1) = 4$.

6. **(D)** Since $\tan\theta = \dfrac{y}{x} = \left(-\dfrac{1}{2}\right) \div \left(\dfrac{\sqrt{3}}{2}\right) = \dfrac{-1}{\sqrt{3}}$, the reference angle is $\dfrac{\pi}{6}$. Because tangent is positive in Quadrant II but negative in Quadrant IV, θ must be in Quadrant IV. Hence, $\theta = 2\pi - \dfrac{\pi}{6} = \dfrac{11\pi}{6}$.

7. **(A)** After a clockwise rotation of $\dfrac{\pi}{6}$ radians, the angle is located in Quadrant IV where y is negative. Since $\tan\dfrac{\pi}{6} = \dfrac{y}{x} = \dfrac{1}{\sqrt{3}}$, the coordinates of image point must produce that ratio. Review the answer choices and choose the answer in which y is negative and $\dfrac{y}{x} = \dfrac{-1}{\sqrt{3}}$, as in choice (A):
$$\frac{y}{x} = \left(\frac{-1}{2}\right) \div \left(\frac{\sqrt{3}}{2}\right) = \frac{-1}{\sqrt{3}}$$

8. **(B)** After a counterclockwise rotation of $\dfrac{3\pi}{4}$ radians, the angle is located in Quadrant II where x is negative and y is negative. Since the reference angle is $\dfrac{\pi}{4}$, the x and y values of the coordinates must have the form $(-k, k)$. Reviewing the answer choices shows that choices (B) and (D) satisfy this condition. Since this is a unit circle, it must also be the case that $x^2 + y^2 = 1$, which eliminates choice (D).

9. **(B)** Since $\sin^2 x + \cos^2 x = 1$, $\sin^2 x = 1 - \cos^2 x$. Hence,
$$\frac{\sin^2 x}{1 + \cos x} = \frac{1 - \cos^2 x}{1 + \cos x}$$
$$= \frac{(1 - \cos x)\,(1 + \cos x)}{1 + \cos x}$$
$$= 1 - \cos x$$

10. **(A)** Since
$$\tan\left(-\frac{\pi}{3}\right) = -\tan\left(\frac{\pi}{3}\right) = \frac{\sqrt{3}}{1} = \frac{\dfrac{-\sqrt{3}}{2}}{\dfrac{1}{2}} = \frac{y}{x}$$
so $y = -\dfrac{\sqrt{3}}{2}$. It would not be correct to stop at $\tan\left(-\dfrac{\pi}{3}\right) = \dfrac{\sqrt{3}}{1} = \dfrac{y}{x}$ since x and y must also satisfy the condition that $x^2 + y^2 = 1$.

11. **(C)** If x is a positive acute angle and $\cos x = a$, then the side adjacent to angle x is a units long, and the hypotenuse is 1 unit long:

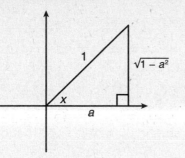

Use the Pythagorean theorem to express the length of the remaining side of the right triangle in terms of a:

$$a^2 + y^2 = 1$$
$$y^2 = 1 - a^2$$
$$y = \pm\sqrt{1 - a^2}$$

Since y is positive in Quadrant I, $y = \sqrt{1 - a^2}$ so

$$\tan x = \frac{y}{a} = \frac{\sqrt{1 - a^2}}{a}$$

TAKING PRACTICE TESTS

Practice Tests 1 and 2

This section provides two full-length practice SAT Math tests. Each test consists of two timed math sections that should be taken under testlike conditions, using the sample SAT answer forms at the beginning of each test. At the end of each practice test you will find an Answer Key, as well as detailed solutions for all questions.

ANSWER SHEET
Practice Test 1

Math (No Calculator)

1. Ⓐ Ⓑ Ⓒ Ⓓ   5. Ⓐ Ⓑ Ⓒ Ⓓ   9. Ⓐ Ⓑ Ⓒ Ⓓ   13. Ⓐ Ⓑ Ⓒ Ⓓ
2. Ⓐ Ⓑ Ⓒ Ⓓ   6. Ⓐ Ⓑ Ⓒ Ⓓ   10. Ⓐ Ⓑ Ⓒ Ⓓ   14. Ⓐ Ⓑ Ⓒ Ⓓ
3. Ⓐ Ⓑ Ⓒ Ⓓ   7. Ⓐ Ⓑ Ⓒ Ⓓ   11. Ⓐ Ⓑ Ⓒ Ⓓ   15. Ⓐ Ⓑ Ⓒ Ⓓ
4. Ⓐ Ⓑ Ⓒ Ⓓ   8. Ⓐ Ⓑ Ⓒ Ⓓ   12. Ⓐ Ⓑ Ⓒ Ⓓ

16.

17.

18.

19.

20.

25 MINUTES, 20 QUESTIONS

Turn to Section 3 of your answer sheet to answer the questions in this section.

Directions: For questions 1–15, solve each problem and choose the best answer from the given options. Fill in the corresponding circle on your answer document. For questions 16–20, solve the problem and fill in the answer on the answer sheet grid.

Notes:

- Calculators are **NOT PERMITTED** in this section.
- All variables and expressions represent real numbers unless indicated otherwise.
- All figures are drawn to scale unless indicated otherwise.
- All figures are in a plane unless indicated otherwise.
- Unless indicated otherwise, the domain of a given function is the set of all real numbers x for which the function has real values.

REFERENCE INFORMATION

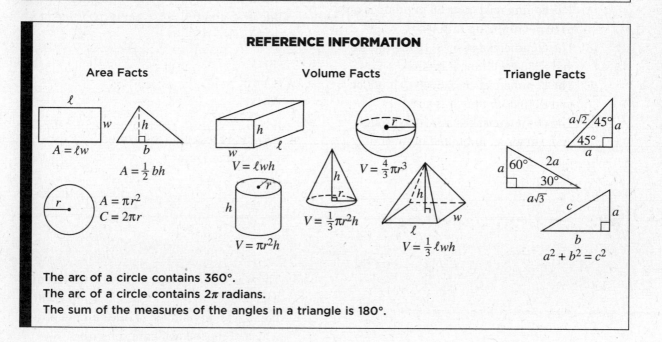

Area Facts

Volume Facts

Triangle Facts

The arc of a circle contains 360°.

The arc of a circle contains 2π radians.

The sum of the measures of the angles in a triangle is 180°.

GO ON TO THE NEXT PAGE

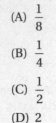

1. If $x^{-2} = 64$, what is the value of $x^{\frac{1}{3}}$?

   (A) $\dfrac{1}{8}$

   (B) $\dfrac{1}{4}$

   (C) $\dfrac{1}{2}$

   (D) 2

$$C(n) = 110n + 900$$

2. The cost of airing a commercial on television, C, is modeled by the function above where n is the number of times the commercial is aired. Based on this model, which statement is true?

   (A) The commercial costs $0 to produce and $110 per airing up to $900.

   (B) The commercial costs $110 to produce and $900 each time it is aired.

   (C) The commercial costs $900 to produce and $110 each time it is aired.

   (D) The commercial costs $110 to produce and can air an unlimited number of times.

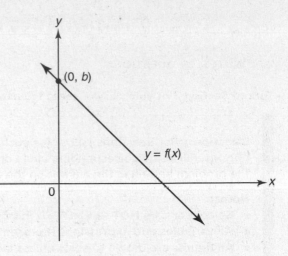

Note: Figure not drawn to scale.

3. The figure above shows the graph of the linear function $y = f(x)$. If the slope of the line is -2 and $f(3) = 4$, what is the value of b?

   (A) 8

   (B) 9

   (C) 10

   (D) 11

4. If $x - 3$ is 1 less than $y + 3$, then $x + 2$ exceeds y by what amount?

   (A) 4

   (B) 5

   (C) 6

   (D) 7

5. The weights of 5 boxes of screws vary from 2.85 pounds to 3.45 pounds. If w represents the weight, in pounds, of one of these boxes, which of the following must be true?

   (A) $|w - 2.85| \le 0.3$

   (B) $|w - 3.15| \le 0.3$

   (C) $|w - 5| \le 0.3$

   (D) $|w - 0.3| \le 3.15$

GO ON TO THE NEXT PAGE

6. Mikala exercises in her gym by jogging on the treadmill at an average rate of 4 miles per hour and then pedaling on a stationary bicycle at an average rate of 8 miles per hour. In her workout, she jogs the equivalent of x miles and bicycles the equivalent of y miles. If Mikala works out for at least 45 minutes, which of the following is true?

(A) $\dfrac{x}{4} + \dfrac{y}{8} \geq \dfrac{3}{4}$

(B) $x + \dfrac{y}{2} \geq \dfrac{3}{4}$

(C) $4x + 8y \geq 45$

(D) $\dfrac{4}{x} + \dfrac{8}{y} \geq 45$

7. If $7^k = 100$, what is the value of $7^{\frac{k}{2}+1}$?

(A) 18
(B) 51
(C) 57
(D) 70

$$3y + 6 = 2x$$

$$2y - 3x = 6$$

8. The system of equations above can best be described as having

(A) no solution.
(B) one solution with the graphs intersecting at right angles in the xy-plane.
(C) one solution with the graphs *not* intersecting at right angles in the xy-plane.
(D) infinitely many solutions.

9. Which of the following statements is true about the parabola whose equation in the xy-plane is $y = (2x - 6)(x + 1)$?

   I. The line $x = 2$ is a vertical line of symmetry.
  II. The minimum value of y is -8.
 III. The y-intercept is -6.

(A) I and III only
(B) II and III only
(C) I and II only
(D) I, II, and III

10. A survey is conducted in which 60% of the individuals who responded indicated that they do *not* support issuing a bond to help raise money to fund the construction of a new sports arena in their city. A statistician calculates the confidence level to be 95% for an interval of 5% below and above the 60% mark. What conclusion is best supported by this information?

(A) 95% of the people surveyed do *not* support the issuing of the bond.
(B) The probability that a person selected at random from the sample does *not* support the issuing of the bond ranges from 0.57 to 0.63.
(C) The probability that a person selected at random from the sample supports the issuing of the bond is 0.4.
(D) If the survey were to be repeated 100 times, 95% of the time the number of people who would *not* support the issuing of the bond would range from 55% to 65% of those surveyed.

GO ON TO THE NEXT PAGE

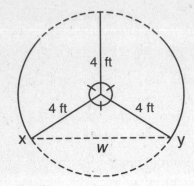

11. The accompanying diagram shows a revolving door with three panels, each of which is 4 feet long. What is the number of feet in the width, w, of the opening between points x and y?

(A) $\dfrac{4}{\sqrt{3}}$

(B) $4\sqrt{3}$

(C) $8\sqrt{2}$

(D) $8\sqrt{3}$

12. Impedance measures the opposition of an electrical circuit to the flow of electricity. The total impedance in a particular circuit is given by the formula $Z_T = \dfrac{Z_1 \cdot Z_2}{Z_1 + Z_2}$. What is the total impedance of a circuit, Z_T, if $Z_1 = 1 + 2i$ and $Z_2 = 1 - 2i$? [Note: $i = \sqrt{-1}$]

(A) $-\dfrac{3}{2}$

(B) $2i$

(C) $\dfrac{1}{2}$

(D) $\dfrac{5}{2}$

13. At 9:00 A.M. Allan began jogging and Bill began walking at constant rates around the same circular $\dfrac{1}{4}$ mile track. The figure above compares their times in minutes and corresponding distances in miles. Which statement or statements must be true?

I. Bill's average rate of walking was 2 miles per hour.

II. At 9:10 A.M., Allan had jogged $\dfrac{3}{5}$ mile more than Bill had walked.

III. At 9:30 A.M., Allan had completed 8 more laps around the track than Bill.

(A) I only

(B) II only

(C) I and II only

(D) I and III only

GO ON TO THE NEXT PAGE

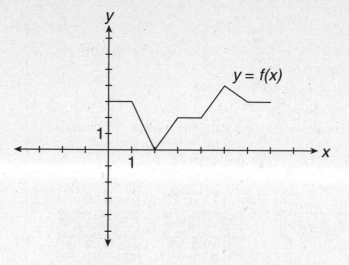

y = f(x)

14. The figure above shows part of the graph of function *f*. If $f(x + 6) = f(x)$ for all values of *x*, what is the value of $f(23)$?

(A) 0
(B) 2
(C) 3
(D) 4

15. Which function could represent the graph above?

(A) $f(x) = (x - 6)(x^2 - 4x + 3)$
(B) $f(x) = (x - 3)(x^2 + x - 2)$
(C) $f(x) = (x - 1)(x^2 - 5x - 6)$
(D) $f(x) = (x + 2)(x^2 - 4x - 12)$

GO ON TO THE NEXT PAGE

Grid-in Response Directions

In questions 16–20, first solve the problem, and then enter your answer on the grid provided on the answer sheet. The instructions for entering your answers follow.

- First, write your answer in the boxes at the top of the grid.
- Second, grid your answer in the columns below the boxes.
- Use the fraction bar in the first row or the decimal point in the second row to enter fractions and decimals.

Write your answer in the boxes

Grid in your answer

Answer: $\frac{8}{15}$

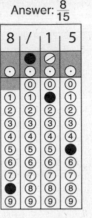

Answer: 1.75

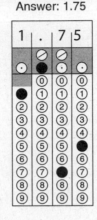

Answer: 100

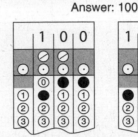

Either position is acceptable

- Grid only one space in each column.
- Entering the answer in the boxes is recommended as an aid in gridding but is not required.
- The machine scoring your exam can read only what you grid, so you **must grid-in your answers correctly to get credit**.
- If a question has more than one correct answer, grid-in only one of them.
- The grid does not have a minus sign; so no answer can be negative.
- A mixed number *must* be converted to an improper fraction or a decimal before it is gridded.

  Enter $1\frac{1}{4}$ as 5/4 or 1.25; the machine will interpret 11/4 as $\frac{11}{4}$ and mark it wrong.

- **All decimals must be entered as accurately as possible.** Here are three acceptable ways of gridding

$$\frac{3}{11} = 0.272727\ldots$$

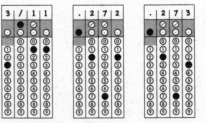

- Note that rounding to .273 is acceptable because you are using the full grid, but you would receive **no credit** for .3 or .27, because they are less accurate.

GO ON TO THE NEXT PAGE

16. On a test that has a normal distribution of scores, a score of 59 falls two standard deviations below the mean, and a score of 74 is one standard deviation above the mean. If x is an integer score that lies between 2.5 and 3.0 standard deviations above the mean, what is a possible value of x?

Hours Worked in a Week	Total Payment
8	$108.00
23	$310.50
17	$229.50

17. Andrew keeps track of his paychecks over the past several weeks, recording the number of hours he worked and his total payments as indicated in the table above. He wants to model the relationship between h hours worked and total payments p, in dollars, using an equation of the form $p = kh$ where k is a constant. Based on the data in the table, what value of k should he use?

18. If $\dfrac{-3}{x} + 4 \le -11$ and $x > 0$, what is the *greatest* possible value for x?

19. The equation of a circle in the xy-plane is $x^2 + 4x + y^2 - 10y = 20$. If the line $x = k$ intersects the circle in exactly one point, what is a possible value of k?

x	1	2	3	4	5
$f(x)$	3	4	5	6	7

x	3	4	5	6	8
$g(x)$	4	6	8	10	7

20. The tables above give the values of functions f and g for several values of x. If $g(f(b)) = 8$, what is the value of b?

GO ON TO THE NEXT PAGE

ANSWER SHEET
Practice Test 1

Math (Calculator)

1. Ⓐ Ⓑ Ⓒ Ⓓ
2. Ⓐ Ⓑ Ⓒ Ⓓ
3. Ⓐ Ⓑ Ⓒ Ⓓ
4. Ⓐ Ⓑ Ⓒ Ⓓ
5. Ⓐ Ⓑ Ⓒ Ⓓ
6. Ⓐ Ⓑ Ⓒ Ⓓ
7. Ⓐ Ⓑ Ⓒ Ⓓ
8. Ⓐ Ⓑ Ⓒ Ⓓ

9. Ⓐ Ⓑ Ⓒ Ⓓ
10. Ⓐ Ⓑ Ⓒ Ⓓ
11. Ⓐ Ⓑ Ⓒ Ⓓ
12. Ⓐ Ⓑ Ⓒ Ⓓ
13. Ⓐ Ⓑ Ⓒ Ⓓ
14. Ⓐ Ⓑ Ⓒ Ⓓ
15. Ⓐ Ⓑ Ⓒ Ⓓ
16. Ⓐ Ⓑ Ⓒ Ⓓ

17. Ⓐ Ⓑ Ⓒ Ⓓ
18. Ⓐ Ⓑ Ⓒ Ⓓ
19. Ⓐ Ⓑ Ⓒ Ⓓ
20. Ⓐ Ⓑ Ⓒ Ⓓ
21. Ⓐ Ⓑ Ⓒ Ⓓ
22. Ⓐ Ⓑ Ⓒ Ⓓ
23. Ⓐ Ⓑ Ⓒ Ⓓ
24. Ⓐ Ⓑ Ⓒ Ⓓ

25. Ⓐ Ⓑ Ⓒ Ⓓ
26. Ⓐ Ⓑ Ⓒ Ⓓ
27. Ⓐ Ⓑ Ⓒ Ⓓ
28. Ⓐ Ⓑ Ⓒ Ⓓ
29. Ⓐ Ⓑ Ⓒ Ⓓ
30. Ⓐ Ⓑ Ⓒ Ⓓ

31. [grid-in answer field]
32. [grid-in answer field]
33. [grid-in answer field]
34. [grid-in answer field]

35. [grid-in answer field]
36. [grid-in answer field]
37. [grid-in answer field]
38. [grid-in answer field]

MATHEMATICS TEST—CALCULATOR

55 MINUTES, 38 QUESTIONS

Turn to Section 4 of your answer sheet to answer the questions in this section.

Directions: For questions 1–30, solve each problem and choose the best answer from the given options. Fill in the corresponding circle on your answer document. For questions 31–38, solve the problem and fill in the answer on the answer sheet grid.

Notes:

- Calculators **ARE PERMITTED** in this section.
- All variables and expressions represent real numbers unless indicated otherwise.
- All figures are drawn to scale unless indicated otherwise.
- All figures are in a plane unless indicated otherwise.
- Unless indicated otherwise, the domain of a given function is the set of all real numbers x for which the function has real values.

REFERENCE INFORMATION

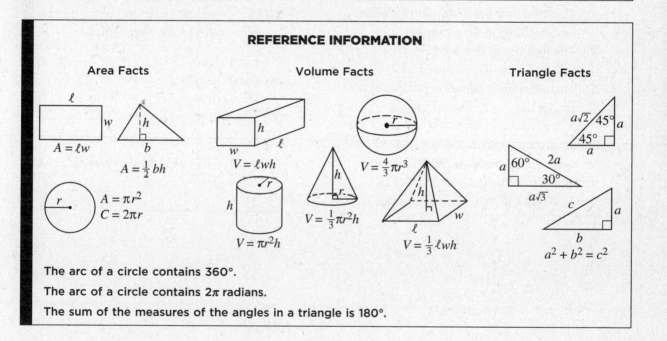

Area Facts

$A = \ell w$

$A = \frac{1}{2} bh$

$A = \pi r^2$

$C = 2\pi r$

Volume Facts

$V = \ell w h$

$V = \pi r^2 h$

$V = \frac{1}{3} \pi r^2 h$

$V = \frac{4}{3} \pi r^3$

$V = \frac{1}{3} \ell w h$

Triangle Facts

$a^2 + b^2 = c^2$

The arc of a circle contains 360°.

The arc of a circle contains 2π radians.

The sum of the measures of the angles in a triangle is 180°.

GO ON TO THE NEXT PAGE

1. If $(2b - 7)(2b + 7) = 1$, what is the value of $2b^2$?

   (A) 15
   (B) 25
   (C) 32
   (D) 50

2. The number of donation pledges, p, made to a charity d days after the charity began a campaign for donations can be approximated by the equation $p = 117 + 32d$. What is the best interpretation of the number 32 in this equation?

   (A) The number of donation pledges received before the campaign for donations started.
   (B) The total number of donation pledges received during the campaign.
   (C) The number of donation pledges received each day of the campaign.
   (D) The number of donation pledges made on the last day of the campaign.

3. A long-distance telephone call costs $1.80 for the first 3 minutes and $0.40 for each additional minute. If the charge for an x-minute long-distance call at this rate was $4.20, then $x =$

   (A) 7
   (B) 8
   (C) 9
   (D) 10

Gender	Type of College				
	4-Year Same State	2-Year Same State	4-Year Out-of-State	None	Total
Male	64	26	22	7	119
Female	41	19	15	6	81
Total	105	45	37	13	200

4. Based on the data in the table above, which of the following statements must be true?

   I. For every 3 men who applied to a same state college, 2 women applied to a same state college.
   II. If a female student is selected at random, the probability that she did not apply to a 2-year college is greater than 75%.
   III. Of the students who applied to a same state college, 40% were females.

   (A) I and II only
   (B) I and III only
   (C) II and III only
   (D) I, II, and III

5. If $3x - 1 = x - \dfrac{7}{9}$, what is the value of $2x + 1$?

   (A) $\dfrac{11}{9}$
   (B) $\dfrac{4}{3}$
   (C) $\dfrac{25}{9}$
   (D) $\dfrac{10}{3}$

GO ON TO THE NEXT PAGE

6. The price of gas increased by 12% per gallon sometime during the first fiscal quarter and then decreased by 25% per gallon by the end of the second fiscal quarter. The final price of gas per gallon at the end of the second quarter decreased by what percent compared to the starting price at the beginning of the first fiscal quarter?

(A) 13%
(B) 16%
(C) 18.5%
(D) 20%

7. A population, $T(x)$, of wild turkeys, in a certain rural area is represented by the function $T(x) = 17(1.15)^{2x}$, where x is the number of years since 2010. According to this model, how many more turkeys are in the population for the year 2015 than were available for 2010?

(A) 46
(B) 49
(C) 51
(D) 68

8. If an equation of a parabola in the xy-plane is $f(x) = -(x + 2)^2 - 1$, what are the coordinates of the vertex of the parabola defined by $g(x) = f(x - 2)$?

(A) (0, −1)
(B) (4, −1)
(C) (−2, −3)
(D) (−2, 1)

9. A city planner estimates that due to lower birth rates and changing demographics, enrollment in city's public schools will decrease at the rate of 16% per year for the next 5 years. If the city planner uses the equation $P = P_0(r)^n$ to estimate the school enrollment, P, after n years, what value should be used for the value of r?

(A) 1.16
(B) 0.84
(C) 0.80
(D) 0.16

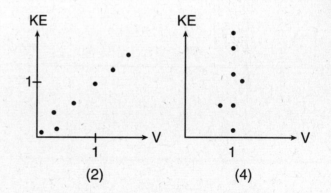

(1)   (3)

(2)   (4)

10. In the physics lab, a student determined the kinetic energy, KE, of an object at various velocities, V, and found a strong positive association between KE and V. Which of the above scatterplots show this relationship?

(A) Graph (1)
(B) Graph (2)
(C) Graph (3)
(D) Graph (4)

GO ON TO THE NEXT PAGE

11. The average (arithmetic mean) of a, b, c, and d is 3 times the median. If $0 < a < b < c < d$, what is a in terms of b, c, and d?

    (A) $5(b + c) - d$
    (B) $3(b + c) + d$
    (C) $5(b + c) + d$
    (D) $3(b + c) - d$

12. A person spent a total of $720 for dress shirts and sport shirts, each priced at $35 and $20, respectively. If the person purchased two $35 dress shirts for each $20 sport shirt, what is the total number of shirts purchased?

    (A) 16
    (B) 21
    (C) 24
    (D) 28

13. If 10 cubic centimeters of blood contains 1.2 grams of hemoglobin, how many grams of hemoglobin are contained in 35 cubic centimeters of the same blood?

    (A) 2.7
    (B) 3.0
    (C) 3.6
    (D) 4.2

Players' Salaries (in millions of dollars)					
0.5	0.5	0.6	0.7	0.75	0.8
1.0	1.0	1.1	1.25	1.3	1.4
1.6	1.8	2.5	3.7	3.8	4.0
4.2	4.6	5.1	6.0	6.3	7.2
Total = 61.7 Million					

14. The table above shows the annual salaries for the 24 members of a professional sports team in terms of millions of dollars. If the team signs an additional player to a contract worth 7.3 million dollars per year, which statement about the median and mean is true?

    (A) The median and mean will increase by the same amount.
    (B) The median will increase by a greater amount.
    (C) The mean will increase by a greater amount.
    (D) Neither will change.

$$m = \frac{M}{\sqrt{1 - \dfrac{v^2}{c^2}}}$$

15. The equation above describes, according to Einstein's theory of relativity, how the mass of an object increases with velocity where m is the mass of a moving object, M is the mass the object when it is not moving, v is the velocity of the object relative to a stationary observer, and c is the speed of light. Which of the following expresses v in terms of m, M, and c?

    (A) $c\sqrt{1 - \left(\dfrac{M}{m}\right)^2}$

    (B) $c\sqrt{1 + \left(\dfrac{M}{m}\right)^2}$

    (C) $\sqrt{c^2 + \left(\dfrac{M}{m}\right)^2}$

    (D) $\sqrt{\left(c + \dfrac{M}{m}\right)^2 - 1}$

GO ON TO THE NEXT PAGE

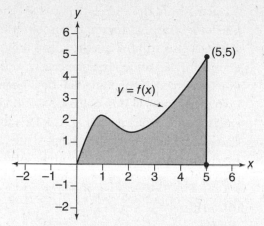

16. Function f is defined for $0 \le x \le 5$, as shown in the accompanying figure. If (r, s) is a point inside the shaded region bounded by the x-axis, the line $x = 5$, and $y = f(x)$, which statement must be true?

    I. $r + s \le 5$
    II. $s \le f(r)$
    III. $r \ne s$

(A) I only
(B) II only
(C) III only
(D) I and III only

17. Natalie is planning a school celebration and wants to have live music and food for everyone who attends. She has found a band that will charge her $750 and a caterer who will provide snacks and drinks for $2.25 per person. If her goal is to keep the average cost per person between $2.75 and $3.25, how many people, p, must attend?

(A) $225 < p < 325$
(B) $325 < p < 750$
(C) $500 < p < 1,000$
(D) $750 < p < 1,500$

18. If $p(x)$ is a polynomial function with $p(3) = 0$, which statement must be true?

(A) $p(x)$ is divisible by 3.
(B) $x - 3$ is a factor of $p(x)$.
(C) $p(x)$ is divisible by $x + 3$.
(D) The highest power of x in $p(x)$ is 3.

19. A group of p people plan to contribute equally to the purchase of a gift that costs d dollars. If n of the p people decide not to contribute, by what amount in dollars does the contribution needed from each of the remaining people increase?

(A) $\dfrac{d}{p - n}$

(B) $\dfrac{pd}{p - n}$

(C) $\dfrac{pd}{n(p - n)}$

(D) $\dfrac{nd}{p(p - n)}$

20. Which of the following statements includes a function divisible by $2x + 1$?

    I. $f(x) = 8x^2 - 2$
    II. $g(x) = 2x^2 - 9x + 4$
    III. $h(x) = 4x^3 + 2x^2 - 6x - 3$

(A) I only
(B) I and II only
(C) I and III only
(D) I, II, and III

21. When Sophie was born her parents invested a sum of $20,000 in her college fund. They invested it at a nominal annual rate of 5% with interest compounded quarterly. Which equation could be used to find the number of dollars, y, in the account, after 18 years assuming no other deposits or withdrawals are made?

(A) $y = 20,000(1.05)^{18}$
(B) $y = 20,000(0.21)^{18 \times 4}$
(C) $y = 20,000(1.0125)^{\frac{18}{4}}$
(D) $y = 20,000(1.0125)^{18 \times 4}$

GO ON TO THE NEXT PAGE

22. If function g is defined by g(x) = x − 1 and
    2g(c) = 10, what is the value of g(3c)?

    (A) 6
    (B) 9
    (C) 15
    (D) 17

Population

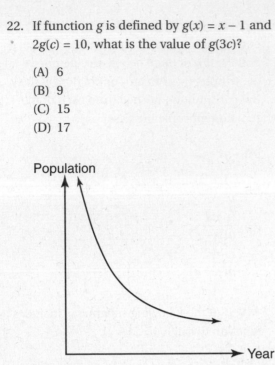

→ Year

23. The graph above shows how the size of a
    country's population has changed over time.
    Which of the following are the most likely
    underlying reasons for the type of graph
    shown?

    (A) A moderate increase in annual birthrates
        and a liberal immigration policy.
    (B) A large increase in annual birthrates and
        increased life expectancy rates.
    (C) A liberal immigration policy and a
        thriving economy with increased job
        opportunities.
    (D) The spread of a highly contagious fatal
        disease and a history of political strife
        and unrest.

24. A teacher from the United States wishes to
    purchase textbooks for her classroom when
    she goes on a trip to Canada, where they are
    on sale for 45 Canadian dollars each. At the
    time of purchase one Canadian dollar can be
    exchanged for 0.76 U.S. dollars. Assuming
    she is able to exchange her U.S. dollars for
    Canadian dollars at no cost, what is the
    exact cost, in U.S. dollars, to purchase 30
    books?

    (A) $849
    (B) $1026
    (C) $1350
    (D) $1776

Age (years)	Average Pupil Diameter (mm)
20	4.7
40	3.9
60	3.1
80	2.3

25. The table above shows the average diameter,
    in millimeters, of a pupil in a person's eye as
    she or he grows older from age 20 to age 80.
    Which equation expresses the relationship
    between pupil diameter, p, and age, a?

    (A) $p = -0.04a + 5.5$
    (B) $p = 0.04a + 3.9$
    (C) $p = -0.04a + 34.3$
    (D) $p = 0.235a$

GO ON TO THE NEXT PAGE

26. A small, open-top packing box, similar to a shoebox without a lid, is three times as long as it is wide, and half as high as it is long. Each square inch of the bottom of the box costs $0.08 to produce, while each square inch of any side costs $0.03 to produce. If x represents the number of inches in the width of the box, which of the following functions represent the cost, C, of producing the box?

(A) $C(x) = 0.42x^2$
(B) $C(x) = 0.60x^2$
(C) $C(x) = 0.72x^2$
(D) $C(x) = 0.96x^2$

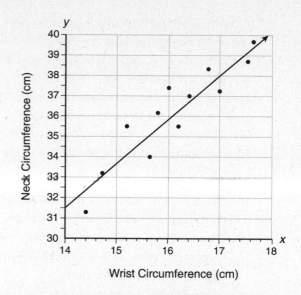

Wrist Circumference (cm)

27. The scatterplot above summarizes the wrist and neck circumference measurements, in centimeters, for 12 people. The line of best fit is drawn. What proportion of the 12 measurements satisfy the inequality $|o - p| \leq d$ where o is the observed measurement, p is corresponding measurement predicted by the line of best fit, and d is 0.5 cm?

(A) $\dfrac{1}{6}$

(B) $\dfrac{1}{4}$

(C) $\dfrac{1}{3}$

(D) $\dfrac{1}{2}$

28. An arch is built so that it has the shape of a parabola with the equation $y = -3x^2 + 24x$ where y represents the height of the arch in meters. How many times greater is the maximum height of the arch than the width of the arch at its base?

(A) 4
(B) 6
(C) 8
(D) 12

GO ON TO THE NEXT PAGE

29. A political strategist wants to conduct a survey to determine how the likely voters in a given state of 10,000,000 people feel about a politician's stand on an infrastructure spending plan. The strategist has a budget to make phone calls to 1,000 people. What would be the most effective approach for him to minimize the margin of error in his survey results?

(A) Place calls to randomly selected phone numbers of residents within the state.
(B) Place calls to residents of the state's largest city who have indicated they are members of a political party.
(C) Place calls to rural residents of the state who have demonstrated political activism.
(D) Place calls to places of business so that people can more likely be reached during the work day.

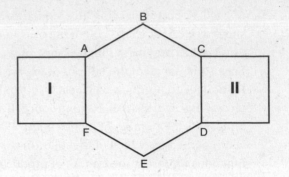

30. A metal belt buckle is being designed so that it has the shape of a regular hexagon in the center and squares at opposite ends as shown in the figure above where *ABCDEF* is a regular hexagon and figures I and II are squares. The hexagon will be gold plated and the two squares silver plated. The length of a side of each square is 6 centimeters. Which of the following is closest to the percent of the total surface area of the buckle that will be silver plated?

(A) 41
(B) 44
(C) 47
(D) 49

GO ON TO THE NEXT PAGE

Grid-in Response Directions

In questions 31–38, first solve the problem, and then enter your answer on the grid provided on the answer sheet. The instructions for entering your answers follow.

- First, write your answer in the boxes at the top of the grid.
- Second, grid your answer in the columns below the boxes.
- Use the fraction bar in the first row or the decimal point in the second row to enter fractions and decimals.

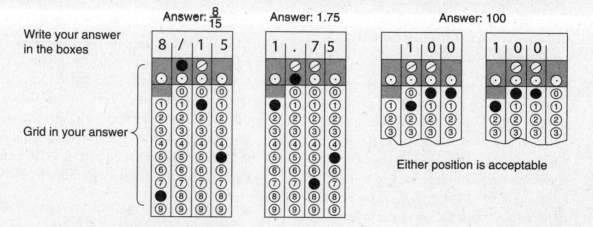

Write your answer in the boxes

Grid in your answer

Answer: $\frac{8}{15}$ Answer: 1.75   Answer: 100   Either position is acceptable

- Grid only one space in each column.
- Entering the answer in the boxes is recommended as an aid in gridding but is not required.
- The machine scoring your exam can read only what you grid, so you **must grid-in your answers correctly to get credit**.
- If a question has more than one correct answer, grid-in only one of them.
- The grid does not have a minus sign; so no answer can be negative.
- A mixed number *must* be converted to an improper fraction or a decimal before it is gridded.

  Enter $1\frac{1}{4}$ as $\frac{5}{4}$ or 1.25; the machine will interpret 11/4 as $\frac{11}{4}$ and mark it wrong.

- **All decimals must be entered as accurately as possible.** Here are three acceptable ways of gridding

$$\frac{3}{11} = 0.272727\ldots$$

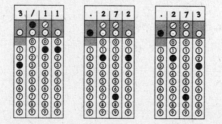

- Note that rounding to .273 is acceptable because you are using the full grid, but you would receive **no credit** for .3 or .27, because they are less accurate.

GO ON TO THE NEXT PAGE

| 3 teaspoons = 1 tablespoon |
| 16 tablespoons = 1 cup |
| 1 cup = 8 ounces |
| 29.6 milliliters = 1 ounce |

31. Using the conversion relationships above, what is the maximum number of 2-teaspoon doses of cough medicine that can be dispensed from a bottle that contains 225 milliliters of cough medicine?

32. NASA's *New Horizons* interplanetary probe has been making its way to Pluto since January 2006. In July 2015, it reached Pluto and sent a radio transmission signal at a speed of 1.86×10^5 miles per second. If the signal traveled a distance back to Earth of approximately 3.06×10^9 miles, how many minutes did it take for the signal to reach Earth, *correct to the nearest 5 minutes?*

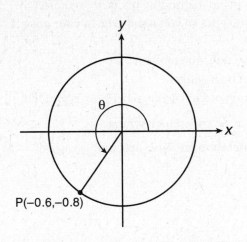

P(−0.6,−0.8)

33. If $P(-0.6, -0.8)$ is a point on the unit circle in the figure above, what is the *exact* value of $\tan\theta + \sin\theta$?

34. If $a + 2b = 13$ and $8a + b = 20$, what is value of $3a + b$?

35. An opinion poll survey was conducted in which 120 sports fans and 75 non-sports fans participated. If the sample size was increased by 65 non-sports fans, how many sports fans should be added so that $\frac{3}{5}$ of those polled are sports fans?

36. The Eye Surgery Institute just purchased a new laser machine for $500,000 to use during eye surgery. The Institute must pay the inventor $550 each time the machine is used. If the Institute charges $2,000 for each laser surgery, what is the *minimum* number of surgeries that must be performed in order for the Institute to make a profit?

GO ON TO THE NEXT PAGE

Questions 37 and 38 refer to the following information

The U.S. Federal Government tracks the Consumer Price Index (CPI)—a comprehensive standard used to estimate the average price change for the typical goods and services purchased by consumers. This measure gives economists a useful way to estimate the rates of inflation or deflation, which reflects the respective general increase or decrease of prices of goods and service in the economy. The accompanying tables summarizes the changes in the CPI for the years 2005 through 2014, which can be assumed to be the corresponding percent rates of inflation.

Yearly Percent Change in Urban Consumer Price Index in the United States

Year	Annual	First Half of Year	Second Half of Year
2005	3.4	3.0	3.8
2006	3.2	3.8	2.6
2007	2.8	2.5	3.1
2008	3.8	4.2	3.4
2009	−0.4	−0.6	−0.1
2010	1.6	2.1	1.2
2011	3.2	2.8	3.5
2012	2.1	2.3	1.8
2013	1.5	1.5	1.4
2014	1.6	1.7	1.5

Source: United States Bureau of Labor and Statistics

37. An economist purchases a kitchen appliance at the beginning of 2014 for $3,000. The salesperson advises him that the only changes in price for the appliance since the beginning of 2012 have been due to inflation. Assuming that is the case, what would have been the purchase price for the appliance at the beginning of 2012 *correct to the nearest dollar*?

38. At the beginning of 2015, a retired person is shopping for a retirement annuity, which is an investment policy that will give him fixed monthly payments for the rest of his life. He would like the amount of his annuity payments to more than keep up with the rate of inflation. He decides that he will choose a policy that issues payments that increase annually at a rate that is at least 1.5% greater than the *average* yearly compounded rate of inflation calculated from the period that extends from the second half of 2005 through the first half of 2008. What should be the minimum annual rate of increase in his monthly annuity payments, *correct to the nearest tenth*?

GO ON TO THE NEXT PAGE

Answer Key
For Practice Test 1

Math (No Calculator)

1. **C**	5. **B**	9. **B**	13. **D**
2. **C**	6. **A**	10. **D**	14. **D**
3. **C**	7. **D**	11. **B**	15. **B**
4. **D**	8. **C**	12. **D**	

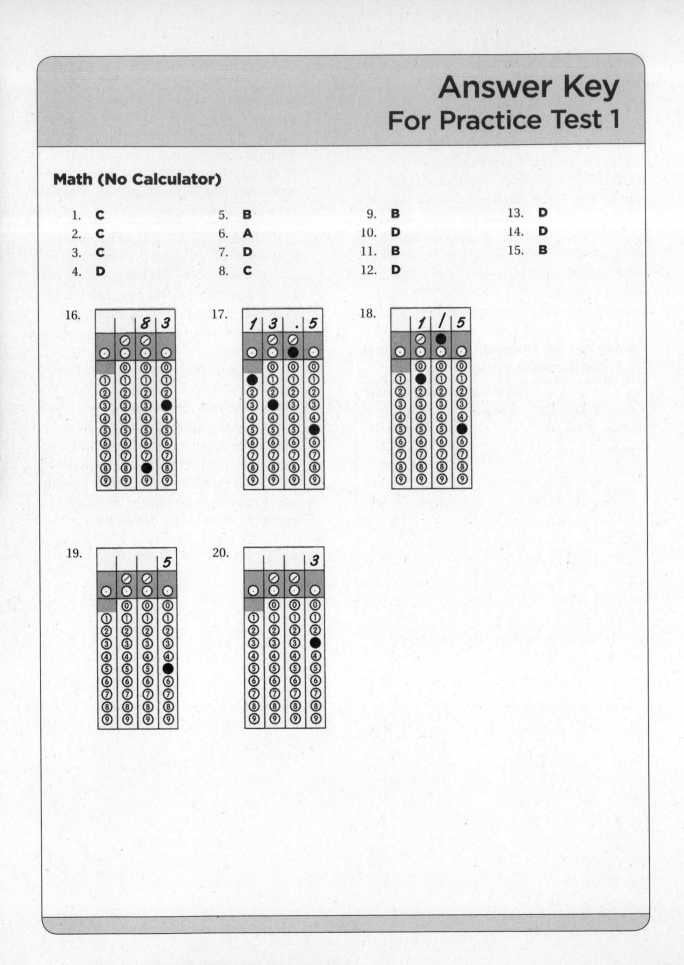

16. 83
17. 13.5
18. 1/5
19. 5
20. 3

Answer Key
For Practice Test 1

Math (Calculator)

1. **B**	9. **B**	17. **D**	25. **A**
2. **C**	10. **B**	18. **B**	26. **B**
3. **C**	11. **A**	19. **D**	27. **C**
4. **D**	12. **C**	20. **C**	28. **B**
5. **A**	13. **D**	21. **D**	29. **A**
6. **B**	14. **C**	22. **D**	30. **B**
7. **C**	15. **A**	23. **D**	
8. **A**	16. **B**	24. **B**	

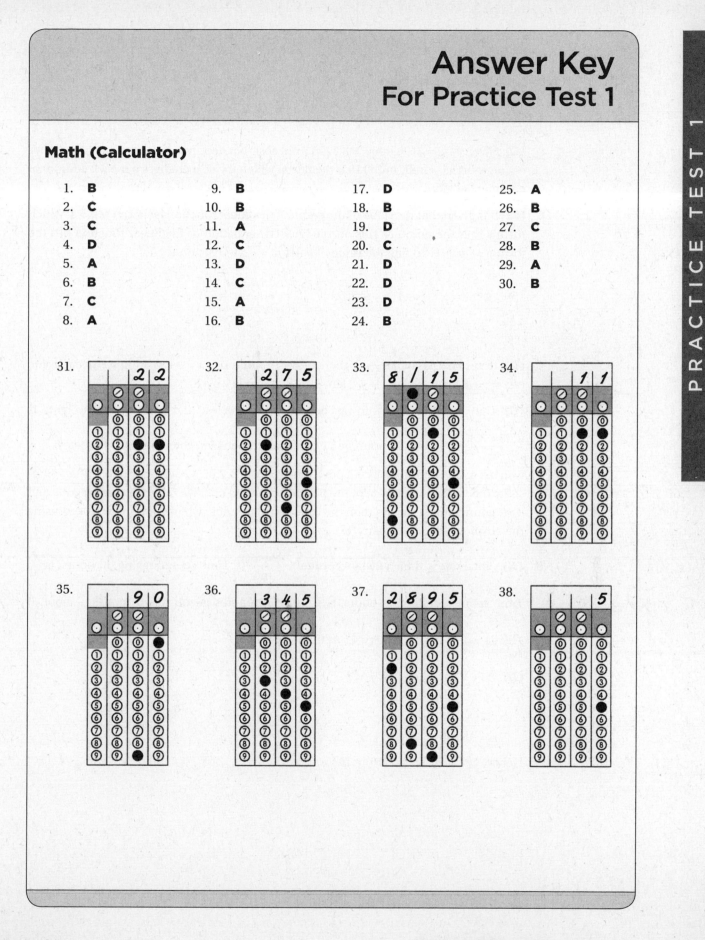

No-Calculator Section

1. **(C)** If $x^{-2} = 64$, then $\dfrac{1}{x^2} = 64$ so $x = \sqrt{\dfrac{1}{64}} = \dfrac{1}{8}$. Then $x^{\frac{1}{3}} = \left(\dfrac{1}{8}\right)^{\frac{1}{3}} = \dfrac{1}{\sqrt[3]{8}} = \dfrac{1}{2}$.

2. **(C)** Given $C(n) = 110n + 900$, $900 is a fixed cost that does not depend on n, the number of commercials aired, and $110 is the rate at which the cost changes per each additional commercial aired.

3. **(C)** It is given that the slope of the graph of the linear function is -2 and $f(3) = 4$, which means that the point $(3, 4)$ is on the line. The equation of the linear function has the form $y = mx + b$. To find the value of b, let $m = -2$, $x = 3$, and $y = 4$:

$$4 = -2(3) + b$$
$$4 = -6 + b$$
$$10 = b$$

4. **(D)** If $x - 3$ is 1 less than $y + 3$, then $x - 3 = (y + 3) - 1$ so $x = y + 5$. Adding 2 to each side of $x = y + 5$ gives $x + 2 = y + 7$. Hence, $x + 2$ exceeds y by 7.

5. **(B)** The weight that is midway between the smallest and greatest box weights is

$\dfrac{2.85 + 3.45}{2} = \dfrac{6.3}{2} = 3.15$ pounds. The difference between the smallest or greatest box weights and 3.15 pounds is 0.3 pounds since $3.15 - 2.85 = 3.45 - 3.15 = 0.3$. If w represents the weight of any one of the 5 boxes, then the positive difference between w and 3.15 pounds must be less than or equal to 0.3 pounds, which can be expressed using the absolute value inequality $|w - 3.15| \le 0.3$.

6. **(A)** Since rate $\times$ time = distance, time $= \dfrac{\text{distance}}{\text{rate}}$. Time spent jogging, in hours, is $\dfrac{x}{4}$.

Time spent bicycling, in hours, is $\dfrac{y}{8}$ and 45 minutes is equivalent to $\dfrac{45}{60} = \dfrac{3}{4}$ hour. If

Mikala works out for *at least* 45 minutes, then $\dfrac{x}{4} + \dfrac{y}{8} \ge \dfrac{3}{4}$.

7. **(D)** If $7^k = 100$, then

$$\left(7^k\right)^{\frac{1}{2}} = (100)^{\frac{1}{2}}$$
$$7^{\frac{k}{2}} = \sqrt{100} = 10$$

Hence, $7^{\frac{k}{2}+1} = 7^1 \cdot 7^{\frac{k}{2}} = 7(10) = 70$.

8. **(C)** Compare the given equations in slope-intercept form:

- If $3y + 6 = 2x$, then $y = \dfrac{2}{3}x - 2$.

- If $2y - 3x = 6$, then $y = \dfrac{3}{2}x + 3$.

- Since the lines have slopes that are different, the graphs of the lines intersect so there is one solution. The slopes are not negative reciprocals, so the lines do not intersect at right angles in the xy-plane.

9. **(B)** If $y = (2x - 6)(x + 1) = 2x^2 - 4x - 6$. The y-intercept of the parabola is -6 since when $x = 0$, $y = -6$. Hence, Statement III is true. Write the parabola equation in vertex form by completing the square:

$$y = 2x^2 - 4x - 6$$
$$= 2(x^2 - 2x + 1) - 6 - 2$$
$$= 2(x - 1)^2 - 8$$

The vertex of the parabola is $(1, -8)$ so the $x = 1$ is a vertical line of symmetry and -8 is the minimum value of y, which means Statement I is false and Statement II is true. Hence, only Statements II and III are true.

10. **(D)** In general, a probability level of $p\%$ associated with a confidence interval does *not* give the probability that the sample statistic is in the interval of a particular sample. Instead, it states that if a large number of samples are repeatedly drawn from a population, then the sample statistic will be in the stated interval of values $p\%$ of the time. In this problem, the range of the interval is $60\% \pm 5\%$. Hence, if the survey were to be repeated 100 times, 95% of the time the number of people who would *not* support the issuing of the bond would fall between 55% and 65% of those surveyed.

11. **(B)** The central angle formed measures $\dfrac{360}{3} = 120$. Draw perpendicular $\overline{PQ}$ from the center of the circle to $\overline{XY}$, which bisects the angle and $\overline{XY}$:

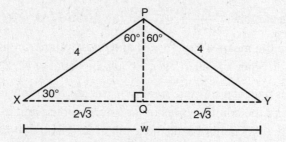

In a right triangle, the length of the side opposite the 60° angle is one-half of the hypotenuse. Hence, $XQ = 2\sqrt{3}$ so $w = 2\sqrt{3} + 2\sqrt{3} = 4\sqrt{3}$.

12. **(D)** $Z_T = \dfrac{(1 + 2i)(1 - 2i)}{(1 + 2i) + (1 - 2i)} = \dfrac{1 - 4i^2}{2} = \dfrac{1 - 4(-1)}{2} = \dfrac{5}{2}$.

13. **(D)** Determine whether each Roman numeral choice is true or false:

- I. From the graph, Bill walks 1 mile in 30 minutes. Since he is walking at a constant rate, he walks 2 miles in 60 minutes. Hence, Bill's average rate of walking was 2 miles per hour. This choice is correct.
- II. At 9:10 A.M., Allan jogged 1 mile. Since Bill was walking at a constant rate of 2 miles per hour, he walked

$$2 \text{ miles} \times \frac{1}{6} \text{ hr} (= 10 \text{ min}) = \frac{1}{3} \text{ mile}.$$

  Hence, at 9:10 A.M. Allan had jogged $1 - \dfrac{1}{3} = \dfrac{2}{3}$ mile more than Bill had walked. This choice is not correct.

- III. At 9:30 A.M., Allan had jogged 3 miles and Bill had walked 1 mile. Hence, Allan jogged 2 miles more than Bill had walked. Four laps around a $\dfrac{1}{4}$-mile track equals 1 mile. Hence, Allan completed 8 more laps around the track than Bill. This choice is correct.

  Since Roman numeral choices I and III are correct, the answer is choice (D).

14. **(D)** Since $f(x + 6) = f(x)$, $f(23) = f(17) = f(11) = f(5)$. Reading from the graph, $f(5) = 4$.

15. **(B)** Since the x-intercepts of the graph of a function correspond to its x-intercepts,

$$\begin{aligned}
f(x) &= (x - (-2))(x - 1)(x - 3) \\
&= (x + 2)(x - 1)(x - 3) \\
&= (x - 3)(x + 2)(x - 1) \\
&= (x - 3)(x^2 + x - 2)
\end{aligned}$$

16. **(83)** The difference between 74 and 59 represents a width of 3 standard deviations so 1 standard deviation is $\dfrac{74 - 59}{3} = 5$ and the mean is $59 + 10 = 69$. Three standard deviations above the mean is $69 + 15 = 84$ and 2.5 standard deviations above the mean is $69 + 12.5 = 81.5$. Since x is an integer, x could be equal to 82, 83, or 84.

17. **(13.5)** Andrew is paid at a constant rate, so to calculate the constant of proportionality, simply take one of the combinations of his hours worked and his total payment. Let's take 8 hours and a payment of \$108. Plug 108 in for p and 8 in for h: $108 = k \times 8$. Then divide both sides by 8 to solve for k, giving a value of 13.5 for k.

18. **(1/5)** Simplify the inequality and isolate x. If $\dfrac{-3}{x} + 4 \le -11$, then

$$\frac{-3}{x} \le -15$$

$$\frac{1}{x} \ge 5$$

$$\frac{1}{5} \ge x \quad \text{so} \quad x \le \frac{1}{5}$$

The greatest value of x is $1/5$.

19. **(5)** Rewrite the equation of the circle in center-radius form by completing the square for both x and y:

$$x^2 + 4x + y^2 - 10y = 20$$
$$\left(x^2 + 4x + 4\right) + \left(y^2 - 10y + 25\right) = 20 + 4 + 25$$
$$(x + 2)^2 + (y - 5)^2 = 49$$

The center of the circle is $(-2, 5)$ with radius 7. The line $x = k$ is a vertical line. Since it is given that it intersects the circle in one point, it intersects the circle at a horizontal distance of 7 units from $x = -2$ so $k = -2 + 7 = 5$ or $-2 - 7 = -9$

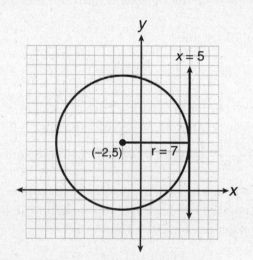

20. **(3)** It is given that $g(f(b)) = 8$. Since $g(5) = 8$, $f(b) = 5$ so $x = 3$.

Calculator Section

1. **(B)** If $(2b - 7)(2b + 7) = 1$, then $4b^2 - 49 = 1$ so $4b^2 = 50$ and $2b^2 = \dfrac{50}{2} = 25$.

2. **(C)** In the linear function $p = 117 + 32d$, 32 is the coefficient of d, the number of days, so it represents the rate of change of p (the number of pledges) received per day.

3. **(C)** If a call lasts x minutes and x is greater than 3, then the charge for the first 3 minutes is \$1.80 and the charge for the next $x - 3$ minutes is $0.40(x - 3)$. Since the total charge was \$4.20,

$$0.40(x - 3) + 1.80 = 4.20$$
$$40(x - 3) + 180 = 420$$
$$40x - 120 + 180 = 420$$
$$40x = 360$$
$$x = \frac{360}{40} = 9$$

4. **(D)** Consider each Roman numeral statement in turn.

- I. Number of men who applied to a same state college to the number of women who did the same is $\frac{64 + 26}{41 + 19} = \frac{90}{60} = \frac{3}{2}$. Hence, Statement I is true. ✔

- II. The probability that a female student selected at random did not apply to a 2 year college is $\frac{41 + 15 + 6}{81} = \frac{62}{81} \approx .765 > 75\%$. Thus, Statement II is true. ✔

- III. Of the students who applied to a same state college, the percent of females is $\frac{41 + 19}{105 + 45} = \frac{60}{150} = \frac{2}{5} = 40\%$. Hence, Statement III is true. ✔

Statements I, II, and III are true.

5. **(A)** If $3x - 1 = x - \frac{7}{9}$, then $2x - 1 = -\frac{7}{9}$ so $(2x - 1) + 2 = -\frac{7}{9} + 2$ and $2x + 1 = \frac{11}{9}$.

6. **(B)** Pick an easy number for the starting price of a gallon of gas, say \$1. After a 12% increase the price is \$1.12. After a 25% decrease the price is $0.75 \times \$1.12 = \$0.84$, which corresponds to a $\frac{1.00 - 0.84}{1.00} \times 100\% = 16\%$ decrease in the price of a gallon of gas.

7. **(C)** The number of wild turkeys in the population in 2010 is calculated by evaluating $T(0)$. Since x represents the number of years since 2010, $T(5)$ represents the number of wild turkeys in the population in 2015. Thus,

$$T(5) - T(0) = 17(1.15)^{10} - 17(1.15)^{0}$$
$$= 17(4.04 - 1)$$
$$= 17(3.04)$$
$$= 51.68$$

Since the answer must be a whole number, round down to 51.

8. **(A)** If $g(x) = f(x - 2)$, the function g is the function that results from shifting function f 2 units to the right in the horizontal direction. The vertex of the parabola $f(x) = -(x + 2)^2 - 1$ is $(-2, -1)$, which shifted 2 units to the right is mapped onto $(0, -1)$.

9. **(B)** Since $1 - 16\% = 1 - 0.16 = 0.84$, the city planner should use .84 for r.

10. **(B)** A strong positive association in a scatterplot is indicated by a cluster of data points that is rising from left to right which, of the four graphs, is best illustrated by graph (2).

11. **(A)** The median of an ordered set of 4 numbers is the average of the two middle values. Hence,

$$\frac{a+b+c+d}{4} = 3\left[\frac{(b+c)}{2}\right]$$
$$a+b+c+d = 6b+6c$$
$$a = 5b+5c-d$$
$$= 5(b+c)-d$$

12. **(C)** If x represents the number of $20 sport shirts purchased, then $2x$ is the number of $35 dress shirts purchased. Since a total of $720 was spent on the shirts, $20(x) + 35(2x)$ = 720 or $20x + 70x = 720$, so $90x = 720$. Hence,

$$x = \frac{720}{90} = 8 \quad \text{and} \quad 2x = 2(8) = 16$$

The total number of shirts purchased is $x + 2x$ or $8 + 16 = 24$.

13. **(D)** If there are 1.2 grams of hemoglobin in 10 cubic centimeters of blood and x represents the number of grams of hemoglobin contained in 35 cubic centimeters of the same blood, then

$$\frac{\text{Blood}}{\text{Hemoglobin}} = \frac{10}{1.2} = \frac{35}{x}$$
$$10x = 42$$
$$x = \frac{42}{10} = 4.2$$

14. **(C)** The previous median $= \dfrac{1.4+1.6}{2} = 1.5$, and the previous mean $= \dfrac{61.7}{24} \approx 2.57$. The new median is the 13th score, which is 1.6, and the new mean $= \dfrac{61.7+7.3}{25} = 2.76$.

The median increases by 0.10, whereas the mean increases by 0.19 so the mean increases by a greater amount.

15. **(A)**

$$m = \frac{M}{\sqrt{1-\dfrac{v^2}{c^2}}}$$

$$\left(\sqrt{1-\frac{v^2}{c^2}}\right)^2 = \left(\frac{M}{m}\right)^2$$

$$1-\frac{v^2}{c^2} = \left(\frac{M}{m}\right)^2$$

$$\frac{v^2}{c^2} = 1-\left(\frac{M}{m}\right)^2$$

$$v^2 = c^2\left(1-\left(\frac{M}{m}\right)^2\right)$$

$$\sqrt{v^2} = \sqrt{c^2\left(1-\left(\frac{M}{m}\right)^2\right)}$$

$$v = c\sqrt{1-\left(\frac{M}{m}\right)^2}$$

16. **(B)** Consider each Roman numeral choice in turn:

- I. Roman numeral choice I is false, since the point $(4, 2)$ lies inside the shaded region and $4 + 2 > 5$. ✗
- II. For any given point (r, s) inside the shaded region, $f(r)$ represents the y-value of the point on the graph directly above it, so $s \leq f(r)$. For example, if $f(r, s) = (3, 1)$, then $1 < f(3)$ since $f(3)$, according to the graph, has a value between 1 and 2. Roman numeral choice II must be true. ✓
- III. Because points such as $(1, 1)$ are contained in the shaded region, it is not always the case that $r \neq s$. Roman numeral choice III is false. ✗

Since only Roman numeral choice II must be true, the correct answer is choice **(B)**.

17. **(D)** The cost for p people attending is $\dfrac{750 + 2.25p}{p} = \dfrac{750}{p} + 2.25$, which must be between $2.75 and $3.25 per person so

$$2.75 < \frac{750}{p} + 2.25 < 3.25$$

$$0.50 < \frac{750}{p} < 1.00$$

$$0.50p < 750 \quad and \quad 750 < 1.00p$$

- If $0.50p < 750$, then $p < \dfrac{750}{0.50}$ so $p < 1{,}500$.
- If $750 < 1.00p$, then $p > 750$.
- Hence, $750 < p < 1{,}500$.

18. **(B)** If $p(x)$ is a polynomial function and $p(r) = 0$, then the remainder when $p(x)$ is divided by $x - r$ is 0 so $p(x)$ is divisible by $x - r$ or, equivalently, $x - r$ is a factor of $p(x)$. Since it is given that $p(3) = 0$, $x - 3$ is a factor of $p(x)$.

19. **(D)** If a group of p people plan to contribute equally to the purchase of a gift that costs d dollars, then each person must contribute $\dfrac{d}{p}$ dollars. If n of the p people decide not to contribute, then each of the $p - n$ people who are left must contribute $\dfrac{d}{p - n}$ dollars.

The difference between the two contribution rates represents the amount of increase for each person who contributes:

$$\frac{d}{p-n} - \frac{d}{p} = \frac{d}{p-n}\left(\frac{p}{p}\right) - \frac{d}{p}\left(\frac{p-n}{p-n}\right)$$

$$= \frac{dp - d(p - n)}{p(p - n)}$$

$$= \frac{dp - dp + nd}{p(p - n)}$$

$$= \frac{nd}{p(p - n)}$$

20. **(C)** To determine if a function is divisible by $2x + 1$, check to see if it is a factor. Consider each statement in turn.

- I. $$f(x) = \frac{8x^2 - 2}{2x + 1} = \frac{2(4x^2 - 1)}{2x + 1} = \frac{2(2x - 1)(2x + 1)}{2x + 1} = 2(2x - 1). \checkmark$$

- II. $$g(x) = \frac{2x^2 - 9x + 4}{2x + 1} = \frac{(2x - 1)(x - 4)}{2x + 1} \ \text{✗}$$

- III. $$h(x) = \frac{4x^3 + 2x^2 - 6x - 3}{2x + 1} = \frac{(4x^3 + 2x^2) - (6x + 3)}{2x + 1}$$

$$= \frac{(2x + 1)(2x^2 - 3)}{2x + 1}$$

$$= 2x^2 - 3 \checkmark$$

Only Statements I and III are true.

21. **(D)** The function $y = A(1 + r)^n$ describes exponential growth of an initial amount, A, with a compound rate of growth of $r\%$ where n represents the number of compounding periods. In this problem, $A = 20{,}000$, $r = \dfrac{5\%}{4} = 0.0125$, and $n = 18 \times 4$ so $y = 20{,}000(1.0125)^{18 \times 4}$.

22. **(D)** If $g(x) = x - 1$ and $2g(c) = 10$, then $g(c) = \dfrac{10}{2} = 5$.

- $g(c) = c - 1 = 5$ so $c = 6$.
- $g(3c) = 3c - 1 = 3(6) - 1 = 17$.

23. **(D)** The graph describes exponential decay, which means that the population will *decrease* at an exponential rate. An increase in birthrate would *increase* the population, making choices A and B incorrect. A liberal immigration policy and thriving economy would contribute to population growth so choice C is not correct. A highly contagious illness would spread rapidly and would be most likely to cause a drastic, geometric decline in population. Political strife and unrest would tend to make more people leave the country further contributing to an ongoing decline in the size of the population.

24. **(B)** First, calculate the total number of Canadian dollars she will need to make her purchase by multiplying the number of books by the price per book in Canadian dollars: $30 \times 45 = 1{,}350$. Then, convert this amount to U.S. dollars by multiplying by the U.S. to Canadian dollar exchange rate: $1{,}350 \times .76 = 1{,}026$.

25. **(A)** Pick any two years and figure out the rate of change (slope). Using the first two rows of the table:

$$\frac{3.9 - 4.7}{40 - 20} = -\frac{0.8}{20} = -0.04.$$

You can eliminate choices (B) and (D). The desired equation has the form $y = -0.04a + b$. To find b, substitute 20 for a and 4.7 for p:

$$4.7 = -0.04(20) + b$$
$$4.7 = -0.8 + b$$
$$b = 5.5$$

The equation $y = -0.04a + 5.5$ expresses the relationship between p and a.

26. **(B)** If x represents the number of inches in the width of the box, then $3x$ is the length and $\frac{1}{2}(3x) = 1.5x$ is the height. Hence,

- The area of the base is $(x)(3x) = 3x^2$. The cost of producing the base is \$0.08 per square inch or $(3x^2) \times (0.08) = 0.24x^2$.
- The area of each of the longer of the vertical sides is $(3x)(1.5x) = 4.5x^2$. Since there are two of these sides and the cost per square inch is \$0.03, the cost of these two sides is $2(4.5x^2)(0.03) = 0.27x^2$.
- The area of each of the shorter vertical sides is $(x)(1.5x) = 1.5x^2$. Since there are two of these sides and the cost per square inch is \$0.03, the cost of these two sides is $2(1.5x^2)(0.03) = 0.09x^2$.

The cost of producing the box is $0.24x^2 + 0.27x^2 + 0.09x^2 = 0.60x^2$.

27. **(C)** The given inequality $|o - p| \le d$ represents the condition that the data points of interest must fall within 0.5 cm either above or below the line. Of the 12 plotted points on the scatterplot, 4 points satisfy this condition so the proportion or ratio is $\frac{4}{12}$ or $\frac{1}{3}$.

28. **(B)** To determine the width of the arch, find the distance between the x-intercepts of the parabola by setting $y = 0$: $0 = -3x(x - 8)$ so the x-intercepts are 0 and 8. The width of the base of the arch is $8 - 0 = 8$. To find the height of the arch, find the y-coordinate of the vertex of the parabola. For the parabola $y = ax^2 + bx + c$, the formula $x = -\frac{b}{2a}$ gives the x-coordinate of the vertex:

$$x = -\frac{24}{2(-3)} = 4$$

To find the y-coordinate of the vertex, substitute 4 for x in the parabola equation:

$$y = -3(4)^2 + 24(4)$$
$$= -48 + 96$$
$$= 48$$

Since the height of the parabola is 48 meters and its width is 8 meters, the height is 6 times the width.

29. **(A)** Randomly placing calls to as many potential residents as possible will result in the most random, representative sample of the population. All of the other options place significant limitations on the range of potential voters surveyed.

30. **(B)**

- The sum of the areas of the two squares is 36 + 36 = 72.
- To find the area, A, of a regular hexagon with side s, multiply the area of one of the equilateral triangles that comprises the hexagon by 6:

$$A = 6 \times \left(\frac{s^2 \sqrt{3}}{4} \right)$$

$$= 6 \times \left(\frac{36\sqrt{3}}{4} \right)$$

$$= 54\sqrt{3}$$

- The percent of the total surface area that will be silver plated is the ratio of the areas of the two squares to the total area:

$$\frac{72}{72 + 54\sqrt{3}} \approx 43.5 \approx 44\%$$

31. **(22)** Change to ounces:

- Find the equivalent number of ounces in the 225 milliliter (ml) bottle:

$$225\,\text{ml} \div 29.6 \, \frac{\text{ml}}{\text{ounce}} = 7.6 \, \text{ounces}$$

- Find the number of ounces equivalent to two teaspoons:
  1 cup = 16 tablespoons = 16×3 teaspoons = 8 ounces. So 48 teaspoons = 8 ounces. Divide each side of the equation by 24, which shows that 2 teaspoons are equivalent to $\frac{8}{24} = \frac{1}{3}$ ounce.

- Since $7.6 \div \frac{1}{3} = 22.8$, the maximum number of 2-teaspoon doses is 22.

32. **(270)** Use the relationship that time = $\frac{\text{distance}}{\text{rate}}$:

$$\text{time} = \frac{3.06 \times 10^9}{1.86 \times 10^5}$$

$$\approx 1.64516 \times 10^4$$

$$\approx 16{,}451.6 \, \text{seconds}$$

$$\approx \frac{16{,}451.6}{60} \, \text{minutes}$$

$$\approx 274.2 \, \text{minutes}$$

Rounding to *the nearest 5 minutes*, the time is 275 minutes.

33. **(8/15)** In the unit circle given,

- $\tan\theta = \dfrac{y}{x} = \dfrac{-0.8}{-0.6} = \dfrac{4}{3}$
- $\sin\theta = y = -0.8$
- $\tan\theta + \sin\theta = \dfrac{4}{3} - 0.8$

$$= \dfrac{4}{3} - \dfrac{4}{5}$$

$$= \dfrac{20}{15} - \dfrac{12}{15}$$

$$= \dfrac{8}{15}$$

34. **(11)** If $a + 2b = 13$ and $8a + b = 20$, then adding corresponding sides of the two equations gives $9a + 3b = 33$. If each member of $9a + 3b = 33$ is divided by 3, then $3a + b = 11$.

35. **(90)** If x represents the number of sports fans that needs to be added, then

$$\dfrac{120 + x}{120 + 75 + 65 + x} = \dfrac{3}{5}$$

$$\dfrac{120 + x}{260 + x} = \dfrac{3}{5}$$

$$5(120 + x) = 3(260 + x)$$

$$600 + 5x = 780 + 3x$$

$$5x - 3x = 780 - 600$$

$$2x = 180$$

$$x = \dfrac{180}{2} = 90$$

36. **(345)** If x represents the minimum number of surgeries that need to be performed to make a profit, then

$$2{,}000x - (500{,}000 + 550x) > 0$$

$$1{,}450x > 500{,}000$$

$$x > \dfrac{500{,}000}{1{,}450}$$

$$x > 344.83$$

Since x must be an integer greater than 344, the minimum value of x is 345.

37. **(2,895)** Use the percent rates of increase in the CPI from the table as the corresponding rates of inflation. Call p the price of the appliance at the beginning of 2012. Multiply p by (1 + inflation rate) for 2012 and multiply the result by (1 + inflation rate) for 2013 to arrive at the final cost of 3,000 in 2014:

$$(p \times 1.021) \times 1.015 = 3{,}000$$

$$p = \dfrac{3{,}000}{1.021 \times 1.015}$$

$$p \approx 2{,}894.87$$

The cost of the appliance in 2012, to the *nearest dollar*, would have been $2,895.

38. **(5)** Use the CPI rate increases as the rates for inflation for the time specified in the problem. In the second half of 2005, the inflation rate is 3.8% for the 6 month period; in 2006 it was 3.2%; in 2007 it was 2.8%; and in the first half of 2008 it was 4.2% for the 6 month period. Calculate the compounded rate of inflation over this three year time period by using 100 as an initial base amount:

- Using the base amount of 100, for the second *half* of 2005, the amount increases to
  $$100 + 0.038 \times 100 \times \frac{1}{2} = 101.9.$$

- The final amount at the end of 2006 is $101.9(1 + 0.032) = 105.16$.
- The final amount at the end of 2007 is $105.16(1 + 0.028) = 108.10$.
- The final amount at the end of the first *half* year in 2008 is
  $$108.10 + 108.10 \times 0.042 \times \frac{1}{2} = 110.37.$$

Since the initial amount increased from 100 to 110.37, the compounded rate of inflation over the three year period was 10.37%. The average yearly compounded rate of inflation over the three years is $\dfrac{10.37\%}{3} \approx 3.46\%$. Since $1.5\% + 3.46\% = 4.96\%$, the minimum annual rate of increase in his monthly annuity payments, *correct to the nearest tenth* should be 5.0%.

ANSWER SHEET
Practice Test 2

Math (No Calculator)

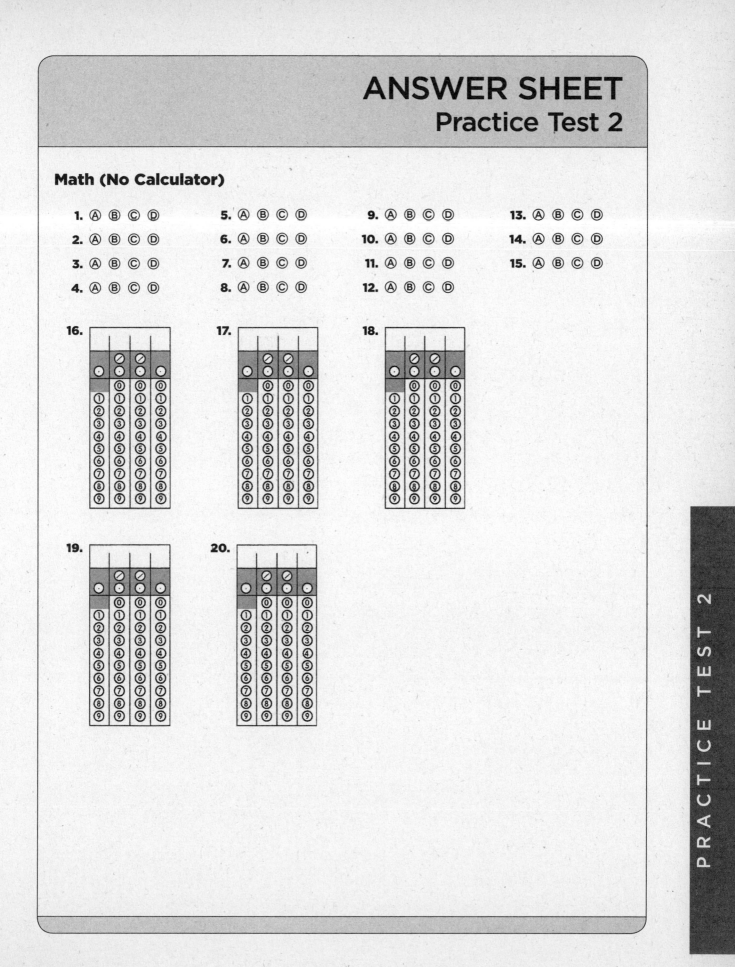

1. Ⓐ Ⓑ Ⓒ Ⓓ
2. Ⓐ Ⓑ Ⓒ Ⓓ
3. Ⓐ Ⓑ Ⓒ Ⓓ
4. Ⓐ Ⓑ Ⓒ Ⓓ

5. Ⓐ Ⓑ Ⓒ Ⓓ
6. Ⓐ Ⓑ Ⓒ Ⓓ
7. Ⓐ Ⓑ Ⓒ Ⓓ
8. Ⓐ Ⓑ Ⓒ Ⓓ

9. Ⓐ Ⓑ Ⓒ Ⓓ
10. Ⓐ Ⓑ Ⓒ Ⓓ
11. Ⓐ Ⓑ Ⓒ Ⓓ
12. Ⓐ Ⓑ Ⓒ Ⓓ

13. Ⓐ Ⓑ Ⓒ Ⓓ
14. Ⓐ Ⓑ Ⓒ Ⓓ
15. Ⓐ Ⓑ Ⓒ Ⓓ

16.
17.
18.
19.
20.

25 MINUTES, 20 QUESTIONS

Turn to Section 3 of your answer sheet to answer the questions in this section.

Directions: For questions 1–15, solve each problem and choose the best answer from the given options. Fill in the corresponding circle on your answer document. For questions 16–20, solve the problem and fill in the answer on the answer sheet grid.

Notes:

- Calculators are **NOT PERMITTED** in this section.
- All variables and expressions represent real numbers unless indicated otherwise.
- All figures are drawn to scale unless indicated otherwise.
- All figures are in a plane unless indicated otherwise.
- Unless indicated otherwise, the domain of a given function is the set of all real numbers x for which the function has real values.

REFERENCE INFORMATION

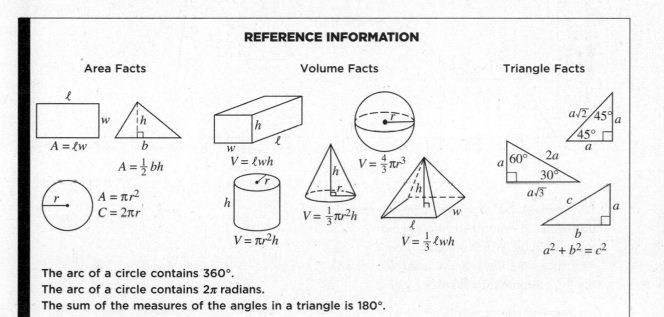

Area Facts

$A = \ell w$

$A = \frac{1}{2} bh$

$A = \pi r^2$
$C = 2\pi r$

Volume Facts

$V = \ell w h$

$V = \pi r^2 h$

$V = \frac{1}{3}\pi r^2 h$

$V = \frac{4}{3}\pi r^3$

$V = \frac{1}{3}\ell w h$

Triangle Facts

$a^2 + b^2 = c^2$

The arc of a circle contains 360°.
The arc of a circle contains 2π radians.
The sum of the measures of the angles in a triangle is 180°.

GO ON TO THE NEXT PAGE

$$\frac{5x - 3y}{3x + 5y} + \frac{2}{3} = 1$$

1. In the equation above, what is the value of $\frac{x}{y}$?

   (A) $\frac{1}{3}$

   (B) $\frac{2}{3}$

   (C) $\frac{5}{6}$

   (D) $\frac{7}{6}$

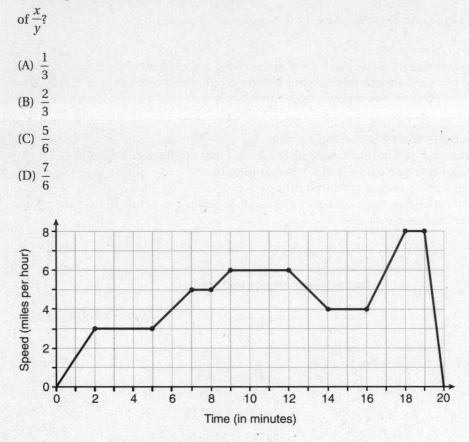

2. The graph above represents a jogger's speed during her 20-minute jog around her neighborhood. Which statement best describes what the jogger was doing during the 9–12 minute interval of her jog?

   (A) She was standing still.
   (B) She was increasing her speed.
   (C) She was decreasing her speed.
   (D) She was jogging at a constant rate.

GO ON TO THE NEXT PAGE

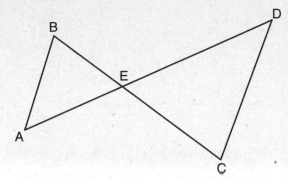

Note: Figure is not drawn to scale.

3. In the figure above, $\overline{AB} \parallel \overline{CD}$, $AD = 42$, $AB = 12$, and $CD = 16$. What is the length of $\overline{DE}$?

(A) 21
(B) 24
(C) 27
(D) 30

$$C = 60 + 0.25d$$

4. The equation above represents the monthly cost of a cell phone that includes up to 1 gigabyte of data after which there is a charge for d gigabytes of any additional data. Which of the following must be true?

  I. The cost of each additional megabyte of data is $60.25.
  II. The y-intercept of the graph of the cost equation represents the charge for each additional megabyte of data used.
  III. If between 5 and 6 megabytes of data are used in a month, the monthly charge is $61.25.

(A) I and II only
(B) I and III only
(C) II only
(D) III only

5. For what set of values of x is the expression $|3x + 4| < 0$ true?

(A) $-\dfrac{4}{3} < 0 < x$

(B) $x < -\dfrac{4}{3}$

(C) No real numbers
(D) All real numbers

6. The distance a free falling object has traveled can be modeled by the equation $d = \dfrac{1}{2}at^2$ where a is acceleration due to gravity and t is the amount of time the object has fallen. What is t in terms of a and d?

(A) $t = \sqrt{\dfrac{da}{2}}$

(B) $t = \sqrt{\dfrac{2d}{a}}$

(C) $t = \left(\dfrac{da}{2}\right)^2$

(D) $t = \left(\dfrac{2d}{a}\right)^2$

7. If $x^2 - y^2 = 24$ and $x - y = 3$, what is the value of y?

(A) $\dfrac{1}{2}$

(B) $\dfrac{3}{2}$

(C) $\dfrac{7}{4}$

(D) $\dfrac{5}{2}$

8. If $\dfrac{z}{2b} = 4$, $\dfrac{z}{2c} = 6$, and $2b + 3c = 12$, what is the value of z?

(A) 16
(B) 20
(C) 24
(D) 48

GO ON TO THE NEXT PAGE

9. A pizza parlor has a fixed initial cost of $180,000, and a variable cost of $4 for each pizza sold. If the pizza parlor charges $10 for each pizza, how many pizzas will it have to sell before it makes a profit?

(A) 24,000
(B) 30,000
(C) 38,000
(D) 42,000

$$(ax + 7)(bx - 1) = 12x^2 + kx + (b - 13)$$

10. If the equation above is true for all values of x where a, b, and k are non-zero constants, what is the value of k?

(A) 40
(B) 25
(C) 17
(D) 8

11. Function f is defined by the equation $f(x) = ax^2 + \dfrac{2}{a}x$. If $f(3) - f(2) = 11$, what is the *smallest* possible value of a?

(A) $\dfrac{1}{6}$

(B) $\dfrac{1}{5}$

(C) $\dfrac{1}{2}$

(D) 2

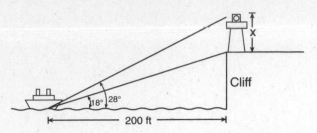

Cliff

18° 28°

200 ft

12. A lighthouse is built on the edge of a cliff near the ocean, as shown in the diagram above. From a boat located 200 feet from the base of the cliff, the angle of elevation to the top of the cliff is 18° and the angle of elevation to the top of the lighthouse is 28°. Which of the following equations could be used to find the height of the lighthouse, x, in feet?

(A) $x = 200 \tan 10°$

(B) $x = 200(\tan 28° - \tan 18°)$

(C) $x = \dfrac{200}{(\tan 28° - \tan 18°)}$

(D) $x = 200\left(\dfrac{\tan 18°}{\tan 28°}\right)$

13. The local deli charges a fee for delivery. On Monday, they delivered two dozen bagels to an office at a total cost of $8. On Tuesday, three dozen bagels were delivered at a total cost of $11. Which system of equations could be used to find the cost of a dozen bagels, b, if the delivery fee is f?

(A) $b + 2f = 8$
$b + 3f = 11$

(B) $2b + f = 8$
$b + 3f = 11$

(C) $b + 2f = 8$
$3b + f = 11$

(D) $2b + f = 8$
$3b + f = 11$

GO ON TO THE NEXT PAGE

14. The equation of a parabola in the xy-plane is $y = 2x^2 - 12x + 7$. What is the distance between the vertex of the parabola and the point $(3, 4)$?

(A) 6

(B) 8

(C) 11

(D) 15

15. When a baseball is hit by a batter, the height of the ball, $h(t)$, at time t, is determined by the equation $h(t) = -16t^2 + 64t + 4$ where $t \geq 0$. For which interval of time, in seconds, is the height of the ball at least 52 feet above the playing field?

(A) $0.5 \leq t \leq 2.5$

(B) $1.0 \leq t \leq 3.0$

(C) $1.5 \leq t \leq 3.5$

(D) $2.0 \leq t \leq 4.0$

GO ON TO THE NEXT PAGE

Grid-in Response Directions

In questions 16–20, first solve the problem, and then enter your answer on the grid provided on the answer sheet. The instructions for entering your answers follow.

- First, write your answer in the boxes at the top of the grid.
- Second, grid your answer in the columns below the boxes.
- Use the fraction bar in the first row or the decimal point in the second row to enter fractions and decimals.

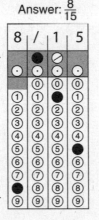

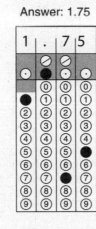

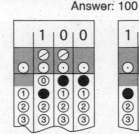

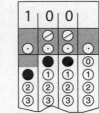

Either position is acceptable

- Grid only one space in each column.
- Entering the answer in the boxes is recommended as an aid in gridding but is not required.
- The machine scoring your exam can read only what you grid, so you **must grid-in your answers correctly to get credit**.
- If a question has more than one correct answer, grid-in only one of them.
- The grid does not have a minus sign; so no answer can be negative.
- A mixed number *must* be converted to an improper fraction or a decimal before it is gridded.

  Enter $1\frac{1}{4}$ as 5/4 or 1.25; the machine will interpret 11/4 as $\frac{11}{4}$ and mark it wrong.

- **All decimals must be entered as accurately as possible.** Here are three acceptable ways of gridding

$$\frac{3}{11} = 0.272727\ldots$$

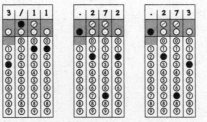

- Note that rounding to .273 is acceptable because you are using the full grid, but you would receive **no credit** for .3 or .27, because they are less accurate.

GO ON TO THE NEXT PAGE

$$\frac{\frac{2}{3}a^2 - \frac{4}{9}a^2}{2a} = 4 \quad \text{where } a \neq 0$$

16. What is the value of a in the expression above?

$$\frac{2}{3}x - \frac{1}{4}y = 6$$

$$kx - \frac{1}{3}y = 8$$

17. If the system of equations above has an infinite number of solutions, what is the value of the constant k?

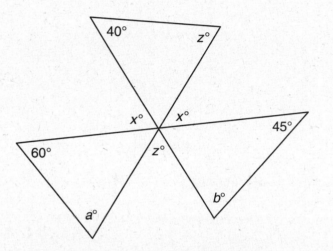

18. In the figure above, the measures of the angles are as marked. What is the value of $a + b$?

19. The equation $W = 120I - 12I^2$ represents the power, W, in watts, of a 120-volt circuit having a resistance of 12 ohms when a current, I, is flowing through the circuit. What is the maximum power, in watts, that can be delivered in this circuit?

20. The graph of a line in the xy-plane passes through the points $(5, -5)$ and $(1, 3)$. The graph of a second line has a slope of 6 and passes through the point $(0, 1)$. If the two lines intersect at (p, q), what is the value of $p + q$?

GO ON TO THE NEXT PAGE

ANSWER SHEET
Practice Test 2

Math (Calculator)

1. Ⓐ Ⓑ Ⓒ Ⓓ
2. Ⓐ Ⓑ Ⓒ Ⓓ
3. Ⓐ Ⓑ Ⓒ Ⓓ
4. Ⓐ Ⓑ Ⓒ Ⓓ
5. Ⓐ Ⓑ Ⓒ Ⓓ
6. Ⓐ Ⓑ Ⓒ Ⓓ
7. Ⓐ Ⓑ Ⓒ Ⓓ
8. Ⓐ Ⓑ Ⓒ Ⓓ

9. Ⓐ Ⓑ Ⓒ Ⓓ
10. Ⓐ Ⓑ Ⓒ Ⓓ
11. Ⓐ Ⓑ Ⓒ Ⓓ
12. Ⓐ Ⓑ Ⓒ Ⓓ
13. Ⓐ Ⓑ Ⓒ Ⓓ
14. Ⓐ Ⓑ Ⓒ Ⓓ
15. Ⓐ Ⓑ Ⓒ Ⓓ
16. Ⓐ Ⓑ Ⓒ Ⓓ

17. Ⓐ Ⓑ Ⓒ Ⓓ
18. Ⓐ Ⓑ Ⓒ Ⓓ
19. Ⓐ Ⓑ Ⓒ Ⓓ
20. Ⓐ Ⓑ Ⓒ Ⓓ
21. Ⓐ Ⓑ Ⓒ Ⓓ
22. Ⓐ Ⓑ Ⓒ Ⓓ
23. Ⓐ Ⓑ Ⓒ Ⓓ
24. Ⓐ Ⓑ Ⓒ Ⓓ

25. Ⓐ Ⓑ Ⓒ Ⓓ
26. Ⓐ Ⓑ Ⓒ Ⓓ
27. Ⓐ Ⓑ Ⓒ Ⓓ
28. Ⓐ Ⓑ Ⓒ Ⓓ
29. Ⓐ Ⓑ Ⓒ Ⓓ
30. Ⓐ Ⓑ Ⓒ Ⓓ

31.
32.
33.
34.

35.
36.
37.
38.

MATHEMATICS TEST—CALCULATOR

55 MINUTES, 38 QUESTIONS

Turn to Section 4 of your answer sheet to answer the questions in this section.

Directions: For questions 1–30, solve each problem and choose the best answer from the given options. Fill in the corresponding circle on your answer document. For questions 31–38, solve the problem and fill in the answer on the answer sheet grid.

Notes:
- Calculators **ARE PERMITTED** in this section.
- All variables and expressions represent real numbers unless indicated otherwise.
- All figures are drawn to scale unless indicated otherwise.
- All figures are in a plane unless indicated otherwise.
- Unless indicated otherwise, the domain of a given function is the set of all real numbers x for which the function has real values.

REFERENCE INFORMATION

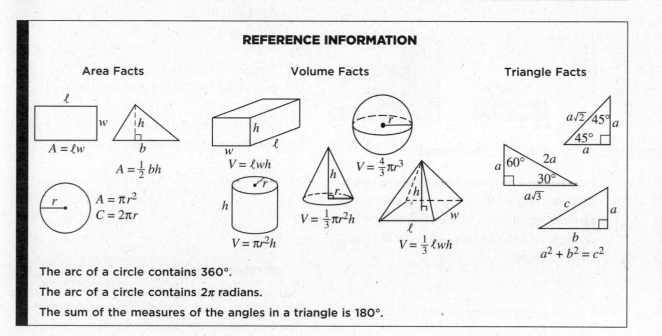

Area Facts

$A = \ell w$

$A = \frac{1}{2}bh$

$A = \pi r^2$
$C = 2\pi r$

Volume Facts

$V = \ell wh$

$V = \pi r^2 h$

$V = \frac{4}{3}\pi r^3$

$V = \frac{1}{3}\pi r^2 h$

$V = \frac{1}{3}\ell wh$

Triangle Facts

$a^2 + b^2 = c^2$

The arc of a circle contains 360°.

The arc of a circle contains 2π radians.

The sum of the measures of the angles in a triangle is 180°.

GO ON TO THE NEXT PAGE

1. If three times 1 less than a number n is the same as two times the number increased by 14, what is the value of n?

   (A) 15
   (B) 17
   (C) 19
   (D) 21

2. George spent 25% of the money he had on lunch and 60% of the remaining money on dinner. If he then had $9.00 left, how much money did he spend on lunch and dinner?

   (A) $19
   (B) $20
   (C) $21
   (D) $27

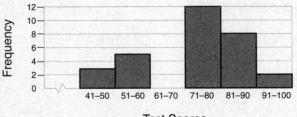

Test Scores

3. The histogram above shows the distribution of 30 test scores. If a test score is selected at random, what is the probability that the score falls in the interval that contains the median score?

   (A) $\dfrac{4}{15}$

   (B) $\dfrac{2}{5}$

   (C) $\dfrac{1}{2}$

   (D) $\dfrac{3}{5}$

4. The breakdown of a 500-milligram sample of a chemical compound in the bloodstream is represented by the function $p(n) = 500(0.8)^n$, where $p(n)$ represents the number of milligrams of the compound that remains at the end of n hours. Which of the following is true?

   I. The amount of the compound present is decreasing by a constant amount.
   II. Each hour the compound gets reduced by 20% of the amount present at the beginning of that hour.
   III. Each hour the compound gets reduced by 80% of 500.

   (A) I only
   (B) II only
   (C) I and III only
   (D) II and III only

5. Maggie's farm stand sold a total of 165 pounds of apples and peaches. She sold apples for $1.75 per pound and peaches for $2.50 per pound. If she made $337.50, how many pounds of peaches did she sell?

   (A) 11
   (B) 18
   (C) 65
   (D) 100

Number of Weeks	1	2	3	4
Number of Downloads	120	180	270	405

6. A computer program application developer released a new game app to be downloaded. The table above gives the number of downloads, y, for the first four weeks after the launch of the app. If w represents the number of weeks after the launch of the app, which equation best models these data?

   (A) $y = 60(w + 1)$
   (B) $y = 96(1.25)^w$
   (C) $y = 80(1.50)^w$
   (D) $y = 90w$

GO ON TO THE NEXT PAGE

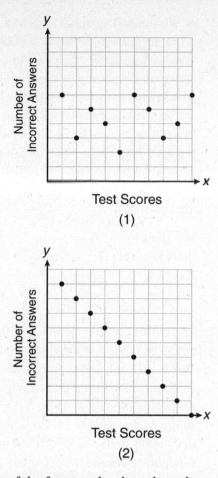

Test Scores

(1)

Test Scores

(2)

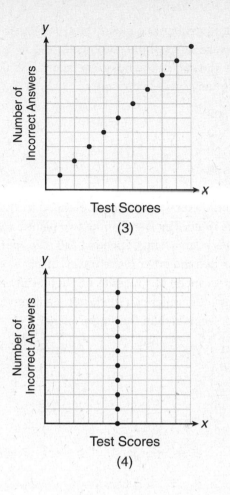

Test Scores

(3)

Test Scores

(4)

7. Which of the four graphs above best shows the relationship between *x* and *y* if *x* represents a student score on a test and *y* represents the number of incorrect answers a student received on the same test?

(A) Graph (1)
(B) Graph (2)
(C) Graph (3)
(D) Graph (4)

GO ON TO THE NEXT PAGE

8. An animal boarding facility houses 3 dogs for every 2 cats. If the combined total of dogs and cats at the boarding facility is 250, how many cats are housed?

(A) 80
(B) 100
(C) 120
(D) 150

9. An airline flies two different planes over the same route. The faster of the two planes travels at an average speed of 540 miles per hour, and the other plane travels at an average speed of 450 miles per hour. How many more miles can the faster plane travel in 12 seconds than the slower plane?

(A) $\dfrac{1}{5}$

(B) $\dfrac{3}{10}$

(C) 9

(D) 18

$$x - 3y = 2y + 7$$
$$x + 2 = 3(y + 1)$$

10. In the above system of equations, what is the value of $\dfrac{x}{y}$?

(A) $\dfrac{8}{3}$

(B) $\dfrac{11}{3}$

(C) 4

(D) 12

11. An Ironman Triathlon consists of swimming 2.4 miles, biking 112 miles, and running a marathon distance of 26.2 miles. Dylan completed an Ironman Triathlon in 12 hours and 30 minutes. He spent approximately half the time biking. He needed about 4 times as much time to run the 26.2 miles as to swim the 2.4 miles. The average rate of minutes per mile at which Dylan ran the marathon part of the Triathlon is closest to which of the following?

(A) 10.6
(B) 11.5
(C) 12.2
(D) 13.4

12. The bottom of a ski slope is 6,500 feet above sea level, the top of the slope is 11,000 feet above sea level, and the slope drops 5 feet vertically for every 11 feet traveled in the horizontal direction. From the top of the slope, Kayla skis down at an average speed of 30 miles per hour. Which of the following functions gives the best estimate for the distance above sea level, d, Kayla is t seconds after she begins her ski run where $6,500 < d < 11,000$? [Note: 5,280 feet = 1 mile]

(A) $d(t) = 11,000 - \left(\dfrac{150}{11}\right)t$

(B) $d(t) = 11,000 - 2.2t$

(C) $d(t) = 11,000 - 20t$

(D) $d(t) = 4,500 - 1,200t$

13. A gardener is planting two types of trees. One type is seven feet tall and grows at a rate of 8 inches per year. The other type is four feet tall and its rate of growth is 50% greater than the rate of the other tree. In how many years will the two trees grow to the same height?

(A) 6
(B) 7
(C) 8
(D) 9

GO ON TO THE NEXT PAGE

	Vaccination and Flu Status				
Age	Unvaccinated No Flu	Unvaccinated Got Flu	Vaccinated No Flu	Vaccinated Got Flu	Total
Under 21	6	4	8	2	20
21–50	17	15	22	14	68
Over 50	2	9	32	19	62

14. The table above summarizes the results of a survey taken at the end of last year's flu season. What fraction of the people who got the flu were unvaccinated?

(A) $\dfrac{2}{3}$

(B) $\dfrac{4}{9}$

(C) $\dfrac{3}{8}$

(D) $\dfrac{1}{12}$

15. The temperature, t, generated by an electrical circuit is represented by $t = f(m) = 0.3m^2$, where m is the number of moving parts. The resistance of the same circuit is represented by $r = g(t) = 150 + 5t$, where t is the temperature. What is the resistance in a circuit that has four moving parts?

(A)   51

(B)   156

(C)   174

(D) 8,670

GO ON TO THE NEXT PAGE

Comparison of Combined State and Local Spending on Education						
	Year					
	2011		2013		2015	
State	Education Spending	Population	Education Spending	Population	Education Spending	Population
California	453,480.7	37.7	447,531.1	38.4	454,003.1	39.2
New York	300,031.9	19.5	306,395.8	19.7	316,104.0	19.8
Texas	221,155.9	25.7	226,805.0	26.5	252,655.5	27.4
Florida	163,070.8	19.1	157,010.2	19.6	162,548.3	20.2
Illinois	129,543.3	12.9	132,848.8	12.9	140,072.6	12.9

Questions 16 and 17 refer to the above table, that shows the population (in millions) and education spending (in millions) and by state for each of the states listed for the years 2011, 2013, and 2015.

16. Which of the following best approximates the average rate of change in education spending in Texas from 2011 to 2015 ?

(A) 3.2 billion per year
(B) 6.3 billion per year
(C) 7.9 billion per year
(D) 10.5 billion per year

17. Based on the data in the table, which of the following must be true?

   I. In 2015 per capita (per person) spending on education in Illinois was greater than per capita spending on education in Texas.
   II. Per capita spending on education in Florida declined in 2015 compared to 2011 spending.

   III. California had the highest per capita spending in education for each year.

(A) I and II only
(B) I and III only
(C) II and III only
(D) I, II, and III

18.

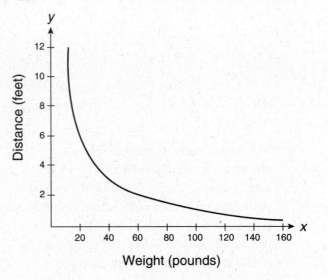

The graph above shows the relationship between a person's weight and the distance that the person must sit from the center of a seesaw to make it balanced. Which of the following best represents the equation of this graph?

(A) $y = 12x^2$
(B) $y = -120x$
(C) $y = 120\left(\dfrac{1}{2}\right)^x$
(D) $y = \dfrac{120}{x}$

GO ON TO THE NEXT PAGE

Average Annual Salary Range By Highest Level of Degree Earned				
	Average Annual Salary			
Highest Degree Earned	Less than $35,000	$35,000 to $70,000	More than $70,000	Total
High School	21	15	3	39
Two Year College	12	24	2	33
Four Year College	18	41	29	93
Graduate School	1	28	46	75
Total	52	108	80	240

19. The table above summarizes the results of a survey taken in which 240 adults were asked about their education level and current annual salary. If a participant who reported earning $35,000 or more per year is selected at random, what is the best estimate of the probability that the person does *not* have a graduate school degree?

(A) 0.31
(B) 0.40
(C) 0.60
(D) 0.69

20. If the sum of 10 dimes, 5 nickels, and x quarters equals $5.25, what is the value of x?

(A) 8
(B) 10
(C) 16
(D) 22

Students at Washington High School	Male	Female	Total
Taking AP Classes	56	72	128
Not Taking AP Classes	23	26	49
Total	79	98	177

21. The table above gives the number of male and female students at Washington High School who are taking Advanced Placement (AP) classes and those who are not. What is the proportion of the total number of students at the school who are both male and NOT taking AP classes?

(A) $\dfrac{23}{177}$

(B) $\dfrac{79}{177}$

(C) $\dfrac{23}{49}$

(D) $\dfrac{23}{56}$

22. A travel agency sells ship cruises for a popular cruise line. Historically, 135 cruises can be sold when the price is $950 per person. If the price drops to the minimum allowed by the cruise line of $725 per person, 180 cruises can be sold. If the number of cruises sold increases at a constant rate as the price p decreases, where $p \geq 725$, which of the following functions best models the situation described?

(A) $f(p) = -\dfrac{1}{29}p + 205$

(B) $f(p) = -\dfrac{1}{19}p + 1{,}135$

(C) $f(p) = -5p + 4{,}885$

(D) $f(p) = -\dfrac{1}{5}p + 325$

GO ON TO THE NEXT PAGE

I. The coordinates of the center are (2, −3).
II. The coordinates of the center are (−2, 3).
III. The length of the radius is $5\sqrt{2}$.
IV. The length of the radius is 50.

23. If an equation of a circle is $x^2 + 4x + y^2 - 6y = 37$, which of the statements above are true?

(A) I and III
(B) I and IV
(C) II and III
(D) II and IV

$$f(x) = \frac{x^4 + 2x^3 - 3x^2 + 4x + 12}{x + 3}$$

24. Which of the following functions is equivalent to the function above for all values of x for which function f is defined?

(A) $g(x) = x^3 - x^2 + 4$
(B) $g(x) = x^2 - x + 4$
(C) $g(x) = x^3 - x^2 + 4x$
(D) $g(x) = x^4 + 2x^3 - 3x^2 + 4$

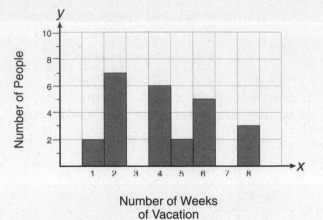

Number of Weeks
of Vacation

25. The histogram above shows the results of a survey taken of 25 individuals who were polled about how many weeks of vacation per year they receive. Which of the following is closest to the average (arithmetic mean) number of weeks of vacation per individual?

(A) 2
(B) 3
(C) 4
(D) 5

26. If $p(x)$ is a polynomial function and $p(-1) = 3$, which statement is true?

(A) The remainder when $p(x)$ is divided by $x - 3$ is −1.
(B) The remainder when $p(x)$ is divided by $x + 3$ is −1.
(C) The remainder when $p(x)$ is divided by $x - 1$ is 3.
(D) The remainder when $p(x)$ is divided by $x + 1$ is 3.

GO ON TO THE NEXT PAGE

$$y = \frac{3}{h-2}x + 5$$
$$hy - 8x = 5$$

27. For what value of h does the system of equations above have no solution?

    (A) $\dfrac{16}{5}$

    (B) $\dfrac{13}{8}$

    (C) $\dfrac{11}{15}$

    (D) $\dfrac{5}{8}$

28. A *troy* ounce is a unit of mass used for precious metals such as gold. There are 12 troy ounces in a troy pound and a troy pound is equivalent to 373.2 grams. If the density of gold is 19.3 grams per cubic centimeter, which of the following is closest to the number of cubic centimeters in the volume of a block of gold with a mass of 5 troy ounces? [Note: density is mass divided by volume]

    (A) 7
    (B) 8
    (C) 9
    (D) 10

29. A researcher is conducting a survey for which she currently has a 93% confidence level. What would be two actions that she could take that would be most likely to increase the confidence level in her survey results?

    (A) Increase the sample size and modify the design of the survey to increase the standard deviation.
    (B) Increase the sample size and modify the design of the survey to decrease the standard deviation.
    (C) Decrease the sample size and increase the randomness of the survey sample.
    (D) Modify the design of the survey to increase the standard deviation and decrease the randomness of the survey sample.

30. The coordinates of the vertex of a parabola in the xy-plane are $(-4, k)$. If the y-intercept of the parabola is 12 and the parabola passes through the point $(-3, 7)$, what is the value of k?

    (A) $\dfrac{20}{3}$

    (B) $\dfrac{16}{5}$

    (C) $\dfrac{14}{3}$

    (D) $\dfrac{12}{5}$

GO ON TO THE NEXT PAGE

Grid-in Response Directions

In questions 31–38, first solve the problem, and then enter your answer on the grid provided on the answer sheet. The instructions for entering your answers follow.

- First, write your answer in the boxes at the top of the grid.
- Second, grid your answer in the columns below the boxes.
- Use the fraction bar in the first row or the decimal point in the second row to enter fractions and decimals.

Write your answer in the boxes

Grid in your answer

Answer: $\frac{8}{15}$

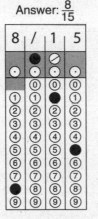

Answer: 1.75

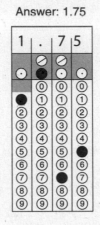

Answer: 100

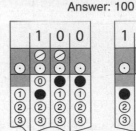 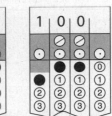

Either position is acceptable

- Grid only one space in each column.
- Entering the answer in the boxes is recommended as an aid in gridding but is not required.
- The machine scoring your exam can read only what you grid, so you **must grid-in your answers correctly to get credit.**
- If a question has more than one correct answer, grid-in only one of them.
- The grid does not have a minus sign; so no answer can be negative.
- A mixed number *must* be converted to an improper fraction or a decimal before it is gridded.

  Enter $1\frac{1}{4}$ as 5/4 or 1.25; the machine will interpret 11/4 as $\frac{11}{4}$ and mark it wrong.

- **All decimals must be entered as accurately as possible.** Here are three acceptable ways of gridding

$$\frac{3}{11} = 0.272727\ldots$$

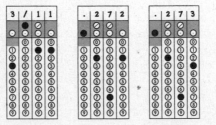

- Note that rounding to .273 is acceptable because you are using the full grid, but you would receive **no credit** for .3 or .27, because they are less accurate.

GO ON TO THE NEXT PAGE

PRACTICE TEST 2

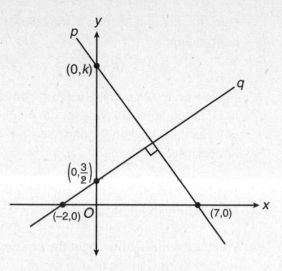

31. In the xy-plane above, line p is perpendicular to line q. What is the value of k?

32. Eleven seconds after a deep sea diver jumps into the ocean he is 69 feet below sea level and 28 seconds later, he is 195 feet below sea level. If he is descending under water at a constant rate, how many feet below sea level will he be 1.5 minutes after his initial descent?

33. What is a possible value of x that satisfies $9 < 4x - |-3| < 10$?

34. One way of estimating a wildlife population of interest is to draw a sample of the population, tag the animals, and then return them to the population. Then, at a later date, draw another sample at random from the same population and compare the results. An ecologist using this methodology captures, tags, and then returns 198 fish to a lake. Three months later the ecologist captures a sample of 135 of the same type of fish, of which 22 were tagged. What would be the ecologist's best estimate for the number of fish of that type that are in the lake?

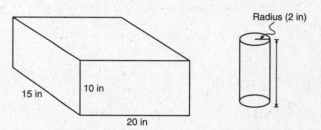

35. In the figure above, a rectangular container with the dimensions 10 inches by 15 inches by 20 inches is to be filled with water, using a cylindrical cup whose radius is 2 inches and whose height is 5 inches. What is the maximum number of full cups of water that can be placed into the container without the water overflowing the container?

GO ON TO THE NEXT PAGE

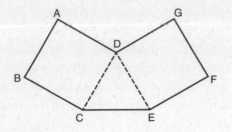

36. A sterling silver pendant is being designed to have the shape of polygon *ABCEFGD* shown above where *ABCD* and *EFGD* are squares and triangle *CDE* is equilateral. If the area of △*CDE* is $\frac{27}{\sqrt{3}}$ square centimeters, what is the total linear distance around the pendant?

Questions 37 and 38 refer to the following information.

$$h(t) = -4.9t^2 + 88.2t$$

When a projectile is launched from ground level, the equation above gives the number of meters in its height, *h*, after *t* seconds have elapsed.

37. How many seconds after the projectile is launched will it hit the ground?

38. What is the maximum height the projectile reaches, correct to the *nearest meter*?

GO ON TO THE NEXT PAGE

Math (No Calculator)

1. **D**	5. **C**	9. **B**	13. **D**
2. **D**	6. **B**	10. **A**	14. **D**
3. **B**	7. **D**	11. **B**	15. **B**
4. **D**	8. **C**	12. **B**	

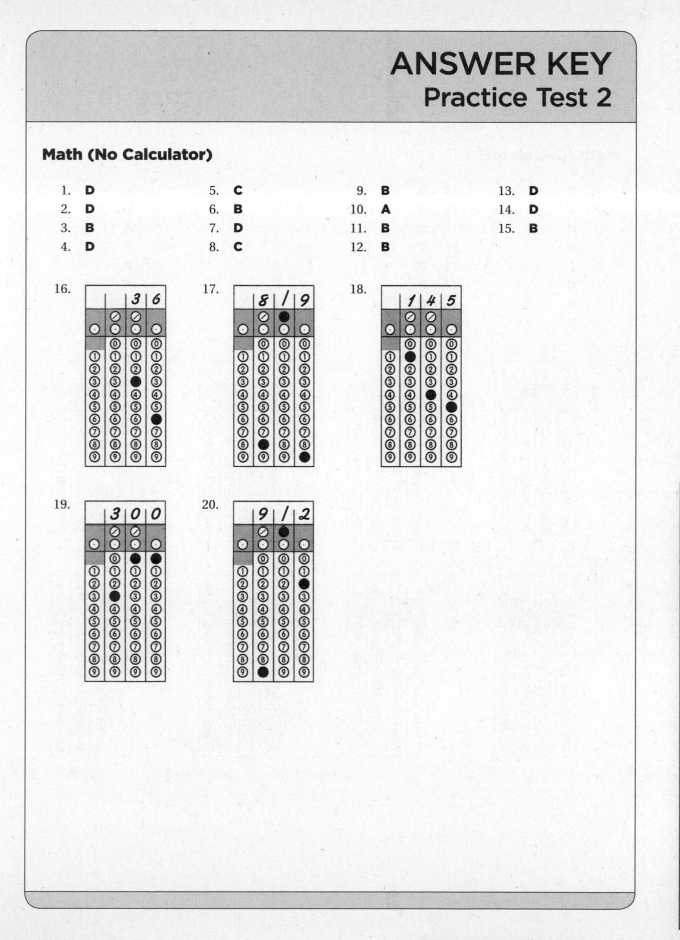

16. 3 6

17. 8 / 9

18. 1 4 5

19. 3 0 0

20. 9 / 2

Math (Calculator)

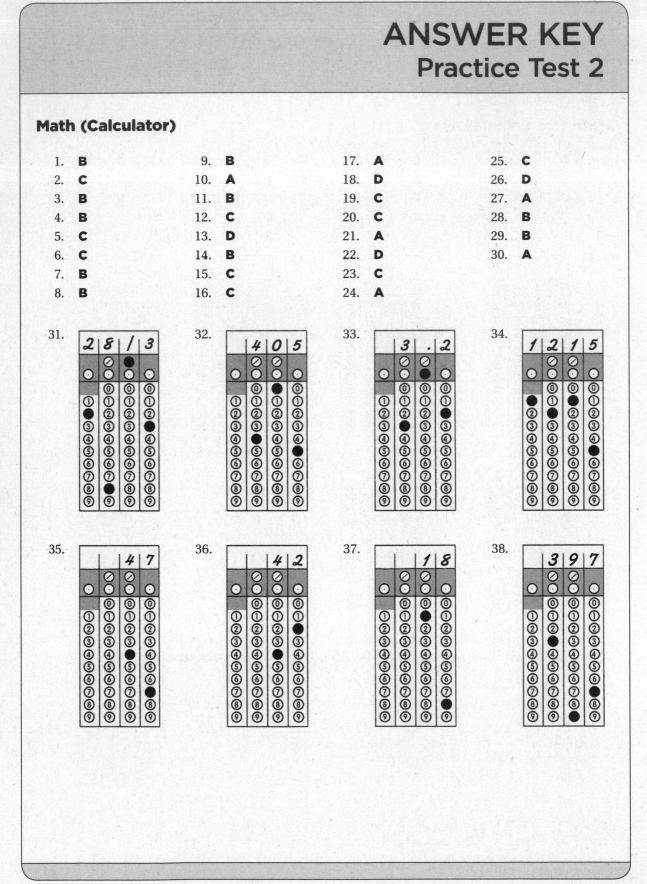

1. **B**
2. **C**
3. **B**
4. **B**
5. **C**
6. **C**
7. **B**
8. **B**

9. **B**
10. **A**
11. **B**
12. **C**
13. **D**
14. **B**
15. **C**
16. **C**

17. **A**
18. **D**
19. **C**
20. **C**
21. **A**
22. **D**
23. **C**
24. **A**

25. **C**
26. **D**
27. **A**
28. **B**
29. **B**
30. **A**

31. 28/3
32. 405
33. 3.2
34. 1215
35. 47
36. 42
37. 18
38. 397

No-Calculator Section

1. **(D)** Subtract $\frac{2}{3}$ from both sides of the equation, and then set the cross-products equal:

$$\frac{5x - 3y}{3x + 5y} = \frac{1}{3}$$
$$3(5x - 3y) = 3x + 5y$$
$$15x - 9y = 3x + 5y \leftarrow \text{Collect like variables}$$
$$12x = 14y \qquad \leftarrow \text{Divide both sides by } 12y$$
$$\frac{x}{y} = \frac{14}{12} = \frac{7}{6}.$$

2. **(D)** During the 9–12 minute interval, the graph is horizontal. Speed is being measured along the vertical axis, and there is no change in speed. As a result, the runner is jogging at a constant rate.

3. **(B)** Since $\triangle AEB$ is similar to $\triangle DEC$, the lengths of their corresponding sides are in proportion. If x represents the length of $\overline{DE}$, then $42 - x$ represents the length of $\overline{AE}$. Thus,

$$\frac{CD}{AB} = \frac{DE}{AE}$$
$$\frac{16}{12} = \frac{x}{42 - x}$$
$$\frac{4}{3} = \frac{x}{42 - x} \leftarrow \text{cross-multiply}$$
$$3x = 4(42 - x)$$
$$3x = 168 - 4x$$
$$7x = 168$$
$$x = \frac{168}{7}$$
$$= 24$$

4. **(D)** In the equation $C = 60 + 0.25d$, d represents the number of gigabytes of data used after the first gigabyte and its coefficient, 0.25, must therefore represent the rate at which the cost, C, changes each time d increases by 1. Consider each statement in turn.

- I. The cost of each additional megabyte of data is not $60.25, it is $0.25. ✗
- II. The y-intercept of the graph of $C = 60 + 0.25d$ is 60, which represents the fixed cost when there are no additional gigabytes of data used so it does not represent the cost of each additional gigabyte of data used. ✗
- III. If between 5 and 6 megabytes of data are used in a month, and since the plan comes with the first megabyte of data free, then there is a charge for 5 rather than 6 gigabytes of data. Hence, $C = 60 + 0.25(5) = \$61.25$. This statement is true. ✔

5. **(C)** The absolute value of a number cannot be less than zero. Therefore, it is impossible for there to be any numbers for which this statement can be true, since the absolute value of an expression cannot be a negative number.

6. **(B)** If $d = \frac{1}{2}at^2$, then $at^2 = 2d$ and $t^2 = \frac{2d}{a}$ so $t = \sqrt{\frac{2d}{a}}$.

7. **(D)** If $x^2 - y^2 = 24$, then $(x - y)(x + y) = 24$. Because it is given that $x - y = 3$, $x + y = 8$. Adding corresponding sides of the two linear equations gives $2x = 11$ so $x = \frac{11}{2}$.

    Substitute $\frac{11}{2}$ for x in $x + y = 8$: $\frac{11}{2} + y = 8$ so $y = 8 - \frac{11}{2} = \frac{16}{2} - \frac{11}{2} = \frac{5}{2}$.

8. **(C)** If $\frac{z}{2b} = 4$ and $\frac{z}{2c} = 6$, then $z = 8b$ and $z = 12c$. Adding corresponding sides of the two equations gives $2z = 8b + 12c$. Dividing each member of the equation by 2 makes $z = 4b + 6c = 2(2b + 3c) = 2(12) = 24$.

9. **(B)** To break even, the pizza parlor must have a profit of at least zero. Profit is calculated as *Revenue – Cost*. The total revenue is 10 dollars for each pizza, and the total cost is 4 dollars for each pizza plus the initial $180,000 fixed cost. Express this algebraically, using p as the number of pizzas sold:

$$10p - (4p + 180,000) = 0$$
$$6p - 180,000 = 0$$
$$6p = 180,000$$
$$p = \frac{180,000}{6} = 30,000$$

10. **(A)** Since the product of the last terms of the binomial factors must be equal to -7: $b - 13 = -7$ so $b = 6$. The product of the first terms of the binomials factors must be 12, the coefficient of the x^2-term so $a = 2$. Hence,

$$(2x + 7)(6x - 1) = 12x^2 - 2x + 42x - 7$$
$$= 12x^2 + 40x - 7$$

    Since k is the coefficient of the x-term, $k = 40$.

11. **(B)** It is given that $f(x) = ax^2 + \frac{2}{a}x$. Then $f(3) = 9a + \frac{6}{a}$ and $f(2) = 4a + \frac{4}{a}$ so

$$f(3) - f(2) = \left(9a + \frac{6}{a}\right) - \left(4a + \frac{4}{a}\right) = 11. \text{ Simplify:}$$

$$5a + \frac{2}{a} = 11 \quad \leftarrow \text{Multiply each term by } a$$
$$5a^2 - 11a + 2 = 0 \quad \leftarrow \text{Solve by factoring}$$
$$(5a - 1)(a - 2) = 0$$
$$5a - 1 = 0 \quad or \quad a - 2 = 0$$
$$a = \frac{1}{5} \quad or \quad a = 2$$

    Hence, the smallest possible value of a is $\frac{1}{5}$.

12. **(B)** The tangent of an acute angle of a right triangle is the length of the side opposite the angle divided by the length of the side adjacent to the angle. Hence,

- $\tan 28° = \dfrac{x + \text{height of cliff}}{200}$ so height of cliff $+ x = 200 \tan 28°$

- $\tan 18° = \dfrac{\text{height of cliff}}{200}$ so height of cliff $= 200 \tan 18°$

- Subtracting corresponding sides of the previous two equations gives
$$x = 200 \tan 28° - 200 \tan 18°$$
$$= 200 \left(\tan 28° - \tan 18°\right)$$

13. **(D)** If b represents the cost of a dozen bagels and f is the delivery fee, then

- The cost of two dozen bagels plus delivery is $2b + f = 8$.
- The cost of three dozen bagels plus delivery is $3b + f = 11$.

14. **(D)** Find the vertex of the parabola $y = 2x^2 - 12x + 7$ by completing the square:

$$y = 2x^2 - 12x + 7$$
$$= 2\left(x^2 - 6x\right) + 7$$
$$= 2\left(x^2 - 6x + 9\right) + 7 - 18$$
$$= 2(x - 3)^2 - 11$$

Hence, the vertex of the parabola is $(3, -11)$. The vertical distance between the vertex and $(3, 4)$ is $4 - (-11) = 4 + 11 = 15$.

15. **(B)** According to the conditions of the problem,

$$-16t^2 + 64t + 4 \geq 52$$
$$-16t^2 + 64t - 48 \geq 0 \quad \leftarrow \text{Divide each member by} - 16$$
$$t^2 - 4t + 3 \leq 0 \quad \leftarrow \text{The direction of the inequality gets reversed}$$
$$(t - 1)(t - 3) \leq 0 \quad \leftarrow \text{Factors must have opposite signs}$$
$$t - 1 \geq 0 \quad and \quad t - 3 \leq 0$$
$$t \geq 1 \quad and \qquad t \leq 3$$

Hence, the solution interval is $1.0 \leq t \leq 3.0$.

16. **36** Factor the numerator and divide out a:

$$\frac{a\left(\dfrac{2}{3} - \dfrac{4}{9}\right)}{2a} = 4$$

$$a\left(\dfrac{2}{3} - \dfrac{4}{9}\right) = 8$$

$$a\left(\dfrac{6}{9} - \dfrac{4}{9}\right) = 8$$

$$a\left(\dfrac{2}{9}\right) = 8$$

$$a = \dfrac{9}{2} \times 8 = 36$$

17. Given $\frac{2}{3}x - \frac{1}{4}y = 6$ and $kx - \frac{1}{3}y = 8$.

If a system of linear equations has an infinite number of solutions, then the two equations can be made to look exactly the same.

- Multiply the first equation by 12 to eliminate the fractions:

$$12\left(\frac{2}{3}x - \frac{1}{4}y\right) = 12(6)$$
$$8x - 3y = 72$$

- Multiply the second equation by 9 so that the constant terms agree:

$$9\left(kx - \frac{1}{3}y\right) = 9(8)$$
$$9kx - 3y = 72$$

- The two equations will look the same when the coefficients of x are the same. Hence, $9k = 8$ so $k = \frac{8}{9}$.

18. **(145)** Since vertical angles have the same measure:

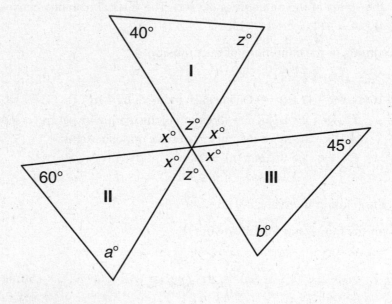

- In $\triangle$I, $2z + 40 = 180$ so $z = 70$.
- $x + z + x = 180$ so $2x + 70 = 180$ and $x = 55$.
- In $\triangle$II, $a + 60 + 55 = 180$ so $a = 65$. In $\triangle$III, $b + 45 + 55 = 180$ so $b = 80$.

Hence, $a + b = 60 + 85 = 145$.

19. **(300)** The graph of the equation $W = 120I - 12I^2$ in the xy-plane is a parabola. To find the maximum value of W, find the y-coordinate of the vertex of the parabola:

$$W = -12I^2 + 120I$$
$$= -12(I^2 - 10I)$$
$$= -12(I^2 - 10I + 25) + 300$$
$$= -12(I - 5)^2 + 300$$

Since the vertex of the parabola is at $(-5, 300)$, the maximum value of W is 300.

20. **(9/2)** The slope of the line through the points $(5, -5)$ and $(1, 3)$ is $\dfrac{3 - (-5)}{1 - 5} = \dfrac{8}{-4} = -2$.

The equation of the line through these two points has the form $y = -2x + b$. To find b, let $x = 1$ and $y = 3$. $3 = -2(1) + b$ so $b = 5$ and the equation of the line is $y = -2x + 5$. The equation of the line with slope 6 and which passes through $(0, 1)$ is $y = 6x + 1$. Find the coordinates of the intersection of the two lines by setting their y-values equal:

$$-2x + 5 = 6x + 1$$
$$4 = 8x$$
$$\frac{1}{2} = x$$

If $x = \dfrac{1}{2}$, then $y = 6\left(\dfrac{1}{2}\right) + 1 = 4$. If the lines intersect at (p, q), then $(p, q) = \left(\dfrac{1}{2}, 4\right)$ so $p + q = \dfrac{1}{2} + 4 = \dfrac{9}{2}$.

Calculator Section

1. **(B)** If three times 1 less than a number n $(3(n - 1))$ is the same as two times the number n increased by 14 $(2n + 14)$, then $3n - 3 = 2n + 14$ so $3n - 2n = 14 + 3$ and $n = 17$.

2. **(C)** If x represents the amount of money George started with, then

   - $0.25x + 0.60(x - 0.25x) = 0.7x$ is the amount of money he spent on lunch and dinner.
   - The amount of money George had left is $x - 0.7x = 9$ so $0.3x = 9$ and $x = \dfrac{9}{0.30} = 30$.
   - The amount George spent on lunch and dinner is $0.7x = 0.7(30) = \$21$.

3. **(B)** Since there are 30 test scores, the median lies between the 15th score and 16th score, which are contained in the interval 71–80. This interval contains 12 of the 30 scores. Hence, the probability that a score picked at random will fall in the interval that contains the median is $\dfrac{12}{30} = \dfrac{2}{5}$.

4. **(B)** Consider each statement in turn.

   I. The amount of the compound present is decreasing by a fixed percent of whatever amount remains, not by a constant amount. ✗
   II. Since the base of the exponential function is 0.8 and $0.8 = 1 - 0.20 = 1 - 20\%$, each hour the compound gets reduced by 20% of the amount present. ✓
   III. Each hour the compound gets reduced by 20% of whatever is present rather than by the fixed amount of 80% of 500 so this statement is false. ✗

5. **(C)** If x represents the number of pounds of peaches sold, then $165 - x$ represents the number of pounds of apples sold. Thus,

$$2.50x + 1.75(165 - x) = 337.50$$
$$2.50x - 1.75x + 288.75 = 337.50$$
$$0.75x = 48.75$$
$$x = \frac{48.75}{0.75} = 65$$

6. **(C)** Test the equation in each of the answer choices for $x = 1$ to $x = 4$ to see if it produces the correct values for y for each of the given values of x. The equation in choice (C), $y = 80(1.50)^x$, is the only equation that works. For example, for $x = 4$, $y = 80(1.50)^4 = 80(5.0625) = 405$.

7. **(B)** Since the test score (x) must increase as the number of incorrect responses decreases (y), the linear pattern of dots must have a negative slope as in graph (2).

8. **(B)** The ratio of cats to the total number of cats and dogs is $\dfrac{2}{2 + 3} = \dfrac{2}{5}$. Set up a proportion to solve for the total number of cats:

$$\frac{2}{5} = \frac{x}{250}$$
$$5x = 500$$
$$x = \frac{500}{5} = 100$$

9. **(B)** Since rate × time = distance,

- Find the distance the faster plane travels:

$$540 \frac{\text{miles}}{\text{hour}} \times 12 \text{ seconds} = 540 \frac{\text{mi}}{\text{hr}} \times 12 \text{ sec} \times \frac{1 \text{ min}}{60 \text{ sec}} \times \frac{1 \text{ hr}}{60 \text{ min}}$$
$$= 1.8 \text{ miles}$$

- Find the distance the slower plane travels:

$$450 \frac{\text{miles}}{\text{hour}} \times 12 \text{ seconds} = 450 \frac{\text{mi}}{\text{hr}} \times 12 \text{ sec} \times \frac{1 \text{ min}}{60 \text{ sec}} \times \frac{1 \text{ hr}}{60 \text{ min}}$$
$$= 1.5 \text{ miles}$$

- The difference in the distances traveled is $1.8 - 1.5 = 0.3$ or $\dfrac{3}{10}$ miles.

10. **(A)** Simplify the equations:

$$x - 3y = 2y + 7 \quad \Rightarrow x = 5y + 7$$
$$x + 2 = 3(y + 1) \quad \Rightarrow x = 3y + 1$$

Hence, $5y + 7 = 3y + 1$ so $2y = -6$ and $y = -3$ If $y = -3$, then $x = 3(-3) + 1 = -8$. The value of $\dfrac{x}{y} = \dfrac{-8}{-3} = \dfrac{8}{3}$.

11. **(B)** The total amount of time swimming and running took one-half of the 12 hours and 30 minutes it took to complete the triathlon or 6 hours and 15 minutes. Let x represent the amount of time swimming and $4x$ represent the amount of time running the marathon. Then

$$x + 4x = 6.25$$
$$5x = 6.25$$
$$x = \frac{6.25}{5} = 1.25$$
$$4x = 4(1.25) = 5$$

The 26.2 mile marathon distance was run in 5 hours or 300 minutes. Since $\frac{300}{26.2} \approx 11.45$, the average rate of minutes per mile at which Dylan ran the marathon is closest to 11.5.

12. **(C)** The rate of change in height when skiing from the top of the ski slope to its bottom is $-\frac{5}{11}$.

   - Convert 30 miles per hour to feet per second:

   $$30\,\frac{\text{mi}}{\text{hr}} \times 5{,}280\,\frac{\text{ft}}{\text{mi}} \times \frac{1\,\text{hr}}{60\,\text{min}} \times \frac{1\,\text{min}}{60\,\text{sec}} = 44\,\frac{\text{ft}}{\text{sec}}$$

   - Multiplying $-\frac{5}{11}$ by $44\,\frac{\text{ft}}{\text{sec}}$ gives the change in height per second of the skier.

   - Thus, $d(t)$ is the difference between the starting height of 11,000 feet and the product formed by multiplying the rate at which height is decreasing per second times the number of seconds that have elapsed in the ski run:

   $$d(t) = 11{,}000 - \overbrace{\left(\frac{5}{11}\right)(44)}^{\substack{\text{change in height}\\\text{per second}}} \times t$$
   $$= 11{,}000 - 20t$$

13. **(D)** If x represents the number of years it will take for the two trees to grow to the same height, then

   $$84 + 8x = 48 + 12x$$

   where each side of the equation repressents the height of the tree in inches. Thus,

   $$84 - 48 = 12x - 8x$$
   $$\frac{36}{4} = \frac{4x}{4}$$
   $$9 = x$$

14. **(B)** From the table, fill in the numbers to form the ratio:

   $$\frac{(4 + 15 + 9)}{(4 + 15 + 9) + (2 + 14 + 19)} = \frac{28}{63} = \frac{4}{9}.$$

15. **(C)** If the circuit has 4 moving parts, evaluate $f(4)$:

$$f(4) = 0.3(4)^2 = 4.8 = t.$$

Next, evaluate $g(4.8)$:

$$\begin{aligned} g(4.8) &= 150 + 5(4.8) \\ &= 150 + 24 \\ &= 174 \end{aligned}$$

16. **(C)** For Texas,

$$\text{Rate of change} = \frac{7,900 - 221,155.9}{4} \approx \text{million}$$

To change from millions to billions, divide by 1,000, which gives 6.3 billion.

17. **(A)** Consider each statement in turn.

■ I. In 2015, the per capita spending for Illinois was $\dfrac{140,072.6}{19.6} \approx 10,858$, which was greater than the per capita spending for Texas of $\dfrac{252,655.5}{27.4} \approx 9,221$. ✔

■ II. Per capita spending in Florida in 2015 was $\dfrac{162,548.3}{20.2} \approx 8,047$, where as in 2011 it was $\dfrac{163,070.8}{19.1} \approx 8,538$ so it declined in 2015. ✔

■ III. New York rather than California had the highest per capita spending for each year. ✗

18. **(D)** For each point along the curve, the product of x and y is 120. Hence, an equation of the graph is $xy = 120$ or, equivalently, $y = \dfrac{120}{x}$.

19. **(C)** Find the ratio of the number of individuals earning at least \$35,000 without a graduate school degree to the total number of individuals who earned at least \$35,000:

$$\begin{aligned} \frac{(108 - 28) + (80 - 46)}{108 + 80} &= \frac{80 + 34}{188} \\ &= \frac{114}{188} \\ &\approx 0.60 \end{aligned}$$

20. **(C)** A dime is .1 dollars, and a nickel is .05 dollars, so the total money from the dimes and nickels is $.1 \times 10 + .05 \times 5 = 1.25$. So, the amount that must come from quarters is $5.25 - 1.25 = 4.00$. Since a quarter is .25 dollars, take 4.00 and divide it by .25: $\dfrac{4}{.25} = 16$.

21. **(A)** Divide the number of males not taking AP classes, 23, by the total number of students in the school, 177, to get $\dfrac{23}{177}$.

22. **(D)** Find the rate of change of cruises sold:

$$\frac{\Delta \text{cruises sold}}{\Delta \text{price}} = \frac{135 - 180}{950 - 725} = \frac{-45}{225} = -\frac{1}{5}$$

You can eliminate choices (A), (B), and (C).

23. **(C)** Write the equation $x^2 + 4x + y^2 - 6y = 37$ in center-radius form by completing the square for both variables:

$$x^2 + 4x + y^2 - 6y = 37$$
$$(x^2 + 4x + 4) + (y^2 - 6y + 9) = 37 + 4 + 9$$
$$(x + 2)^2 + (y - 3)^2 = 50$$

Hence, the center of the circle is at $(-2, 3)$. Since $r^2 = 50$, $r = \sqrt{50} = \sqrt{25} \cdot \sqrt{2} = 5\sqrt{2}$. The correct combination of statements is Statements II and III.

24. **(A)** Simplify the given function by factoring the numerator:

$$f(x) = \frac{(x^4 + 2x^3 - 3x^2) + (4x + 12)}{x + 3}$$
$$= \frac{x^2(x^2 + 2x - 3) + (4x + 12)}{x + 3}$$
$$= \frac{x^2(x + 3)(x - 1) + 4(x + 3)}{x + 3}$$
$$= \frac{\cancel{(x + 3)}[x^2(x - 1) + 4]}{\cancel{x + 3}}$$
$$= x^3 - x^2 + 4$$

25. **(C)** Find the weighted average by calculating the sum of the products of the number of weeks of vacation and the number of people who receive the vacation, and then dividing the sum of the products by 25:

$$\frac{(1 \times 2) + (2 \times 7) + (4 \times 6) + (5 \times 2) + (6 \times 5) + (8 \times 3)}{25} = \frac{104}{25}$$
$$= 4.16$$
$$\approx 4$$

26. **(D)** If $p(x)$ is a polynomial function and $p(-1) = 3$, then when x is divided by $x - (-1)$ or, equivalently, $x + 1$, the remainder is 3.

27. **(A)** A system of linear equations has no solution if the graphs of the lines are parallel. Two lines are parallel when they have the same slope.

- If $hy - 8x = 5$, then $y = \frac{8}{h}x + \frac{5}{h}$ so the slope is $\frac{8}{h}$.

- If $y = \frac{3}{h - 2}x + 5$, the slope is $\frac{3}{h - 2}$.

- Find the value of h that makes the system of equations have no solution by setting the slopes equal:

$$\frac{3}{h-2} = \frac{8}{h}$$
$$8(h-2) = 3h$$
$$8h - 16 = 3h$$
$$5h = 16$$
$$h = \frac{16}{5}$$

28. **(B)** Five troy ounces is equivalent to $\frac{5}{12} \times 373.2 = 155.5$ grams. Since density is mass divided by volume, V:

$$19.3 = \frac{155.5}{V}$$
$$V = \frac{155.5}{19.3} \approx 8.06$$

The volume is closest to 8 cubic centimeters.

29. **(B)** To have an increase in the confidence level for survey results, the sample size should be as large as possible and the standard deviation should be as small as possible. A larger sample size will result in a data set that more completely mirrors the population, and if the standard deviation is low, the surveyor can be confident that there will not be great variation among the survey results.

30. **(A)** It is given that the vertex of a parabola is $(-4, k)$ so its equation is $y = a(x + 4)^2 + k$. Since it is also given that the parabola contains the points $(0, 12)$ and $(-3, 7)$, the coordinates of these points must satisfy the parabola equation.

- For $(0, 12)$:

$$12 = a(0 + 4)^2 + k$$
$$12 = 16a + k$$

- For $(-3, 7)$:

$$7 = a(-3 + 4)^2 + k$$
$$7 = a + k$$

- Solve the two equations simultaneously for k. From the second equation, $a = 7 - k$. Substitute $7 - k$ for a in the first equation:

$$12 = 16(7 - k) + k$$
$$12 = 112 - 16k + k$$
$$100 = 15k$$
$$\frac{100}{15} = k$$
$$\frac{20}{3} = k$$

31. **(28/3)** Find the slope of each line.

- The slope of line p is $\dfrac{k-0}{0-7} = -\dfrac{k}{7}$.

- The slope of line q is $\dfrac{\frac{3}{2}-0}{0-(-2)} = \dfrac{3}{4}$.

- Since perpendicular lines have slopes that are negative reciprocals:

$$-\frac{k}{7} = -\frac{4}{3}$$
$$k = \frac{28}{3}$$

32. **(405)** Rate of change in depth in feet per second is

$$\frac{195-69}{39-11} = \frac{126}{28} = \frac{9}{2}\frac{\text{feet}}{\text{second}}$$

Since 1.5 minutes is equivalent to 90 seconds, 1.5 minutes after his initial descent the diver will be

$$\frac{9}{2}\frac{\text{feet}}{\text{second}} \times 90\,\text{seconds} = 405\,\text{feet}$$

below sea level.

33. **(3.2)** Since $|-3| = 3$, rewrite the inequality as $9 < 4x - 3 < 10$. Adding 3 to each member of the inequality gives $12 < 4x < 13$. Dividing each member of the inequality by 4 makes $3 < x < \dfrac{13}{4}$ or, equivalently, $3 < x < 3.25$. Hence, x can be any number between 3 and 3.25, such as 3.2.

34. **(1,215)** Assume that the ratio of the tagged animals drawn to its sample size is approximately the same. If x represents the total number of fish of interest in the lake, then

$$\frac{198}{x} = \frac{22}{135}$$
$$22x = (198)(135)$$
$$x = \frac{\overset{9}{\cancel{(198)}}(135)}{\cancel{22}}$$
$$= 1{,}215$$

35. **(47)** The volume of the rectangular container is $10 \times 15 \times 20 = 3{,}000$ cubic inches. The volume of the cylinder is $\pi r^2 h = \pi(2)^2(5) = 20\pi$. To find the maximum number of full cylindrical cups of water that can be placed into the container with no overflow, divide the volume of the rectangular container by the volume of the cylindrical cup and round the answer down:

$$\frac{3{,}000}{20\pi} \approx 47.7$$

so, the maximum number of full cups of water without overflowing is 47.

36. **(42)** The area of an equilateral triangle with side s is $\frac{s^2}{4}\sqrt{3}$. Hence,

$$\frac{s^2}{4}\sqrt{3} = \frac{27}{\sqrt{3}}$$

$$s^2 = \frac{27 \times 4}{\sqrt{3} \times \sqrt{3}}$$

$$= \frac{108}{3}$$

$$s = \sqrt{36} = 6$$

Since the pendant has 7 sides each of which has a length of 6 centimeters, the distance around the pendant is $7 \times 6 = 42$ centimeters.

37–38. It is given that the function $h(t) = -4.9t^2 + 88.2t$ describes the height of a projectile, in meters, after t seconds have elapsed.

37. **(18)** The projectile will hit the ground when $h(t) = 0$:

$$-4.9t^2 + 88.2t = 0$$

$$4.9t(t - 18) = 0 \Rightarrow t = 18$$

38. **(397)** To find the maximum height, write the parabola equation in vertex form:

$$h(t) = -4.9t^2 + 88.2t$$

$$= -4.9(t^2 - 18t)$$

$$= -4.9(t^2 - 18t + 81) + (4.9)(81)$$

$$= -4.9(t - 9)^2 + 396.9$$

The y-coordinate of the vertex, to the *nearest meter*, is 397.

HOW TO EVALUATE YOUR PERFORMANCE ON A PRACTICE TEST

To estimate where your practice SAT math test scores fall on a 200–800 SAT scale, follow the steps in the Self Scoring Conversion Chart that follows. Keep in mind that the conversion table actually used by the College Board will be different than the one presented here. The College Board develops its conversion table using a large database of student test score data.

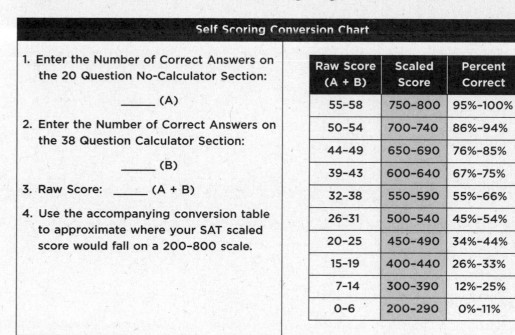

Self Scoring Conversion Chart

1. Enter the Number of Correct Answers on the 20 Question No-Calculator Section:

   _____ (A)

2. Enter the Number of Correct Answers on the 38 Question Calculator Section:

   _____ (B)

3. Raw Score: _____ (A + B)

4. Use the accompanying conversion table to approximate where your SAT scaled score would fall on a 200–800 scale.

Raw Score (A + B)	Scaled Score	Percent Correct
55–58	750–800	95%–100%
50–54	700–740	86%–94%
44–49	650–690	76%–85%
39–43	600–640	67%–75%
32–38	550–590	55%–66%
26–31	500–540	45%–54%
20–25	450–490	34%–44%
15–19	400–440	26%–33%
7–14	300–390	12%–25%
0–6	200–290	0%–11%

NOTES

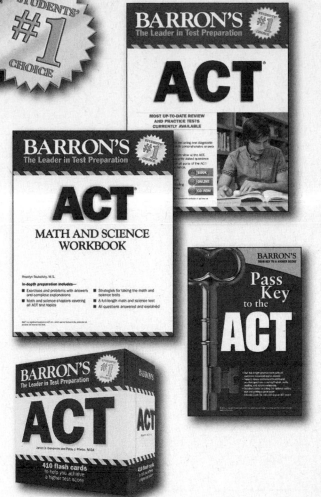

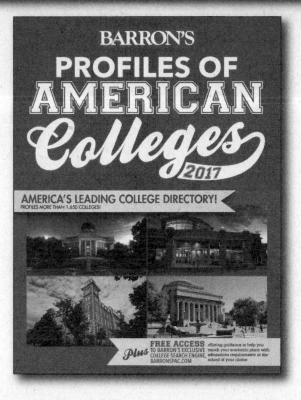